Butterflies
of Maine and the Canadian Maritime Provinces

Phillip G. deMaynadier

John Klymko

Ronald G. Butler

W. Herbert Wilson Jr.

John V. Calhoun

With a Foreword by Ernest H. Williams

Comstock Publishing Associates an imprint of
Cornell University Press | Ithaca and London

Publication of this book was made possible by a generous grant from the Maine Department of Inland Fisheries and Wildlife.

First published 2023 by Cornell University Press

Printed in China

Design and composition by Chris Crochetière, BW&A Books, Inc.

Library of Congress Cataloging-in-Publication Data
Names: DeMaynadier, Phillip G., author. | Klymko, John, author.
   | Butler, Ronald G. (Ronald George), 1951– author. | Wilson,
   W. Herbert, Jr., author. | Calhoun, John V., author.
Title: Butterflies of Maine and the Canadian maritime provinces
   / Phillip G. deMaynadier, John Klymko, Ronald G. Butler,
   W. Herbert Wilson Jr., and John V. Calhoun ; with a foreword by
   Ernest H. Williams.
Description: Ithaca : Comstock Publishing Associates, an imprint
   of Cornell University Press, 2023. | Includes bibliographical
   references and index.
Identifiers: LCCN 2022041254 | ISBN 9781501768941 (paperback)
Subjects: LCSH: Butterflies—Maine. | Butterflies—Maritime
   Provinces.
Classification: LCC QL551.M2 D46 2023 | DDC 595.78/9—dc23
   /eng/20230106
LC record available at https://lccn.loc.gov/2022041254

Butterflies of Maine and the Canadian Maritime Provinces

To Molly, who shares an affinity for all things winged and wild. I am grateful for your love and our journey.

Phillip

To my mother and late father, who fostered my interest in the natural world and tolerated my bringing home pet salamanders and boiling down raccoon skeletons on the stovetop.

John K.

To Ann, for her patience and support over the years that it took to bring this project to fruition, and to Devon, for understanding his nature-geek dad's fascination and respect for the "little things that run the world."

Ron

To my wife, Bets Brown, for her love, shared interest in nature, and infinite patience with this project.

Herb

To Laurel, my wife from Maine, for her unwavering patience and for embracing me as an honorary Mainer.

John C.

CONTENTS

FOREWORD

Butterflies are more than just enchanting examples of the diversity of nature, appealing to young and old alike. As herbivores, pollinators, and prey for other animals, butterflies are integral components of our native ecosystems, and learning more about them opens a window into a better understanding and appreciation of the world in which we live. With maps, flight graphics, historical information, conservation details, and much more, *Butterflies of Maine and the Canadian Maritime Provinces* provides a timely and comprehensive description of the butterflies present in a biogeographically unique region of North America.

Biological communities and their component species exist within natural boundaries, not political ones. This new book appropriately focuses on an important international biological area, the Acadian region encompassing Maine and the Canadian Maritime Provinces. Inhabiting coastal margins, inland areas, wetlands, and mountainous interiors, the species of this region form a natural association. To the task of compiling current knowledge about the Acadian region's butterflies, a talented team of authors has combined well-established expertise and long-term field experience with information from museum specimens and records from volunteers. The species included range from the Monarch, the most recognized butterfly in North America, to the Katahdin Arctic and other rare specialties of the region, to more than one hundred other species, each with its own ecological story.

Two decades into the twenty-first century, we are in a conservation crisis, with the diversity of our biological world being diminished by altered land use, climate change, and pesticides. This new book is important and useful because it provides an understanding of the current state of the butterfly fauna in the region of Maine and the Maritimes. Butterflies and other insects are essential parts of our biological heritage, and monitoring them contributes to an understanding of the pace, nature, and effects of anthropogenic changes. With *Butterflies of Maine and the Canadian Maritime Provinces* as a guide, observers can find new distributional records and increase public understanding of the species of this region. Community science has grown tremendously in importance over the past 20 years, and with good reason; it allows the accumulation of a wide range of extensive field observations for the benefit of professional biologists as well as curious naturalists.

The natural history of butterflies is one of the few areas in science in which such community contributions are especially important. In the field years ago, I encountered L. Paul Grey of Lincoln, Maine, a naturalist and early contributor to understanding the butterflies of Maine and other parts of North America. I spent an afternoon with him chasing butterflies, inspired by his knowledge and enthusiasm. Some of Grey's records are incorporated here, and he is profiled in the historical section. Users of this book, whether community scientists or professionals, can continue the tradition of contributing to what we know. What species do you see? Where are they? What are their behaviors?

The information in this beautifully prepared volume is a foundation for continued observation and learning about a significant group of organisms. It is a book for nature watchers as well as science professionals, an identification guide with in-depth information on the natural history and conservation status of each species. *Butterflies of Maine and the Canadian Maritime Provinces* is an important and much-needed contribution that will help all of us develop a deeper understanding of the butterflies of a unique region with a very special fauna.

Ernest H. Williams
9 January 2022

ACKNOWLEDGMENTS

This book was made possible by the contribution of many individuals. First and foremost, a biological survey of this magnitude requires an army of volunteers. Over 850 community scientists submitted tens of thousands of photo and specimen vouchers for the Maine and Maritimes atlas efforts in the modern era. Recognized individually in Appendix C, their enthusiasm and dedication were boundless.

The Maritimes Butterfly Atlas (MBA) could not have been completed without the helpful advice of the MBA Steering Committee: Rosemary Curley (Prince Edward Island Department of Natural Resources, retired); Mark Elderkin (Nova Scotia Department of Agriculture and Forestry, retired); Donald McAlpine (New Brunswick Museum); Martin Raillard (Environment and Climate Change Canada); and Reginald Webster. During the early years, the Maine Butterfly Survey (MBS) also benefited from the input of several advisors, including Donald Cameron (Maine Natural Areas Program), Richard Dearborn (Maine Entomological Society), Richard Hildreth, Donald Katnik (Maine Department of Inland Fisheries and Wildlife, MDIFW), Mark McCollough (U.S. Fish and Wildlife Service), Kent McFarland (Vermont Center for Ecostudies, VCE), Dale Schweitzer (NatureServe), Nancy Sferra (The Nature Conservancy in Maine), Beth Swartz (MDIFW), Reginald Webster, and Paula Work (Maine State Museum).

In addition to our atlas volunteers, we are grateful to the following individuals for providing private and institutional butterfly data as well as other forms of technical assistance: John Albright; Laura Bennett, Andrew Hebda, Kathy Ogden, and Stephanie Smith (Nova Scotia Museum); Paul Bentzen, Roger Gillis, and James Kho (Dalhousie University); Sean

Blaney, R. A. Lautenschlager, and Sarah L. Robinson (Atlantic Canada Conservation Data Centre, AC CDC); Richard Boscoe; Pamela Bryer, Patsy Dickinson, Robert Stevens, and Nathaniel Wheelwright (Bowdoin College); Kevin Byron; John Burger and Donald S. Chandler (University of New Hampshire Collection); Rick Cech; Brooke Childrey (Acadia National Park Collection); Charles Covell Jr. and Andrew D. Warren (McGuire Center for Lepidoptera and Biodiversity, Florida Museum of Natural History); Richard Dearborn (Maine Forest Service Insect Collection); Robert Dirig, Jason Dombroskie, and Richard Hoebeke (Cornell University); Charlene Donahue and Paula Work (Maine State Museum); Gail Everett; Richard Folsom; Lawrence Gall and Raymond J. Pupedis (Peabody Museum of Natural History, Yale University); Suzanne Green and Eric Quinter (American Museum of Natural History); Alex Grkovich; Chuck Harp; Rachael Hawkins and Philip Perkins (Museum of Comparative Zoology, Harvard University); Brian Harris, Michael Pogue, and Robert Robbins (National Museum of Natural History, Smithsonian Institution); Henry Hensel; Blanca Huertas (Natural History Museum, London); Richard Hildreth; Kirk Hillier (Acadia University); Don Katnik (MDIFW); Boris Kondratieff (Colorado State University); J. Donald Lafontaine and B. Christian Schmidt (Canadian National Collection of Insects, Arachnids, and Nematodes); Maxim Larrivée (Insectarium de Montréal); Ross Layberry; Christopher Livesay; Chrystal Maier and Rebekah Baquiran (Field Museum of Natural History); Christopher Marshall (Oregon State University); Donald McAlpine (New Brunswick Museum); David McCorquodale (Cape Breton University); Kent McFarland (VCE); James Monroe; Randall Mooi (Manitoba Museum); Gaétan Moreau (Université de Moncton); Glenn Morrell; Robert R. Muller; Christine Noronha (Agriculture and Agri-Food Canada); Jane O'Donnell and David Wagner (University of Connecticut); Jeffrey Ogden (Nova Scotia Department of Natural Resources and Renewables); Paul Opler; Dale Pasino (Boston University); Harry Pavulaan; Jonathan Pelham; Edward Perry and Mary Beth Prondzinski (Alabama Museum of Natural History); John Rawlins (Carnegie Museum of Natural History); Michael A. Roberts; Richard Romeyn; Brian Scholtens (College of Charleston); Dale Schweitzer (NatureServe); Debra Staber (L. C. Bates Museum); Sharon Stichter; Kristine Wallstrom; Steve Woods (University of Maine Collection); and Xinbao Zhang (eButterfly). In addition to supplying data from their personal collections, special recognition is due Jim Edsall and Reginald Webster for the exceptional volume of specimen and image vouchers confirmed from the region. Ernest Williams (Hamilton College) submitted numerous valuable records from Monhegan Island, Maine, and generously accepted our invitation to write the Foreword.

For help in locating literature and unpublished manuscripts, we thank Armand Esai (Field Museum of Natural History); Eva Grizzard and Barbara Harvey (Boston Museum of Science); Kirstin Kay and Anne Moore (University of Massachusetts); Kara Kugelmeyer (Colby College); Jeffers Lennox (Wesleyan University); Kent McFarland (VCE); Sophia Mendoza and Kat Stefko (Bowdoin College); Evan Peugh (Academy of Natural Sciences of Drexel University); Lorraine Portch (Natural History Museum, London); Mai Reitmeyer (American Museum of Natural History); Suzanne Smailes (Wittenberg University); and Robert Young (Museum of Comparative Zoology, Harvard University).

Several individuals graciously provided historical images and permission to reproduce them, including Daniel Barbiero (National Academy of Sciences); Paul Bentzen and Roger Gillis (Dalhousie University); Leslie Caseley and Matthew McRae (Prince Edward Island Museum and Heritage Foundation); Andrew Cormack and Peter Howson (Society for Army History Research); William Cromartie (American Entomological Society); Todd Gilligan and Keith Summerville (The Lepidopterists' Society); Pippa Hansen (Mount Desert Island Biological Laboratory); Marcia Kazmierczak (Museum of Comparative Zoology, Harvard University); Jennifer Longon (New Brunswick Museum); Barbara Murphy (National Academies Press); Robert B. Schwartz (Bowdoin College); and Ben Sechrist.

Many colleagues and institutions assisted with specialized questions and tasks that improved the final product. In particular, we are grateful to Sean Basquill (Nova Scotia Department of Natural Resources and Renewables) and Justin Schlawin (Maine Natural Areas Program, MNAP) for advice on ecoregional classifications; Donald Cameron (MNAP) for his patient botanical assistance; Andrew Couturier (Bird Studies Canada) for providing GIS layers for the Maritime Provinces; Jason Czapiga (MDIFW) for producing our study area map; Charlene Donahue and Paula Work (Maine State Museum) for organizing and curating the Maine specimen collection; Denis Doucet (Parks Canada) and R. A. Lautenschlager (AC CDC) for conceptualizing the Maritimes Butterfly

Atlas; Laura Dunn, Melissa Rafuse, Keegan Smith, and Miranda Weigensberg (Mount Allison University) for volunteering their time to enter Maritime Butterfly Atlas data; Manny Gimond (Colby College) for assistance with the R code used to generate flight histograms; Alan MacNaughton for creating interactive maps that helped Maritime volunteers target their atlas efforts; Kendra Mills for index preparation; Glen Mittelhauser (Maine Natural History Observatory) for early species account formatting assistance; Suzanne Nash (Princeton University) for extensive editorial assistance; Charity Robichaud and James Churchill (AC CDC) for help preparing the species richness heat maps; Lisa St. Hilaire (MNAP) for technical assistance with updating NatureServe S-ranks for Maine butterflies; and Beth Swartz (MDIFW) for keeping Maine volunteers well supplied and for advice on flight histograms and maps.

Facilities for Maine Butterfly Survey workshops were graciously provided by the Department of Biology at Colby College and the Delta Institute of Natural History. The following groups generously hosted presentations and workshops promoting the Maritimes Butterfly Atlas: Fundy National Park, Kouchibouguac National Park, Prince Edward Island National Park, Cape Breton Highlands National Park, Kejimkujik National Park, Chignecto Naturalist Club, Club des Ami(e)s de la Nature du Sud-Est, Club de naturalistes de la Péninsule acadienne, Fredericton Nature Club, Nature Miramichi, Nature Moncton, Nature Sussex, Saint John Naturalists Club, Daly Point Nature Preserve, River Valley Garden Club, New Brunswick Museum, Natural History Society of Prince Edward Island, South Shore Naturalists Club, Blomidon Naturalists' Society, Pictou County Naturalists Club, Nature Nova Scotia, Eastern Shore Garden Club, Mersey Tobeatic Research Institute, Nova Scotia Department of Lands and Forestry, and the Nova Scotia Nature Trust.

In a herculean effort, Bryan Pfeiffer captured an exceptional number of stunning images for the project, involving multiple excursions from Vermont to Maine in pursuit of elusive species otherwise lacking quality imagery. He also generously helped to manage the vetting, organizing, and editing of most butterfly imagery in the book. Bryan and the many other expert photographers whose images are included herein are recognized by name in the photo captions. In addition to their beauty, many of the published images also constitute valuable new flight or geographic records for the Acadian region.

Funding for the collection of data and its synthesis toward this volume came from a diversity of sources. In particular, the Maine Butterfly Survey was supported through financial contributions by the Maine Outdoor Heritage Fund, The Nature Conservancy in Maine, a State Wildlife Grant bestowed by the U.S. Fish and Wildlife Service, and proceeds from the Endangered and Nongame Wildlife Fund administered by MDIFW. It is unlikely the project could have culminated in a work of this depth without the support of senior management at MDIFW, all of whom recognize the value of applied research and scholarship in support of the agency's mission to conserve Maine's wildlife heritage; thank you especially to Judy Camuso, James Connolly, and Nathan Webb. The University of Maine at Farmington provided internship funding to support field surveys by students in Maine's Western Mountains for several years of the project. Additionally, Colby College Department of Biology provided research assistantships for more than twenty students who helped prepare Maine vouchers to be deposited as museum specimens. For funding the Maritime Butterfly Atlas, we recognize Environment Canada's EcoAction Community Funding Program, the New Brunswick Wildlife Trust Fund, the Gosling Foundation, New Brunswick Power Corporation, the New Brunswick Department of Natural Resources and Energy Development, Toronto Dominion Friends of the Environment Foundation, and the Prince Edward Island Department of Environment, Energy, and Forestry. BioQuip Products generously offered survey equipment at a discount for the Maritimes Butterfly Atlas.

Finally, we are grateful for the diverse expertise provided by the editorial team at Cornell University Press (CUP), especially Kitty Liu, Allegra Martschenko, Susan Specter, and Jacqulyn Teoh. It was Kitty who originally encouraged our team to consider CUP; her wit, patience, and thoughtful feedback helped to shape a better product than could otherwise have been achieved. Any factual errors or omissions are solely our own.

Butterflies of Maine and the Canadian Maritime Provinces

INTRODUCTION AND HISTORY OF BUTTERFLY STUDY IN THE REGION

BUTTERFLIES OF THE ACADIAN REGION

Encompassing over half of the planet's 1.8 million described species, insects are unquestionably one of the dominant forms of multicellular life on earth. Among the more diverse groups of insects are moths and butterflies, which together compose the order Lepidoptera (from the Greek *lepidos*, meaning scale, and *pteron*, meaning wing). Even casual observers of the biota of Maine and the Maritime Provinces are likely to be familiar with some of the more colorful and conspicuous species in the order, for example Monarchs and Luna Moths (*Actias luna*) (Figure 1). However, the richness of Lepidoptera that lies within and beyond our home gardens and meadows is more stunning and surprising than many realize. To put moth and butterfly diversity into perspective for wildlife enthusiasts of the Acadian region—defined here as Maine, New Brunswick, Nova Scotia, and Prince Edward Island (Figure 2)—it is helpful to compare their species richness with that of more familiar vertebrate groups. Based on published sources to date, approximately 3000 species of Lepidoptera are known from Maine and the Maritimes, a diversity fully eight times that of the region's breeding birds, mammals, freshwater fishes, reptiles, and amphibians combined! Clearly, moths and butterflies constitute a rich and important component of Acadian natural heritage, and yet a comprehensive assessment of their status and distribution in the region was mostly lacking, until now.

The term butterfly refers to members of the superfamily Papilionoidea, a branch on the evolutionary tree within the order Lepidoptera

Figure 1. The Monarch and Luna Moth are colorful and familiar examples of a butterfly and moth (Order: Lepidoptera) from the Acadian region (photos: Bryan Pfeiffer).

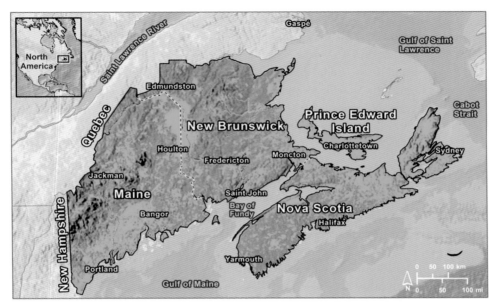

Figure 2. Highlighted in green is the Acadian region of North America covered by this book, including the jurisdictions of Maine, New Brunswick, Nova Scotia, and Prince Edward Island. Note that higher elevations are indicated with darker green tones.

(all other branches on the tree are collectively considered to represent moths). While the Papilionoidea is well-defined in the taxonomic sense, the criteria generally used to separate butterflies and moths are relatively subjective, as some moths display characters of butterflies, and vice versa. In general, butterflies are diurnal (day fliers), have clubbed antennae, hold their wings more vertically at rest, and do not make cocoons in which to pupate.

As is the case for Lepidoptera everywhere on earth, the diversity of butterflies is dwarfed by that of moths in the Acadian region. Still, with at least 121 documented resident and visiting species (approximately 14% of the American butterfly fauna north of Mexico), students of Acadian butterflies have much to pursue and enjoy. Colorful, diurnal, and mostly active on warm, sunny days, butterflies are an excellent gateway group for budding entomologists and amateur naturalists, some of whom will move on to study moths and other less conspicuous insect orders. While there is a rich history of butterfly research by expert lepidopterists in the Acadian region, much remains to be learned about the distribution and biology of this popular insect group, even by those less formally trained as evidenced by the many significant contributions of community scientists to this atlas.

In addition to their beauty and accessibility for study, butterflies play an important ecological role in our region's terrestrial and wetland ecosystems. Butterflies contribute to the pollination of native shrubs and wildflowers, and they help fuel the food chain (both as caterpillars and adults) for myriad insectivorous species ranging from spiders to birds. Additionally, caterpillars are important herbivores of the region's flora. Some can feed on a variety of plants across a wide range of habitats, such as the Canadian Tiger Swallowtail and White Admiral. Others are highly specialized to feed on just one or a few related plant species, often restricted to a single ecosystem, such as the cranberry-feeding Bog Copper, uniquely associated with bogs and fens. Their potential economic contribution in terms of watchable wildlife is difficult to estimate, but clearly no other insects have attracted as much attention from amateur naturalists and ecotourists, a group whose ranks increasingly include former bird-watchers now armed with close-focusing binoculars. Finally, butterflies are important for their scientific value as ecological indicators of ecosystem change due to factors such as habitat loss, succession, pollution, and climate change. Indeed, by studying trends in the abundance and distribution of butterfly communities we stand to learn about subtle changes in the natural environment important to the health of other organisms, including ourselves.

In this book, we introduce the reader to each of the Acadian region's confirmed resident and visiting butterfly species, as well as several undocumented species considered "possible" based on habitat, range, and migratory behavior. Additionally, while much of our volume is focused on documenting detailed butterfly species distributions, we embrace an expansive role for the modern biological "atlas" by also exploring questions of species taxonomy, life history, and conservation that are potentially unique to the geography of Maine and the Maritime Provinces. Most of the data summarized in this atlas were contributed by participants in the Maine Butterfly Survey (2006–2015) and the Maritimes Butterfly Atlas (2010–2015). While coordinated separately, both projects shared similar and simple goals: (1) improved scientific understanding of our region's butterfly fauna, and (2) increased public appreciation of the biology and conservation needs of butterflies. In support of these goals, we have tried to write this book in a manner appealing to a broad audience, from professional entomologists and seasoned naturalists to interested members of the general public. An example of this balance is our prioritization of common (English) names for plants and animals in the text, accompanied by the scientific (Latin) names only when first referenced. Scientific names for butterflies of the Acadian region are omitted entirely from much of the text, but they can be found in both the formal species accounts and regional checklist (Appendix A).

Having amassed tens of thousands of new records for the Acadian region—including novel state, province, and even national species additions—and having trained and marshaled the efforts of hundreds of new community scientists (also known as citizen scientists), we are pleased with the project's accomplishments and indebted to its contributors. Completing a comprehensive summary of the state of knowledge of Acadian butterfly fauna (or any taxon) is a daunting task and one that is partially dated the moment it is published. As such, it is our hope that this atlas serves not only as a measure of what we presently know about the region's butterfly assemblage but also as an inspiration for identifying and filling future gaps in our understanding of this most fascinating group of animals.

HISTORICAL BUTTERFLY STUDY IN MAINE AND THE MARITIMES

The first known published reference to the butterflies of the Acadian region occurred over 270 years ago, when the northern boundaries of Maine were still disputed by the British colonies of New Brunswick and Nova Scotia. At that time, an engraved representation of a White Admiral, identified as "The White Admirable," appeared on a map titled *A Plan of the Harbour of Chebucto and Town of Halifax* (Figure 3). In addition to the butterfly, the map figured an indeterminate species of longhorn beetle, a St. Lawrence Tiger Moth (*Arctia parthenos*), and a dreadful representation of a North American Porcupine (*Erethizon dorsatum*), which was copied in reverse from a figure in Edwards (1743). Nicknamed the "Porcupine Map," it was published in 1750 in the British periodical *The Gentleman's Magazine*. Although there is some disagreement, it is generally attributed to the English artist and entomologist Moses Harris (1730–ca.1788) (Kershaw 1993; Layberry et al. 1998; Lennox 2007;

The White Admirable

Figure 3. Detail of White Admiral from 1750 map of Halifax, Nova Scotia (library of John V. Calhoun).

Bushell et al. 2020). A notation in an earlier issue of the publication indicated that "the real butterflies brought from Halifax" (possibly including those portrayed on the map) could be viewed at the printer's shop in London (Anonymous 1750). These are the first butterflies known to have been captured by Europeans in Canada.

The first known reference to butterflies in Maine was by William D. Williamson (1832), who identified two species using imaginative names of his own invention: "Papilio magnus; Great Butterfly" and "Papilio Communis; Common Butterfly." Unfortunately, there is no way to know to what species he was referring. Nathaniel Brown (1772–ca.1825), a baker from Ipswich, Massachusetts, moved to Hallowell, Maine, in 1798 and presented specimens of ten species of Maine Lepidoptera to the collection of the newly founded Boston Society of Natural History in 1834 (Anonymous 1837). Possibly including butterflies, Brown's specimens probably no longer exist.

It was not until the 1860s that the study of the region's butterflies began in earnest. Many people, mostly amateur naturalists, collected specimens and published their discoveries. A number of earlier collectors and researchers were mentioned by Ferguson (1954), Brower (1974), and Webster and deMaynadier (2005). A list of publications about butterflies of Maine, New Brunswick, and Nova Scotia was compiled by Field et al. (1974). Regrettably, we cannot mention everyone who has assembled collections or shared their findings over the years, but we offer brief biographies of some of the more noteworthy individuals who helped lay the groundwork for our scientific knowledge of the region's butterflies.

Pioneers

Arriving at Pictou, Nova Scotia, from Scotland in 1803, **Thomas McCulloch** (1776–1843) (Figure 4) was a Presbyterian minister and controversial political figure. He founded the Pictou Academy in 1816 and later became the first president of Dalhousie College (now Dalhousie University) in Halifax (McCulloch 1920; Buggey and Davies 1988). In 1822, McCulloch sent 4000 insect specimens to the University of Glasgow (his alma mater) and the University of Edinburgh (Raine 1857). The following year, he sent several boxes of pinned butterflies and moths to Edinburgh (Young 1824). He established a natural history museum at Pictou Academy, which was visited by the celebrated naturalist John James Audubon

Figure 4. Thomas McCulloch (courtesy of Dalhousie University Archives).

in 1833 (Audubon and Coues 1897). Several of McCulloch's moths were mentioned by Kirby (1837), who described the species *Alypia maccullochii* (now *Androloma maccullochii*) in McCulloch's honor. Some of McCulloch's butterflies may still exist, but they are no longer recognizable as having been collected by him.

After serving in the West Indies from 1807 to 1816, **Richard S. Redman** (?–1831) (Figure 5), a lieutenant in the British Army, was stationed in Halifax, Nova Scotia, from at least 1820 until his death (Young 1952). He pursued the study of entomology during his idle hours, collecting many species of insects around Halifax and perhaps beyond. In 1821, Redman sent five or six cases of insects and other natural history objects to the British Museum in London, England (Hawes 1835; Piers 1918). Forty-one butterflies from Nova Scotia, representing fifteen different species, were listed in the museum's collection by Doubleday (1844, 1847), who explicitly credited "Lt. Redman" in one instance. At least some of Redman's specimens survive in the Natural

Figure 5. Richard S. Redman (reprinted from Young 1952, courtesy of Society for Army Historical Research).

Figure 6. Alpheus S. Packard (reprinted from Cockerell 1920).

History Museum, London, representing the oldest known extant butterflies from Canada. They include a Mustard White labeled "Nova Scotia Redman," and a Harvester labeled "N. Scotia Redman."

Alpheus S. Packard (1839–1905) (Figure 6) was born to an accomplished family in Brunswick, Maine. His father was a professor at Bowdoin College in Brunswick, which afforded the young Packard access to the school's library and its natural history books (Cockerell 1920). He started collecting insects as a teenager during the 1850s and enrolled at Bowdoin in 1857. It was then that he developed a particular fondness for Lepidoptera. He collected butterflies around Brunswick, and some of his specimens are deposited at the Museum of Comparative Zoology (Harvard University) (Figure 7). Among his many publications about insects, Packard (1862) authored the first overview of Maine's butterflies as well as a popular book, *Guide to the Study of Insects* (Packard 1869), in which he mentioned several butterfly records from Maine. In 1878, Packard was elected to serve as professor of zoology

Figure 7. Butterflies collected by A. S. Packard (Baltimore Checkerspot, top) and S. I. Smith (Compton Tortoiseshell, bottom), probably during the 1860s (courtesy of Museum of Comparative Zoology, Harvard University).

and geology at Brown University (Providence, Rhode Island), where he worked for the remainder of his career.

Sidney I. Smith (1843–1926) (Figure 8) was born in the small town of Norway in western Maine. He developed interests in botany and entomology at a young age (Coe 1929b). He was strongly influenced by **Addison E. Verrill** (1839–1926) (Figure 9), who also grew up studying natural history in Maine. Verrill was born in Greenwood, Maine, and moved with his family to the town of Norway when he was 14 years old (Coe 1929a). Smith and Verrill frequently collected insects together around Norway, and they forged a lifelong friendship. Scudder (1863, 1868) listed a number of their butterfly records from Norway. Verrill would go on to become the

Figure 8. Sidney I. Smith (reprinted from Coe 1929b, courtesy of National Academy of Sciences Archives).

first professor of zoology at Yale University (New Haven, Connecticut), and in 1865 he married Smith's older sister. In 1864, Smith followed in Verrill's footsteps and was admitted to Yale, where he later accepted a professorship in comparative zoology. Smith was previously credited with collecting the only known Maine specimens of four rare butterflies: Persius Duskywing, Frosted Elfin, Karner Blue, and Tawny Crescent. A recent investigation, however, revealed that these records were based on misidentifications and misinterpretations of specimen data (Calhoun 2017a, 2017b). Smith was, however, the first to record many other butterfly species in Maine, including the imported Cabbage White, which he found at Norway in 1865. This was also the first record of this butterfly from the United States (Scudder 1887a). Most of the insects that Smith collected prior to 1864 were acquired by the Museum of Comparative Zoology (Harvard University) (Figure 7). His remaining specimens are deposited at the Peabody Museum of Natural History (Yale University) (Uhler 1865; Verrill 1926; Coe 1929b).

Figure 9. Addison E. Verrill (reprinted from Verrill 1914).

Figure 10. Thomas Belt (reprinted from Belt 1888).

Thomas Belt (1832–1878) (Figure 10) was born in Newcastle, England, where he became interested in natural history, especially botany and entomology (Wright 1878; Anonymous 1888). He later studied geology and became a mining engineer. He arrived in Nova Scotia in 1863 to oversee the mining operations of the Nova-Scotian Gold Company. While there, he published several articles on geology as well as a list of the butterflies that he observed in the vicinity of Halifax in 1862 and 1863 (Belt 1864), representing the first publication on the butterflies of the Maritimes. Belt lived in Nicaragua from 1868 to 1874, resulting in his popular book *The Naturalist in Nicaragua* (Belt 1874), which is still considered one of the most interesting works on natural history travel. He perhaps borrowed the title of his book from a similar work by his friend, John M. Jones (see below). In 1878, Belt journeyed to the United States on business, where he was stricken with rheumatic fever in Colorado and died at the age of 45. The fate of the butterflies that Belt collected in Nova Scotia is unknown.

Early Contributors

From 1868 to 1876, **Henry H. Lyman** (1854–1914) (Figure 11) spent six to eight weeks each summer collecting butterflies at Cape Elizabeth, Maine (Lyman 1874, 1880). He also visited there in 1882 (Lyman 1910). Of particular interest, he observed that the now-extirpated Regal Fritillary was occasionally "somewhat plentiful" at Cape Elizabeth (Lyman 1876). Lyman was born and lived in Montreal, Quebec, where he managed a lucrative drug company. He became interested in Lepidoptera at an early age (Bethune 1900, 1914). He developed a hearing disability before his death, and an ear-trumpet became his constant companion (Vickery and Moore 1964). Tragically, he and his wife, Florence, died with over 1000 other passengers when the ocean liner they were aboard, the *Empress of Ireland*, collided with another vessel and sank in heavy fog near the mouth of the St. Lawrence River. They had departed from Quebec City with the hope of enjoying a belated honeymoon trip to England (Renaud

2010). Lyman bequeathed his insect collection of about 20,000 specimens, mostly Lepidoptera, to McGill University in Montreal. In 1961, when the McGill insect collection was moved to the MacDonald Campus of the university, it became known as the Lyman Entomological Museum (Vickery and Moore 1964).

Figure 11. Henry H. Lyman (reprinted from Bethune 1900).

Born in Wales and trained as a lawyer, **John M. (J. Matthew) Jones** (1828–1888) came to New York around 1854 with the intention of traveling to the Rocky Mountains. In 1860, a cholera outbreak forced him to go to Halifax, Nova Scotia, where a relative was serving as the governor (Piers 1903; 1915). He lived there the rest of his life, serving as a founder and president of the Nova Scotian Institute of Science. He was interested in many aspects of natural history, including entomology, leading to his nickname "Bug Jones" (Piers 1915). Of wealthy means, he maintained a large private natural history museum in a separate building on his property. Jones authored and coauthored several publications on the Lepidoptera of Nova Scotia, including the first faunal list of the butterflies of the province (Bethune and Jones 1870; Jones 1870, 1872a, 1872b). In addition to his many publications about the natural history of Nova Scotia, he wrote *The Naturalist in Bermuda* (Jones 1859), based on his experiences on the island. An enthusiastic collector, his insect specimens are deposited in many institutions, including the Natural History Museum, London, the National Museum of Natural History (Smithsonian Institution, Washington, D.C.), and the Nova Scotia Museum (Piers 1903).

Considered the pioneer entomologist of the Natural History Society of New Brunswick, **Caroline E. Heustis** (fl. 1874–1904) lived in Saint John, New Brunswick. Around 1882 she relocated up the Bay of Fundy to Parrsboro, Nova Scotia. Heustis collected insects, minerals, and plants in both New Brunswick and Nova Scotia during the 1870s and 1880s (McTavish 2008; Creese and Creese 2010) and published several observations about butterflies (Heustis 1879a, 1879b, 1880). Her insects from the vicinity of Saint John were presented to the New Brunswick Museum, but they suffered from neglect and no longer exist (Brittain 1918; Fairweather and McAlpine 2011).

Figure 12. Charles Fish (courtesy George J. Mitchell Department of Special Collections and Archives, Bowdoin College Library).

Like several other early Maine entomologists, **Charles Fish** (1832–1915) (Figure 12) graduated from Bowdoin College. Originally from Lincoln, Maine, he worked as a school instructor and administrator in several Maine towns, teaching courses in mathematics and natural sciences (Calhoun 2017a). From 1875 to 1881 he lived in Old Town, Maine, after which he moved to Brunswick, where he resided for the remainder of his life. Fish developed an interest in Lepidoptera during the 1870s, ultimately turning his attention toward moths of the family Pterophoridae. He donated his "valuable collection of mounted butterflies and moths" to Bowdoin College (Anonymous 1886). Totaling 2538 specimens, it was described as being particularly rich in New England species. Although Brower (1974) examined Fish's collection, recent attempts to locate it at Bowdoin College were unsuccessful, and it is believed to be lost.

From Southwest Harbor on Mount Desert Island, Maine, **Charles H. Fernald** (1838–1921) (Figure 13) developed an interest in entomology around the year 1871, when he received a master's degree from Bowdoin College. He was hired as professor of natural history at Maine State College (now the University of Maine, Orono), where he worked until 1886 (Anonymous 1911; Braun 1921). Fernald gave more attention to the study of insects during this period, authoring *The Butterflies of Maine* (Fernald 1884a, 1884b), which was the first account of all the known butterflies of the state. Assisted by his entomologist wife, Maria E. Fernald née Smith (1833–1919), Fernald recorded many species of butterflies in the vicinity of Orono. In 1886, he accepted the position of professor of zoology at the Massachusetts Agricultural College (now University of Massachusetts, Amherst), where he worked until his retirement in 1910. Fernald's butterfly specimens were incorporated into the collection of the University of Maine, which was transferred in 2012

Figure 13. Charles H. Fernald (reprinted from Anonymous 1911).

to the Maine State Museum in Augusta. Other specimens, particularly those that he collected on Mount Desert Island, Maine, are deposited at the University of Massachusetts, Amherst.

Known as the "farmer naturalist," **Francis Bain** (1842–1894) (Figure 14) was born on the family farm in North River, west of Charlottetown, Prince Edward Island. He developed an interest in the natural sciences at a young age (Martin 1979, 1990). In the 1860s he started exploring the island by train and horse-drawn wagon, becoming a self-taught authority on local animals, plants, rocks, and fossils. Bain authored over fifty articles on natural history topics in the local newspaper. He compiled a brief list of butterfly species on the island, which he recorded in

Figure 14. Francis Bain (courtesy of Prince Edward Island Museum and Heritage Foundation, HF.2016.03.1).

1884 (Bain 1885a), and noted the abundance of a few butterflies in 1885 (Bain 1885b). He also published a book on the natural history of Prince Edward Island, which mentioned several species of butterflies (Bain 1890). There is no evidence that any of his butterfly specimens survived (Martin 1983).

Born in Manchester, Maine, **Mattie (Martha) Wadsworth** (1862–1943) (Figure 15) spent her entire life in the same home, working the family farm. She received mail in the nearby town of Hallowell, which led to some confusion about her place of residence (White and Calhoun 2009). In her twenties she became interested in entomology, particularly beetles, dragonflies, and butterflies. Wadsworth collected and reared butterflies during the 1880s and 1890s, but poor health restricted her travels to within about two miles of her home. She corresponded with a number of prominent entomologists of her day, including Samuel H. Scudder, who included many of her butterfly records in his three-volume treatise *The Butterflies of Eastern United States and Canada, with Special Reference to New England* (Scudder 1888–1989). After her death, most of Wadsworth's insect collection was donated to

Figure 15. Mattie Wadsworth (courtesy of Ben Sechrist).

Figure 16. Wadsworth collection in the L. C. Bates Museum, showing drawer with Regal Fritillaries at left (John V. Calhoun).

the L. C. Bates Museum (Hinckley, Maine). It is contained in a thirty-drawer wooden cabinet, which she reportedly constructed (White and Calhoun 2009). Among her surviving specimens are six Regal Fritillaries from Manchester (Figure 16).

Philip Laurent (1857–1942) was born in Philadelphia, Pennsylvania. He began building an insect collection during the 1880s (Champlain 1945). He worked as a candy manufacturer and traveled widely in his spare time in search of specimens. Laurent authored several articles on insects, including a summary of his trip to the neighborhood of King and Bartlett Lake, within the mountainous portion of northern Somerset County, Maine (Laurent 1895). Spending two weeks in the area during August 1895, he called Maine "anything but an 'entomologist's paradise.'" Nonetheless, he collected nine specimens of the rare Hoary Comma. Prior to his death, Laurent was the oldest living member of the American Entomological Society (Anonymous 1941). His insect collection is deposited at the Academy of Natural Sciences, Philadelphia, for which he served as the director of its entomological section for a number of years.

A renowned botanist and mycologist, **Roland Thaxter** (1858–1932) (Figure 17) was born in Newtonville, Massachusetts. His mother was Celia L. Thaxter (1835–1894), one of America's most popular late-nineteenth-century authors. Thaxter attended Harvard University, just as his father, grandfather, and great-grandfather had done (Clinton 1935). Best known for studying mycology, he was initially interested in entomology (Lloyd 1917; Horsfall 1979) and collected butterflies on Cape Breton Island, Nova Scotia,

Figure 17. Roland Thaxter (reprinted from Lloyd 1917).

during the 1870s (Scudder 1877b). For many years, Thaxter maintained a summer home on Kittery Point, Maine. Some of the butterflies he collected there, and on Mount Desert Island, Maine, are deposited at the Museum of Comparative Zoology (Harvard University) and Boston University. At Kittery Point in 1930, Thaxter collected the only known Maine specimen of the Orange-barred Sulphur, which for 80 years was misidentified as a Large Orange Sulphur.

Little is known about **Lucy C. Eaton** (1867–?), an amateur entomologist with a passion for Lepidoptera. From Truro, Nova Scotia, she collected butterflies and other insects in the vicinity of her hometown. In 1895, the distinguished historian and naturalist Harry Piers read a paper by Eaton at a meeting of the Nova Scotian Institute of Science titled "The Butterflies of Truro, N.S.," which was subsequently published (Eaton 1896). Many of Eaton's Lepidoptera records were mentioned by Bethune (1897) and Perrin and Russel (1912). Her modest insect collection was purchased by the Provincial Museum of Nova Scotia (now the Nova Scotia Museum) (Piers 1906), but it was largely destroyed by insect pests during the ensuing decades (Ferguson 1954).

Twentieth Century

Born in Scotland, **William McIntosh (MacIntosh)** (1867–1950) (Figure 18) moved to Saint John, New Brunswick, when he was a boy. He

Figure 18. William McIntosh (courtesy of New Brunswick Museum—Musée du Nouveau-Brunswick; X10721).

worked as a landscape gardener and florist, becoming honorary curator of the Museum of the Natural History Society of New Brunswick in 1898. Spelling his name as both McIntosh and MacIntosh, he served as the first director of the New Brunswick Museum, which opened in 1934 (McTavish and Dickison 2007). He acted as the first provincial entomologist of New Brunswick from 1913 to 1923, and he spent a great deal of time developing nature lessons for children, who affectionately called him "Mr. Mac." Entirely self trained in natural history, McIntosh received an honorary doctorate from the University of New Brunswick in 1934. He authored the first faunal lists of the butterflies of New Brunswick (McIntosh 1899a, 1899b, 1904), which mentioned the collections of Caroline E. Heustis and Lucy C. Eaton (see above). Insects that McIntosh collected, including Lepidoptera, are deposited in the New Brunswick Museum, though most of the material prior to 1914 was destroyed (Fairweather and McAlpine 2011; McAlpine 2018).

   Joseph Perrin (1865–1936) came to Nova Scotia from England in 1887, when his father purchased McNabs Island, Halifax. Perrin lived the remainder of his life on the island, which is situated in the mouth of Halifax Harbour (McLaine 1937). An amateur entomologist, he worked for many years as an inspector of imported nursery stock in the Division of Foreign Pest Suppression at the port of Halifax (Gibson 1924). Perrin was most interested in Lepidoptera, combining his efforts with **John Russell** (1884–1948) to publish a detailed list of the butterflies and moths of Halifax and Digby, Nova Scotia (Perrin and Russell 1912). A few years later, Perrin (1919) published a supplement to this work based mostly on his own continuing studies. Census records reveal that Russell was a surveyor who was born in Digby. He moved to Hope Station, British Columbia, around 1910, then to Edmonton, Alberta, by 1916. Before leaving Digby, he sold much of his Lepidoptera collection to the Provincial Museum of Nova Scotia (now the Nova Scotia Museum) (Brittain 1918), where Perrin's collection is also deposited. Some of Russell's specimens are in the Canadian National Collection, Ottawa, Ontario, and other institutions (Ferguson 1954).

From Wilmington, Delaware, the naturalist **Frank M. Jones** (1869–1962) (Figure 19) was employed in the iron works industry. He made numerous collecting trips to obtain insect specimens, especially Lepidoptera, visiting almost every state in the United States over the course of his lifetime (McDermott 1963). Jones published extensively and was awarded an honorary degree in 1931 by the University of Delaware for his entomological contributions. Between 1905 and 1923, he made several trips to Maine, during which he collected many butterflies. The bulk of his collection, containing approximately 2700 butterflies, was donated to the Peabody Museum of Natural History (Yale University) (Remington 1954).

Figure 19. Frank M. Jones (reprinted from McDermott 1963, courtesy of American Entomological Society).

John C. Parlin (1863–1949) was born near Trap Corner, a small community in western Maine. For many years he worked as a school teacher in various towns in Maine. He was a widely recognized amateur botanist and received an honorary degree in 1947 from the University of Maine (Bean 1948). In addition to a number of botanical articles, he published lists of the butterflies that he collected in several areas of Maine during the early 1920s (Parlin 1922, 1923). Regrettably, his natural history collections were destroyed, along with his home, in the devastating flood of March 1936, which inundated much of New England.

Charles W. Johnson (1863–1932) (Figure 20) was born in Morris Plains, New Jersey, and later moved to St. Augustine, Florida, where he developed an interest in insects around the age of 20 (Brues 1933). He became a highly respected entomologist and published numerous papers on the subject (Melander 1932). He served as curator for the Boston Society of Natural History from 1903 to 1932, during which time he made frequent

Figure 20. Charles W. Johnson (reprinted from Creed 1930).

collecting trips to localities in New England. Between 1918 and 1926, Johnson spent up to two weeks each summer on Mount Desert Island, Maine. Although he was most interested in Diptera (flies), he collected a large number of Lepidoptera and included many butterfly records in his study of the insect fauna of Mount Desert Island (Johnson 1927), which was continued by **William Procter** (1872–1951). Originally from Cincinnati, Ohio, Procter worked for the Procter and Gamble Company, which his grandfather founded (Alexander 1951). He later turned away from business to study biology. When Procter was a boy, his family spent their summers on Mount Desert Island. This inspired him to establish a research laboratory there in 1921. Three years earlier, the Boston Society of Natural History had selected Mount Desert Island for a detailed study of its insect fauna. It was then that Johnson became acquainted with Procter, who continued publishing on the insects of the island after Johnson's death (Procter 1938, 1946). Another prominent contributor to this study was the Maine entomologist Auburn E. Brower (see below). Most of Johnson's butterfly specimens are deposited at Boston University, which houses the former Boston Society of Natural History collection. The bulk of Procter's collection is at the University of Massachusetts, Amherst. Many of Johnson's and Procter's butterflies from Mount Desert Island are contained in the collection of Acadia National Park, located on the island. A catalog of the Lepidoptera in this collection was published by Mittelhauser et al. (2014).

A taxidermist from Lincoln, Maine, who was born to English parents, **Walter J. Clayton** (1876–1942) was interested in birds, mammals, and Lepidoptera. Between 1929 and 1941, he recorded many species of butterflies in Penobscot County, where he lived. The regionally endemic Clayton's Copper was named in his honor by Brower (1940), who described Clayton as "one of Maine's best field naturalists." Clayton's butterflies are deposited in several institutions, including Boston University, the Carnegie Museum of Natural History (Pittsburgh, Pennsylvania), and the National Museum of Natural History (Smithsonian Institution, Washington, D.C.).

James H. McDunnough (1877–1962) (Figure 21) is legendary in the annals of entomology. From Toronto, Ontario, he became interested in butterflies and moths as a youth (Freeman 1962). Described as a linguist, musician, entomological taxonomist, and builder of insect collections, he earned a doctorate in entomology in 1909 from Berlin University in

Germany. The next year, McDunnough was hired by the Illinois physician William Barnes to curate his immense private Lepidoptera collection, which was the largest in existence at that time (Ferguson 1962). During their 9-year association, Barnes and McDunnough jointly published dozens of works, including the popular *Check List of the Lepidoptera of Boreal America* (Barnes and McDunnough 1917). On his own two decades later, McDunnough (1938, 1939a) authored his *Check List of the Lepidoptera of Canada and the United States of America*, which served as a standard reference for 35 years. From 1934–1946, he conducted extensive surveys of the Lepidoptera of Nova Scotia (Ferguson 1954), and authored three articles about butterflies in the province, in which he described the subspecies *Argynnis cybele novascotiae*, *Feniseca tarquinius novascotiae*,

Figure 21. James H. McDunnough (reprinted from Ferguson 1962, courtesy of The Lepidopterists' Society).

and *Papilio brevicauda bretonensis* (McDunnough 1935a, 1935b, 1939b). In 1950, he permanently moved to Halifax, Nova Scotia, where he became a research associate of the Nova Scotia Museum of Science (now the Nova Scotia Museum). The bulk of his collection is deposited at that museum, though many of his specimens are also in other institutions, including the Canadian National Collection in Ottawa.

One of the most distinguished amateur North American lepidopterists of the twentieth century was **Cyril F. dos Passos** (1887–1986) (Figure 22). Born in New York, he was the cousin of the famous American novelist John Dos Passos (1896–1970). Cyril, who used the traditional spelling of his last name (with the lowercase *d*), became a successful attorney within his family law firm, and for a time he was also involved in the railroad business (Wilkinson 1988). The wealth he acquired through these enterprises allowed him to retire at an early age, after which he took up

Figure 22. Cyril F. dos Passos at his camp in Rangeley, Maine (reprinted from Wilkinson 1988, courtesy of The Lepidopterists' Society).

Figure 23. L. Paul Grey (reprinted from Kendall 1977, courtesy of The Lepidopterists' Society).

the study of butterflies at the urging of his first wife, Viola H. dos Passos née Van Hise (1891–1944), who decided to study moths. For many years, dos Passos maintained a camp at Rangeley, Maine, and it was there that he began collecting butterflies in 1929. Soon after, he started corresponding about Maine butterflies with **Lionel P. (L. Paul) Grey** (1909–1994) (Figure 23) (Grey 1988). Born in Houlton, Maine, Grey was then a young man working at his family's farm in Lincoln, Maine, after having recently graduated from Western Maine State Normal School (now University of Maine at Farmington) (Kendall 1977; Wilkinson 1988). While farming and doing carpentry work, he authored several publications about butterflies in Maine (Grey 1932, 1934, 1956) and New Brunswick (Grey 1965). Grey and dos Passos developed a lifelong friendship, and they collaborated on several projects, including an updated list of the butterflies of Maine (dos Passos and Grey 1934a, 1934b). Their most important contribution was a taxonomic revision of the North American greater fritillary butterflies (dos Passos and Grey 1947). In the ensuing years, dos Passos authored systematic and faunal papers on butterflies, including *A Synonymic List of the Nearctic Rhopalocera* (dos Passos 1964), which replaced the earlier check list by McDunnough (1938). The scientific name of the Salt Marsh Copper, *Lycaena dospassosi*, was named in honor of dos Passos (McDunnough 1940). He and Grey collected numerous butterflies in Maine from the early 1920s until the mid-1980s. The dos Passos butterfly collection, containing over 65,000 specimens, was donated to the American Museum of Natural History (New York). Grey's collection is deposited at the same museum, but many of his specimens are also preserved in other institutions as a result of his generous exchanges with fellow entomologists.

Born in Halifax, Nova Scotia, **Douglas C. Ferguson** (1926–2002) (Figure 24) was drawn to the study of natural history during childhood. His full name, a testament to his Scottish roots, was Alexander Douglas Campbell Ferguson, though his friends simply called him "Doug." He became interested in Lepidoptera during the early 1940s and worked as a field assistant to J. H. McDunnough (see above) in 1946. He subsequently

served as curatorial assistant, curator of entomology, and chief curator at the Nova Scotia
Museum from 1949 to 1963 (Hodges 2003). Ferguson earned a doctorate in entomology from
Cornell University and became a research entomologist at the National Museum of Natural
History (Smithsonian Institution, Washington,
D.C.) until his retirement in 1996. His interest
in butterflies and moths resulted in the comprehensive publication *The Lepidoptera of Nova
Scotia* (Ferguson 1954), which was reissued with
additions the following year (Ferguson 1955).
Although his attention was primarily on moths,
he would often collect butterflies during the day.

Figure 24. Douglas C. Ferguson
(reprinted from Hodges 2003,
courtesy of The Lepidopterists'
Society).

His Lepidoptera specimens, totaling more than
200,000, are deposited at the Nova Scotia Museum, the National Museum of Natural History,
and the Peabody Museum of Natural History (Yale University).

   Although not originally from Maine, **Auburn E. Brower** (1898–1994)
(Figure 25) was the undisputed dean of Maine lepidopterists. Born in
St. Louis, Missouri, he grew up on the family farm in the town of Wil-
lard, just outside Springfield, Missouri (Anonymous 1995). Known as
"Doc" or "Ed" to his friends (his middle name was Edmond), he became
interested in natural history as a boy. He briefly taught high school science in Missouri (Anonymous 1938) and earned
a doctorate degree from Cornell University in
1931 with a dissertation on the biology of un-
derwing moths of the genus *Catocala*. He subsequently took a position with the Maine Forest
Service, where he spent 37 years as the forest
entomologist. He attempted to document the
occurrence of all Lepidoptera species in Maine,
explaining, "I found that state lists are of a very
definite value to workers," adding, "When I came
to Maine in May 1931 I set out to gather the specimens, records, proper determinations, etc. of the
Lep[idoptera] here. . . . I have spent some money
every year towards this, and major expenditures

Figure 25. Auburn E. Brower
(reprinted from Kendall 1977,
courtesy of The Lepidopterists'
Society).

each year toward securing authoritative determinations" (letter to J. V. Calhoun, 8 November 1982). He contributed valuable information to several important projects, including the dissertation on the Lepidoptera of New England by Farquhar (1934) and the insect surveys of Mount Desert Island by Procter (1938, 1946). During the 1950s, Brower assisted the Portland Society of Natural History in coauthoring a list of Maine butterflies with Rinda-Mary Payne, then a student in Portland, Maine (Brower and Payne 1956). Brower is best known for his comprehensive *List of the Lepidoptera of Maine*, which was issued in three parts (Brower 1974, 1983, 1984). The first part, on the macrolepidoptera (including butterflies), listed 104 species of butterflies. He accumulated perhaps the largest private entomological collection in New England, with over 130,000 specimens—including more than 115,000 Lepidoptera—which was bequeathed to the National Museum of Natural History (Smithsonian Institution, Washington, D.C.) (Davis and Hevel 1995). Brower was described during his lifetime as Maine's foremost naturalist (Vanderweide 1977).

Today and Beyond

Since the 1970s, research on the region's butterflies has continued at a brisk pace. Valuable contributions, such as *A Preliminary Atlas of the Butterflies of New Brunswick* (Thomas 1996) and *A Baseline Atlas and Conservation Assessment of the Butterflies of Maine* (Webster and de-Maynadier 2005), have greatly enhanced our knowledge of the fauna. Numerous community scientist volunteers participated in the remarkably successful butterfly surveys of Maine and the Maritimes; without their selfless dedication this book would not have been possible. Over 150 years of devoted study, by individuals from every walk of life, is reflected in the species accounts that follow. This book is not a means to an end. Rather, we hope it acts as a solid foundation from which to cultivate a more complete understanding of the extraordinary butterflies that inhabit the Acadian region.

METHODS AND BIOGEOGRAPHICAL FINDINGS

PROJECT METHODS AND DATA SOURCES

The information compiled for this atlas is a result of three separate and complementary approaches at data collection conducted over a period of 15 years (2003–2017): (1) the mining of existing records from museum and private collections, scientific publications, and online databases, (2) targeted professional surveys for regional species of conservation concern, and (3) broad community science surveys for all species across Maine and the Maritime Provinces. This multifaceted approach afforded a comprehensive investigation of the region's historical and current butterfly fauna.

An effort to recognize and summarize the extensive contributions by previous researchers and students in Maine was launched in 2003. Information on the occurrence of butterflies was reviewed from a variety of sources (Table 1), including published references containing records of Maine butterflies, specimens contained in most major northeastern museums and public institutions as well as many larger private collections, and database records shared by the Maine Department of Inland Fisheries and Wildlife (MDIFW), amassed from ecoregional surveys directed primarily at state rare and imperiled species. There is a rich history of Lepidoptera study in Maine. The work of Brower (1974) and many other early Maine Lepidopterists, from S. I. Smith (collected primarily during the 1860s) to Paul Grey (collected from the 1930s to the 1980s), contributed significantly to the historical record and is summarized further in the chapter Introduction and History of Butterfly Study in the Region. As a result of these initial research efforts, a database of nearly 9000 records for 114 Maine species was compiled and ultimately reported in *A Baseline Atlas and Conservation Assessment of the Butterflies of Maine* (Webster and deMaynadier 2005). That project provided the foundation for subsequent survey efforts, which were accomplished through the help of hundreds of dedicated participants.

Records from the Maritime Provinces that predate 2010 were compiled during the Maritimes Butterfly Atlas project from a variety of sources (Table 1), including many that had been amassed previously in the Atlantic Canada Conservation Data Centre (AC CDC) species observation database. Of great importance was a database of Canadian butterfly records, primarily based on museum specimens, compiled by Ross Layberry. It contains records from collections across Canada, and an earlier version was essential to the seminal book *The Butterflies of Canada* (Layberry et al. 1998). Other important sources of published records were *A Preliminary Atlas of the Butterflies of New Brunswick* (Thomas 1996), and *The Lepidoptera of Nova Scotia. Part 1: Macrolepidoptera* (Ferguson 1954). Additional specimen records came from the Nova Scotia Museum, the New Brunswick Museum, several smaller institutions, and private collections, and butterfly surveys that had been conducted by the AC CDC prior to the launch of the Maritime Butterfly Atlas in 2010. These sources collectively yielded a database of over 12,000 records for ninety-one species.

Targeted surveys for butterfly species of regional conservation concern were a priority throughout the Maine Butterfly Survey and parts of the Maritime Butterfly Atlas period. A critical role of the MDIFW and AC CDC is tracking the status of rare, Threatened, and Endangered species in the region. In so doing, these initiatives facilitate conservation efforts by informing strategic land conservation, habitat management, public outreach, and regulatory protections. To this end, Maine and the Maritime Provinces maintain lists of species with official conservation status (Appendix A) and direct resources toward understanding their biology, threats, range, and population status. Many of the region's most vulnerable species have exceptionally restricted geographic ranges (e.g., Katahdin Arctic and Maritime Ringlet) and occupy remote or highly specialized habitats (e.g., Frigga Fritillary and Hessel's Hairstreak). Additionally, a number of species are officially listed as Endangered or Threatened, requiring that special permits be obtained to collect specimens. For all of these reasons, many of the region's most vulnerable

Table 1. Secondary sources used to populate databases for the Maine Butterfly Survey and Maritimes Butterfly Atlas

Source description	Database	Notes
Data source type: Museum collections		
Acadia National Park, Bar Harbor, ME	ME	Contains specimens from W. Proctor, C.W. Johnson, A.E. Brower, and others; also includes BioBlitz data from the Schoodic Peninsula unit of Acadia National Park
Agriculture and Agri-Food Canada Insect Collection, Charlottetown, PE	Maritimes	Contains the University of Prince Edward Island insect collection
Agriculture and Agri-Food Canada Insect Collection, Fredericton, NB[a]	Maritimes	
American Museum of Natural History, New York, NY	ME	Contains specimens collected by L.P. Grey, A.E. Brower, C.F. dos Passos, and others
Atlantic Forestry Centre Insect Collection, Fredericton, NB	Maritimes	
Boston University, Biology Department, Boston, MA (Formerly Boston Society of Natural History)	ME	Contains some valuable older specimens collected by C.W. Johnson, A.E. Brower, C.A. Frost, S.I. Smith, H.H. Newcomb, W.J. Clayton, C.S. Minot, and others
Carnegie Museum of Natural History, Pittsburg, PA	ME	Contains specimens from R.F. Rockwell and others
Canadian National Collection of Insects, Ottawa, ON[a]	ME, Maritimes	Only a few specimens from ME, including from L.P. Grey and A.E. Brower; contains Maritimes specimens from R.A. Layberry, J.H. McDunnough, and others
Cornell University, Ithaca, NY	ME	Only a few specimens from ME, including from L.P. Grey and A.E. Brower
L. C. Bates Museum, Hinckley, ME	ME	Contains a significant collection from M. Wadsworth
Lyman Entomological Museum, McGill University, Ste. Anne de Belliveau, QC	Maritimes	Contains specimens from D.C. Ferguson, V.R. Vickery, and others
Maine Department of Inland Fisheries and Wildlife (MDIFW), Bangor, ME[b]	ME	Contains specimens from J.J. Albright and material from MDIFW surveys
Maine Forest Service Insect Laboratory, Augusta, ME	ME	Contains specimens from A.E. Brower, forest insect surveys, and some preserves owned by The Nature Conservancy
McGuire Center for Lepidoptera and Biodiversity, Florida Museum of Natural History, Gainesville, FL	ME	Contains specimens from L.P. Grey and others
Museum of Comparative Zoology, Harvard University, Cambridge, MA	ME, Maritimes	Contains specimens from W.D. Winter Jr., P.H.H. Gray, D.C. Ferguson, J.H. McDunnough, S. H. Scudder, and others

Table 1 *continued*

Source description	Database	Notes
National Museum of Natural History, Smithsonian Institution, Washington, D.C.	ME	Contains specimens from A.E. Brower, L.P. Grey, and others
Natural History Museum of Utah, Salt Lake City, UT	Maritimes	Contains specimens from R.P. Webster, H. Hensel, and others
New Brunswick Museum, Saint John, NB[a]	Maritimes	Contains specimens from W. McIntosh, A.W. Thomas, R.P. Webster, and many others
Nova Scotia Museum, Halifax, NS[a]	Maritimes	Contains specimens from D.C. Ferguson, J.H. McDunnough, B. Wright, K. Neil, J. Perrin, and many others
Nova Scotia Department of Natural Resources Reference Insect Collection, Shubenacadie, NS[a]	Maritimes	
Peabody Museum of Natural History, Yale University, New Haven, CT	ME, Maritimes	Contains specimens from A.E. Brower, D.C. Ferguson, J.H. McDunnough, L.P. Grey, F.M. Jones, S.I. Smith, J. Russell, and others
Royal British Columbia Museum, Victoria, BC[a]	Maritimes	Contains specimens from D.C. Ferguson and others
Royal Ontario Museum, Toronto, ON[a]	Maritimes	Contains specimens from J. Russell.
University of Connecticut, Storrs, CT	ME	Contains a small number of specimens from S. Sibley
University of Maine, Orono, ME[c]	ME	Contains specimens from M. Copeland, C.B. Hamilton, C.H. Fernald, and others
Université de Moncton collection entomologique, Moncton, NB	Maritimes	
University of New Brunswick Insect Collection, Fredericton, NB	Maritimes	
University of New Hampshire, Durham, NH	ME	Contains specimens from A.E. Brower, L.P. Grey, D.J. Lennox, C.F. dos Passos, and others
Private collections[d]		
John J. Albright, Brunswick, ME	ME	Collected many specimens for MDIFW during the 1980s and early 1990s; collection now housed at the Maine State Museum
Richard Boscoe, Lafayette Hill, PA	ME	Provided voucher records and other data on numerous species reared from ME
Kevin Byron, Kennebunk, ME	ME	Supplied photographs of several species from the area of Wells, ME

Source description	Database	Notes
John V. Calhoun, FL	ME	Supplied data on many specimens from ME
Louis-Emil Cormier, NB	Maritimes	Provided video records, primarily from Kent County, NB
Robert Dirig, Ithaca, NY	ME	Supplied data on all his records from ME
Denis A. Doucet, Riverside-Albert, NB	Maritimes	Supplied data on images and specimens, including records of many rare species
Jim Edsall, Dartmouth, NS	Maritimes	Supplied data on specimens and images, including records of many rare species
Gail Everett, South Portland, ME	ME	Provided many records from southern and central ME
Richard W. Folsom, Pittston, ME	ME	Supplied data on all his records from ME
Alex Grkovich, Boston, MA	ME	Supplied data on all his records from ME
Robert W. Harding, Summerville, PE	Maritimes	Supplied data on specimens collected in PE (now at the New Brunswick Museum), including many new provincial records
Henry Hensel, Edmundston, NB	Maritimes	Supplied data on specimens, most from the Edmundston area
Richard Hildreth, Holliston, MA	ME	Supplied data on all his records from ME, mostly Washington and Hancock Counties
Ross Layberry, Ottawa, ON	Maritimes	Supplied data on all his Maritimes records
Christopher Livesay, Brunswick, ME	ME	Supplied data on all his records from ME, including valuable material from undersurveyed northwestern regions
Robert R. Muller, Milford, CT	ME	Supplied data on his records from Piscataquis County
Jeffrey Ogden, Truro, NS	Maritimes	Supplied data on specimens from Nova Scotia
Michael A. Roberts, Steuben, ME	ME	Supplied data on all his records, mostly from Steuben and vicinity
Dwayne Sabine, Douglas, NB	Maritimes	Supplied data on specimens, mostly from NB (now housed at the New Brunswick Museum)
Phil Schappert, Halifax, NS	Maritimes	Provided photographic records from Nova Scotia
Dale Schweitzer, Port Norris, NJ	ME	Supplied data on all his records from ME, including some notable records from southern barren habitats
Anthony W. Thomas, Nashwaaksis, NB	Maritimes	Supplied data on rare specimens collected in NB after the release of *A Preliminary Atlas of the Butterflies of New Brunswick* (Thomas 1996)

Table 1 *continued*

Source description	Database	Notes
Martin Turgeon, Saint-Basile, NB	Maritimes	Supplied data on specimens from north-western New Brunswick
Kristine Wallstrom, NY	ME	Supplied photographic records from southern and central regions
Reginald Webster, Charters Settlement, NB	ME, Maritimes	Supplied data on all his records from many years of collecting throughout the Maritimes and ME
Existing databases		
MDIFW Rare Species Database	ME	Included record details for ~10 rare butterfly species tracked by MDIFW
AC CDC Species Observation Database	Maritimes	Included records from AC CDC surveys
iNaturalist	ME, Maritimes	Records with verifiable photos were included
eButterfly	ME, Maritimes	Records with verifiable photos were included; many Maritimes atlas participants submitted their records via eButterfly
Butterflies and Moths of North America	ME, Maritimes	Novel and rare species records with verifiable photos were included
Published literature		
See References[e]	ME, Maritimes	An attempt was made to consult all published references containing records from ME or the Maritimes

Note: CT = Connecticut, DC = District of Columbia, FL = Florida, MA = Massachusetts, ME = Maine, NB = New Brunswick, NH = New Hampshire, NJ = New Jersey, NS = Nova Scotia, NY = New York, ON = Ontario, PA = Pennsylvania, PE = Prince Edward Island, QC = Quebec, UT = Utah.

[a] Most or all Maritimes records from these museums were databased by Ross Layberry.

[b] Most of the Maine Department of Inland Fisheries and Wildlife (MDIFW) collection is now housed at the Maine State Museum, Augusta, ME.

[c] The University of Maine collection is now housed at the Maine Forest Service Insect Laboratory, Augusta, ME.

[d] Only private collections with significant numbers of specimen records are listed and generally those amassed prior to the start of the volunteer atlas period (before 2006).

[e] Full citations are given for each literature record in the Maine butterfly database and Maritimes butterfly data set.

species are also among the most challenging to survey, as they are less accessible to many community scientists. As such, targeted surveys by professional biologists were employed throughout the atlas period to improve our knowledge of some of the region's rarest species. With that said, it must be noted that significant contributions were made by community scientists to our knowledge of the status of several regional species of conservation concern, including the discovery of some important novel state and provincial records.

The majority of contributions to this atlas were community science surveys, which were independently organized for Maine and the Maritime Provinces. The Maine Butterfly Survey was conducted by the MDIFW (partnering with Colby College and University of Maine at Farmington [UMF]) from 2006 to 2015. The Maritime Butterfly Atlas, modeled closely after the Maine project, was conducted by the AC CDC from 2010 to 2015. The Maritime Butterfly Atlas also benefited immensely from the second Maritimes Breeding Bird Atlas, which concluded in 2010. Because of the breeding bird atlas, most Maritime naturalists were already acquainted with the notion of an atlas project and many were keen to continue documenting species for the purposes of science and conservation. Additionally, the Maritime Butterfly Atlas adopted the atlas square mapping units used by the breeding bird atlas. While coordinated separately, both the Maine and Maritime butterfly atlas projects shared similar and simple goals: (1) increased scientific understanding of the Acadian region's butterflies, and (2) increased public appreciation of the biology and conservation needs of these popular insects.

Core to the success of any volunteer atlas project is an enthusiastic cadre of community scientists trained in the use of standardized survey methodologies. To this end, both the Maine and Maritimes efforts recruited and instructed new volunteers throughout much of the project period. Workshops introduced participants to basic butterfly biology and taxonomy, survey site selection and documentation, specimen and photo voucher collection and preservation, and project communication tools (e.g., email distribution lists, annual newsletters, blogs, and website updates). Survey supplies were provided to workshop volunteers, including nets, collecting jars, glassine envelopes, field forms, voucher data cards, and participant instructional manuals. Materials for the Maritimes were also made available in French to help engage French-speaking participants in the region. Incidental training courses were also offered to individuals who could not attend a workshop, particularly those with previous experience studying Lepidoptera.

Volunteers were encouraged to survey with few habitat or geographic constraints. In Maine, a scientific collection permit is required to collect one of the many state-listed Endangered, Threatened, or Special Concern species and to collect any species in state parks or national wildlife refuges. In the Maritimes, a special permit is required to collect the Endangered Maritime Ringlet. The geographic unit employed for species distribution mapping of survey results in Maine was the township, of which there are 923 statewide, mostly uniform in shape and size (approximately 25,000 acres, or 100 km^2). The Maritime atlas employed the Universal Transverse Mercator grid, which consists of 10 km by 10 km squares. Both Maine and Maritime volunteers were asked to survey anywhere in their respective regions, though they were encouraged to place emphasis on remote townships and grid cells with low previous survey effort. Maritime volunteers were also solicited to adopt priority squares, a uniformly distributed subset of atlas squares, in hopes of distributing survey effort more evenly.

Volunteer participants presented a wide range of natural history skills and experience. Therefore, to help improve the scientific accuracy of project data sets, only verifiable records supported by specimen vouchers or close-up photographs were included in the atlas. Most voucher specimens collected in Maine are housed in the zoological collection of the Maine State Museum (Augusta), and those collected in the Maritimes are housed in the New Brunswick Museum (Saint John) and the Nova Scotia Museum (Halifax).

To help ensure standardized and comprehensive data collection, field forms were developed for both the Maine and Maritime atlas projects. Site Visit Forms (Figure 26) served as a tool for associating specimen and photo vouchers with specific site characteristics, including observer, date, time, weather, locality, habitat, species name, and abundance. A voucher number was assigned to each specimen or photo accompanying the Site Visit Form. Voucher numbers were assigned from a series of unique values that were preprinted on Voucher Data Cards (Figure 27), one of which was required to accompany every specimen or photo submission. In addition to assigning a unique specimen number, Voucher Data Cards also served as an abbreviated tool for recording site visit details when volunteers collected only an incidental observation of a

maine butterfly survey

SITE VISIT FORM

Return To:

Maine Butterfly Survey
Dr Ron Butler
173 High Street
Preble Hall - UMF
Farmington, ME 04938
(butler@maine.edu)

ID CODES

V = Voucher Specimen
R = Road-killed
P = Photo or Image

HABITAT CODES

Old Fields & Barrens
- ☐ Barren (little vegetation)
- ☐ Grass/Herbs (no shrubs)
- ☐ Few Shrubs
- ☐ Shrubland (shrubs > grass)

Agricultural
- ☐ Active Crop Land
- ☐ Pasture or Hayfield
- ☐ Orchard or Vineyard

Other Openings
- ☐ Power Line Right of Way
- ☐ Rock Outcrop
- ☐ Alpine Meadow

Forest Gaps
- ☐ Natural (blowdown, etc)
- ☐ Forestry (log landing, etc)
- ☐ Regenerating Area
- ☐ Woods Road

Upland Forest Type
- ☐ Deciduous Dominated
- ☐ Coniferous Dominated
- ☐ Mixed (<80% Dominated)
- ☐ Scrub Oak Woodland
- ☐ Pitch Pine Woodland

Wetlands
- ☐ Wet Meadow
- ☐ Fresh Marsh
- ☐ Salt Marsh
- ☐ Shrub Swamp
- ☐ Forested Swamp
  - ☐ Deciduous Dom.
  - ☐ Coniferous Dom.
  - ☐ Mixed (<80% Dom.)
- ☐ Sphagnum Bog/Fen

Road/Roadside
- ☐ Dirt
- ☐ Paved

Residential
- ☐ Lawn
- ☐ Garden
- ☐ Urban

Shoreline
- ☐ Lake (>10 acres)
- ☐ Pond (<10 acres)
- ☐ River (>10 ft. wide)
- ☐ Stream (<10 ft. wide)
- ☐ Ocean

- ☐ **Other** (use Habitat Desc.)

SITE NAME (Use same name for each visit.)

SITE CODE (Your 3 initials & 3 digit #. e.g. RGB001)

DELORME MAP PAGE & GRID (e.g., 03C2)

QUAD (if known)

TOWNSHIP

COUNTY

PARK OR PROTECTED AREA NAME

LANDOWNER (if known)

SITE LOCATION (Give Detailed Location w/ Landmarks or Attach Map)

New site for observer? ☐ Yes   ☐ No

LAT dec. degrees	LONG dec. degrees	ELEV ☐ m ☐ ft

OBSERVERS

CLOUDCOVER ☐ <10%   ☐ 10-50%   ☐ 50-90%   ☐ > 90%

WIND ☐ Still   ☐ Light   ☐ Mod   ☐ Strong

TEMP ☐F or ☐C

DATE (e.g., 4 July 2007)	START TIME	END TIME

SPECIES (USE "L" FOR LARVAE)	#	ID CODE	VOUCHER NUMBER

HABITAT DESCRIPTION (e.g., Dominant Vegetation and Details)

Additional notes on back?   ☐ Yes   ☐ No

Figure 26. The Site Visit Form used to document observer, site locality, and voucher (specimen or photo) details for the Maine Butterfly Survey. The Maritimes Butterfly Atlas used a nearly identical form.

Maritimes Butterfly Atlas – Voucher Data Card

Atlantic Canada Conservation Data Centre—Sackville, New Brunswick

Observer:		MBA Voucher #:	1380
Confirmer:		Site Name:	Province:
Date: (Format 24 May 2010)	Site Code:	Site Location:	MBA Square #:
Common Name:			
Scientific Name:			
Voucher Type: ☐ Specimen ☐ Photo ☐ Roadkill		Lat:	Long:
# of Vouchers:	# Observed:	RETURN TO John Klymko	
Voucher condition (F,S,W,E):		Atlantic Canada Conservation Data Centre Box 6416	
F = Fresh S = Slightly Worn W = Worn E = Extreme		Sackville NB E4L 1G6 jklymko@mta.ca (506) 364-2660	

HABITAT

Old or Regenerating Fields	Forest Gaps	Wetlands	Shoreline
○ Barren (little vegetation)	○ Natural (blowdown, etc.)	○ Sedge/Grass	○ Natural Lake (>5 ha)
○ Grass/Herbs (no shrubs)	○ Forestry (log landing, etc.)	○ Cattail/Reed	○ Pond (<5 ha)
○ Few Shrubs	○ Regenerating Area	○ Shrub	○ Reservoir
○ Shrubland (more shrubs than grass)	○ Woods Road or Trail	○ Beaver Pond	○ Ocean
○ Brush-hogged	○ Ski Slope	○ Wooded	**River or Stream**
Cuts/Year _____ Last Cut _____	**Forest Type**	☐ Deciduous	○ River (>10 feet wide)
Agricultural	○ Deciduous Dominated	☐ Coniferous	○ Stream (<10 feet wide)
○ Active Crop Land	○ Coniferous Dominated	☐ Mixed	
○ Pasture or Hayfield	○ Mixed Woods	○ Bog/Fen	
○ Orchard or Vineyard			

Other Openings	**Road/Roadside**	**Residential**	○ Other (please describe)
○ Power Line Right-of-Way	○ Dirt	○ Lawn	_____
○ Rock Outcrop	○ Paved	○ Garden	_____
○ Coastal Barren		○ Urban	_____

NECTAR PLANT(S)	HOST PLANT(S)

NOTES

Figure 27. The Voucher Data Card used to accompany all specimen vouchers submitted to the Maritimes Butterfly Atlas. The Maine Butterfly Survey used a nearly identical card. Each card was printed with a unique number that is associated with specimen and locality details in the Site Voucher Form and master database.

butterfly (e.g., road-killed specimen) without populating a more detailed Site Visit Form. Photographs were submitted as digital files or prints, often by inserting digital images, appended to a Site Visit Form. All specimen vouchers were dried and stored flat inside of a glassine envelope, accompanied by the uniquely associated Voucher Data Card. Once submitted, all project data were entered into an electronic database, maintained for Maine by MDIFW and UMF, and for the Maritimes by the AC CDC.

The voucher identifications made by participants were subject to expert review by one or more atlas project coordinators. For Maine, most identifications were confirmed by Reginald Webster, while most vouchers for the Maritimes were confirmed by Jim Edsall and John Klymko, with additional assistance from R. Webster. Some difficult genera often precluded the ability to verify species determinations with photographs alone. Maine vouchers of both the *Celastrina* blues (two recognized species) and the *Phyciodes* crescents (two species) were reviewed twice to improve the accuracy of species determinations.

Naturalists are increasingly submitting observations of biota, including butterflies, to online data portals designed to engage the next generation of community scientists. eButterfly (e-Butterfly 2020) is an online database of butterfly observations, populated with records contributed by the public. The Maritime Butterfly Atlas was an early collaborator with eButterfly. When eButterfly was launched in 2011, Maritime participants were encouraged to submit their digital photo records through that database. Over the next 5 years, eButterfly became an important data source for the Maritime Butterfly Atlas and, to a lesser extent, the Maine Butterfly Survey. iNaturalist (iNaturalist 2020) is a similar online database for all taxa, and many community scientists now use it to submit observations captured with smartphone cameras. It, too, has become a significant source of photo-supported butterfly records for the Acadian region, and its records were mined for this project. Finally, selected novel and rare species records were included from the Butterflies and Moths of North America, an online effort to host and manage data for all North American butterflies and moths from Canada to Panama.

For the purposes of this atlas, records were separated into two periods: historical and modern. Following NatureServe conservation guidelines (NatureServe 2002), records more than 20 years old are considered historical. The modern era starts in 1996, which is 20 years before 2015, the last official year for community science contributions to the atlas. Several species that were previously known only from historical records in Maine, such as the Greenish Blue and the Satyr Comma, are now supported by modern documentation as a direct result of the atlas.

SPATIAL AND TEMPORAL PATTERNS OF ACADIAN BUTTERFLIES

Historical and Modern Contributions

Both the Maine and the Maritimes butterfly databases contain records from diverse sources, including historical published accounts, private and institutional collections, online databases, and professional surveys. However, the majority of records in both regions were a result of submissions by community scientist volunteers during the Maine Butterfly Survey (2006–2015) and the Maritimes Butterfly Atlas (2010–2015), hereafter referred to as the Maine Survey and the Maritimes Atlas respectively (Figure 28). The Maine database presently contains 32,534 unique, confirmed specimen, image, or published records, of which 16% are historical and 84% are modern records (post-1995) (Figures 29, 30). The Maritimes database contains 48,295 unique, confirmed records of which 19% are historical and 81% are modern (Figures 29, 30). The

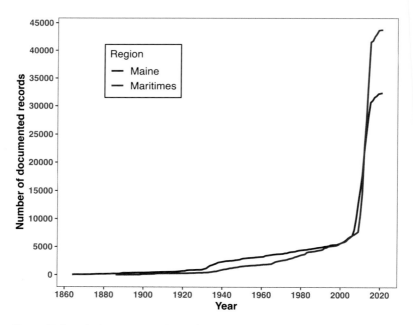

Figure 28. Cumulative increase in butterfly records in Maine and the Maritimes indicating the recent dramatic increases associated with both community science projects.

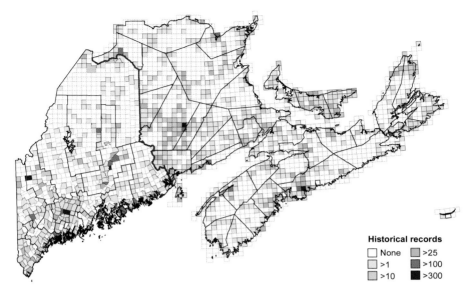

Figure 29. Historical (pre-1996) butterfly records in Maine (by township) and the Maritimes (by grid square).

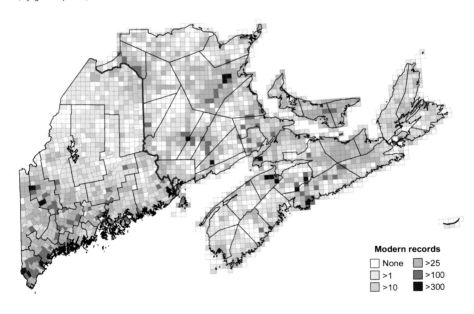

Figure 30. Modern (post-1995) butterfly records in Maine (by township) and the Maritimes (by grid square).

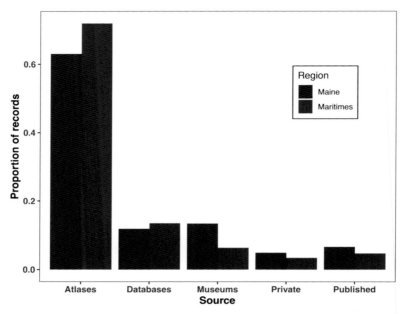

Figure 31. Proportion of butterfly records from five data sources for Maine and the Maritimes. Recent volunteer survey efforts in both regions contributed the majority of the records. Note that not all data sources are exclusive; for example, some existing databases were populated with records from museum collections.

Maine Survey contributed 23,950 (74%) of total confirmable records in the Maine database, of which 12% were new county records and 64% were new township records. Moreover, eleven species were added to the Maine state butterfly list, including a notable new United States record: the Short-tailed Swallowtail (Table 2). The Maritimes Atlas contributed 34,323 (71%) of total records in the Maritimes database, contributing thirteen new provincial records and three new species records for the region (Table 2). Private and institutional collections, online databases (such as eButterfly and iNaturalist), other existing databases and published accounts provided the remainder of the records (Figure 31). The Maine Survey had a higher proportion of records from collections and publications than did the Maritimes Atlas, while the latter included more records from existing databases.

During the Maine Survey, 364 observers contributed specimen or photo vouchers. One-third of observers contributed in more than 1 year, 18% submitted records in 3 or more years, and fourteen individuals

Table 2. New state and provincial butterfly species and subspecies records documented during the Maine (2006–2015) and Maritimes (2010–2015) project periods

Common name	Scientific name	Maine	Maritimes
Long-tailed Skipper	*Urbanus proteus proteus*	ME	
Wild Indigo Duskywing	*Erynnis baptisiae*	ME	
Two-spotted Skipper	*Euphyes bimacula bimacula*		NS
Ocola Skipper	*Panoquina ocola ocola*	ME	NB[a]
Fiery Skipper	*Hylephila phyleus phyleus*	ME	NB
Crossline Skipper	*Polites origenes origenes*		NB[a]
Sassacus Skipper	*Hesperia sassacus sassacus*		PE
Mulberry Wing	*Poanes massasoit massasoit*	ME	
Black Dash	*Euphyes conspicua orono*	ME	
Dusted Skipper	*Atrytonopsis hianna hianna*	ME	
Short-tailed Swallowtail	*Papilio brevicauda gaspeensis*	ME[b]	NB[a]
Clayton's Copper	*Tharsalea dorcas claytoni*		NS
White M Hairstreak	*Parrhasius m-album*	ME[c]	
Eastern Tailed-Blue	*Cupido comyntas comyntas*		NS, PE
American Snout	*Libytheana carinenta bachmanii*		NB
Variegated Fritillary	*Euptoieta claudia*		NS
European Peacock	*Aglais io io*		NS[a]
Eastern Comma	*Polygonia comma*		PE

Note: ME = Maine, NB = New Brunswick, NS = Nova Scotia, PE = Prince Edward Island.
[a] New Maritimes record.
[b] New U.S. record.
[c] Record added after Maine survey period.

participated in 8 or more years of the project. The mean number of records submitted per observer throughout the 10-year survey was 63.2 (range = 1–3437). The Maritimes Atlas had 497 contributors; 42% participated in more than 1 year, 24% participated in more than 2 years, and thirty-nine individuals participated in all 6 years (mean records per observer = 75.6; range = 1–2699). About 28% of Maine Survey records and 53% of Maritimes Atlas records are images. In the future, the proportion of verifiable butterfly records composed of images is expected to rise significantly given the availability of digital cameras and the increasing popularity of crowd science platforms such as iNaturalist and eButterfly.

The distribution of records per township or grid square reveals some interesting demographic patterns, with generally heavier sampling effort in areas of higher human population density, greater road access, and (especially in Maine) proximity to research institutions (Figures 29, 30). This pattern is less pronounced in the modern era, likely because of greater participation in butterfly observation by the general public and efforts to encourage survey participants to target poorly represented areas.

Abundance and Richness

Patterns of Abundance—For most higher taxa, species abundance patterns vary widely with typically a few dominant species and a greater number of rare species (Fisher et al. 1943). This pattern is apparent in the modern records (post-1995) for butterfly fauna of both Maine and the Maritimes (Figure 32). Maine has 27,012 modern records for 102 butterfly species that are considered residents (ninety-five species) or frequent colonists (seven species). The fifteen most abundant of these species constitute 52% of these records (Table 3), and 75% of these records can

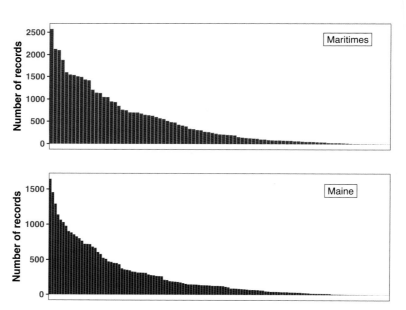

Figure 32. Number of records for butterfly species in Maine and the Maritimes, ordered in decreasing abundance; each bar represents a single species.

Table 3. Fifteen most often documented modern (post-1995) resident and frequent colonist butterfly species occurring in Maine (27,012 records) and the Maritimes (39,187 records)

Maine	Number of records	Maritimes	Number of records
Common Ringlet	1601	Northern Crescent	2277
Clouded Sulphur	1325	Northern Azure	2014
Northern Crescent	1174	Clouded Sulphur	1829
Cabbage White	1078	Cabbage White	1755
Northern Azure	961	Common Ringlet	1579
Canadian Tiger Swallowtail	933	European Skipper	1399
Great Spangled Fritillary	880	Canadian Tiger Swallowtail	1388
White Admiral	850	Great Spangled Fritillary	1352
American Lady	846	Hobomok Skipper	1326
European Skipper	822	Atlantis Fritillary	1307
Atlantis Fritillary	760	Common Wood-Nymph	1215
Monarch	723	Red Admiral	1176
Dun Skipper	717	Long Dash	1044
Long Dash	710	Silvery Blue	1033
Peck's Skipper	670	Viceroy	978

be attributed to only the twenty-eight most abundant species. The least common fifty species account for only 7% of Maine records. A similar pattern of commonness and rarity is evident for the Maritimes, where seventy-six resident and six frequent colonist species are confirmed (39,187 modern records). There, the fifteen most abundant species compose 55% of modern records (Table 3), and the most abundant twenty-six species account for about 75% of the records. The fifty least common species account for only 15% of Maritimes records. The two regions share ten of the fifteen most common butterfly species, including two of North America's most prolific introduced species—the European Skipper and Cabbage White (Table 3). Only four species are shared in common among the least frequently documented residents in Maine and the Maritimes: Hoary Comma, Acadian Hairstreak, Greenish Blue, and Western Pine Elfin (Table 4).

Table 4. Fifteen least often documented modern (post-1995) resident and frequent colonist butterfly species in Maine (27,012 records) and the Maritimes (39,187 records)

Maine	Number of records	Maritimes	Number lof records
Acadian Hairstreak	15	Hoary Comma	39
Edwards' Hairstreak	12	Acadian Hairstreak	36
Hoary Elfin	12	Compton Tortoiseshell	32
Frigga Fritillary	8	Arctic Fritillary	30
Hessel's Hairstreak	8	Bog Fritillary	28
Western Pine Elfin	8	Maritime Ringlet	28
Hoary Comma	6	Sassacus Skipper	26
Western Tailed-Blue	6	Clayton's Copper	25
Arctic Fritillary	6	Silver-spotted Skipper	22
Katahdin Arctic	5	Gray Hairstreak	20
Juniper Hairstreak	4	Greenish Blue	9
Satyr Comma	2	Western Pine Elfin	9
Greenish Blue	1	Crossline Skipper	8
Short-tailed Swallowtail	1	Early Hairstreak	6
Southern Cloudywing[a]	1	European Peacock	6

[a] It is unclear if the Southern Cloudywing is a resident species (see the species account for details).

The most often reported resident butterfly species and frequent colonists to the Acadian region share characteristics that make them easily documented. They are all widespread in the region and common in a variety of habitats, most notably old fields and gardens, favorite and accessible haunts of butterfly enthusiasts. Additionally, many are common in July and August, when people generally have the most time to explore nature. For example, while the Canadian Tiger Swallowtail and Hobomok Skipper peak in abundance in June, both are relatively common in July, and the former is especially spectacular. In contrast, our least often documented species are generally restricted to specialized habitats (e.g., Hessel's Hairstreak in Atlantic White Cedar swamps) or remote regions (e.g., Western Pine Elfin in subboreal woodlands), which are relatively inaccessible to community scientists. Additionally, many infrequently documented species are found early in the season before some butterfly watchers are active (e.g., all seven elfin species), or they fly only for a very brief period (e.g., one or two weeks for Bog and Frigga fritillaries).

Patterns of Richness—Butterflies are found only where you search for them and, not surprisingly, patterns of modern butterfly species richness in our region tend to reflect modern sampling effort (Figure 30). Nonetheless, there is a clear pattern of higher richness in southern and coastal Maine than in the rest of the Acadian region (Figure 33). Temporally, a review of butterfly species accumulation curves for both Maine and the Maritimes reveals that total species richness has increased in a gradual, nearly linear fashion over the past 120 years (Figure 34). This pattern suggests that while our coverage of the region's butterfly fauna is relatively complete, new species records continue to be added in the modern era. Observers should be vigilant for yet additional discoveries, especially those species highlighted in the chapter Butterflies of Possible Occurrence.

North America can be divided into ecologically distinct regions based on patterns of geology, hydrology, biology, and climate. The diversity and composition of butterfly communities may be related to these ecoregional divisions (Andrew et al. 2011). We examined the Acadian region's butterfly assemblage based on the ecological subregional map scheme developed by The Nature Conservancy and Nature Conservancy Canada (Anderson et al. 2006) (Figure 35). The highest species richness occurs in the Acadian "Uplands," while the lowest diversity occurs in the Temiscouata Hills—St. John Uplands. Interestingly, when accounting for differences in area among subregions, the North Atlantic Coast of southwestern Maine signals a diversity hotspot, hosting over three-fourths of the modern Acadian butterfly fauna in a geographic area that is less than 2% of the total region. In terms of geopolitical richness, Maine has 113 confirmed modern-era butterfly species, followed by New Brunswick (87), Nova Scotia (74), and Prince Edward Island (60). Lower species richness in the Maritimes also exists in other well-studied biota, including vertebrates (NatureServe Explorer 2021) and vascular plants (Angelo and Boufford 2016; AC CDC, unpublished data). A general pattern of lower species richness in the Maritimes is likely attributable to southern

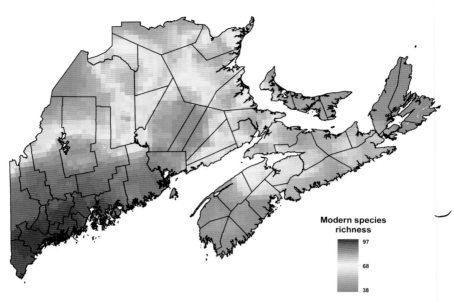

Figure 33. Modern (post-1995) butterfly species richness heat map for Maine and the Maritimes (including county boundaries).

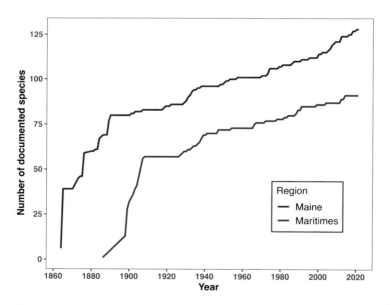

Figure 34. Species accumulation curves for butterfly species recorded for Maine (118 total species) and the Maritimes (93 total species). The value for documented species is counted as one in cases where there is more than a single subspecies in the region (e.g., *Plebejus idas empetri* and *P. i. scudderii*).

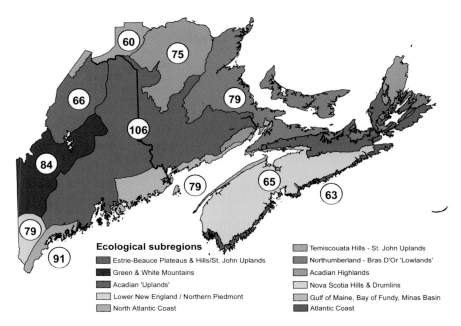

Figure 35. Modern (post-1995) butterfly species richness (in circles) in eleven ecological sub-regions (as delineated by Anderson et al. 2006) for Maine and the Canadian Maritimes.

Maine's warmer climate, as well as the peninsular and insular nature of the Maritime Provinces, a geographic pattern associated with lower butterfly species richness elsewhere in North America (Brown 1987; Brown and Opler 1990).

Notable Species-Specific Patterns

Phenology—Temperature and day length are the chief drivers of natural phenology (i.e., the timing of life-cycle events) for both plants and animals. Given the close association between the phenology of larval food plant growing season and butterfly reproduction (Toftegaard et al. 2019), taken together with the fact that more southern subregions generally experience earlier spring leaf out, it is reasonable to expect some differences in butterfly flight periods across the Acadian region. To test for the presence and magnitude of any subregional differences, we analyzed data for the six most abundant univoltine species (one generation per year; Table 5).

Table 5. Mean flight date of the six most common univoltine species (arranged phylogenetically) for each of the eleven Acadian subregions

Subregion	European Skipper	Canadian Tiger Swallowtail	Silvery Blue	Great Spangled Fritillary	Atlantis Fritillary	Common Wood Nymph
Lower NE	6/29	6/17	5/25	7/24	7/06	8/01
Green Mts.	7/07	6/15	6/13	7/24	7/17	8/04
Temisc. Hills	7/11	6/16	6/17	7/25	7/18	7/26
Acad. Uplands	7/09	6/17	6/16	7/25	7/19	8/05
N. Atlantic Coast	6/28	6/19	5/31	7/31	7/28	8/07
Nova Scotia Hills	7/11	6/16	6/19	8/01	7/30	8/05
Acad. Highlands	7/19	6/22	6/13	7/28	7/26	8/10
Northumb.	7/14	6/20	6/23	7/26	7/23	8/06
Gulf of Maine	7/12	6/20	6/17	7/31	7/22	8/13
Estrie Plateau	7/14	6/17	6/21	7/28	7/24	8/13
Atlantic Coast	7/19	6/20	6/19	8/07	8/01	8/12

Notes: See Figure 10 for the full name of each subregion. The subregions are arranged with earlier mean dates at the top to later dates at the bottom to facilitate interpretation of temporal patterns. ANOVA and post-hoc Tukey HSD tests were used to test each species for differences in mean flight dates among subregions. A Bonferroni correction for multiple comparisons was applied to produce an experiment-wise critical value of $p = 0.05$. Each ANOVA indicated significant differences among subregions ($p < 0.01$ in all cases). A summary of the post hoc pair-wise tests is provided in the text.

All six butterfly species showed significant differences in mean flight dates (1864–2021) among the subregions (ANOVA, $p < 0.01$ in all cases), although the number of differences varied widely among species. For the European Skipper, thirty-one of the fifty-five pair-wise comparisons were significant. Most of the nonsignificant flight date comparisons were among the Nova Scotia Hills, Gulf of Maine, Northumberland, and Estrie Plateau. The Canadian Tiger Swallowtail showed much less variation in mean flight date (only six significant pair-wise contrasts), all involving differences between the latest means with some of the earliest means (e.g., Lower New England Piedmont with the Acadian Highlands). Forty percent of the pair-wise contrasts were significant for Silvery Blue. Most of these significant differences involved the early emerging populations in Lower New England Piedmont and the North Atlantic Coast with later-emerging populations. The remaining nine subregions showed no differences.

The three brushfoots showed similar patterns in mean flight date differences. For the fifty-five pair-wise contrasts for each species, the Great Spangled Fritillary had only eight significant contrasts while the Atlantis Fritillary had twelve and Common Wood Nymph had eleven. In all cases, the differences were between the earliest and latest emerging populations.

A couple of broad phenological trends are evident by looking for patterns among subregional mean butterfly flight dates. The expected trend of earlier mean flight dates in the southern portion of our region is evident, for example in the Lower New England and Green and White Mountains subregions. Perhaps a less expected trend is that coastal butterfly populations emerge later than inland populations. All six species had later mean flight seasons along the coastally situated Gulf of Maine/Bay of Fundy, Atlantic Coast, and Northumberland subregions compared with adjacent upland areas (Table 5). This pattern is likely attributed to the moderating influence of the ocean on the Acadian region's coastal climate, whereby near-coastal areas are slower to warm in the spring and often attain lower maximum summer temperatures than areas farther inland (Raymond and Mankin 2019).

Incontrovertible evidence demonstrates that the earth's climate is warming (IPCC 2019), and numerous studies have documented earlier spring plant germination, leaf out, and flowering dates as well as extended fall photosynthesis and delayed leaf senescence (e.g., Miller-Rushing and Primack 2008; Amano et al. 2010; Gallinat et al. 2015). Similarly, climate-associated changes in butterfly phenology and ranges have been widely reported in the literature (e.g., Roy and Sparks 2000; Polgar et al. 2013; Macgregor et al. 2019). Changes in distribution and abundance to some of our region's most climate-vulnerable butterfly species and other biota are predicted (Whitman et al. 2013) and already documented in nearby Massachusetts (Breed et al. 2013).

In addition to examining spatial differences in butterfly phenology across our region, we investigated potential temporal changes in phenology. Specifically, we examined the relationship between annual median flight date and year for the ten most abundant univoltine species in both Maine and the Maritimes (as listed in Table 6). These species had at least 500 records in each area. Medians were employed in this analysis because of the generally low sample sizes available for flight data in the historical era and because medians are less subject to skew from low or

Table 6. Results of regression analyses to test for changes in median date over time for the first and last deciles of flight records for univoltine Acadian region butterflies with at least 500 records

Species	Maine		Maritimes	
	First decile	Last decile	First decile	Last decile
European Skipper	Negative	NS	NS	Positive
Long Dash	NS	NS	NS	Positive
Dun Skipper	NS	Positive	NS	NS
Canadian Tiger Swallowtail	NS	Positive	Negative	Positive
Silvery Blue	Negative	NS	Negative	NS
Great Spangled Fritillary	NS	Positive	Negative	Positive
Atlantis Fritillary	NS	Positive	Negative	NS
Harris' Checkerspot	NS	NS	Negative	NS
Northern Pearly-Eye	Negative	Positive	Negative	Positive
Common Wood-Nymph	NS	Positive	Negative	Positive

Notes: Negative refers to a statistically significant slope indicating earlier median date in modern years (post-1995); positive refers to a statistically significant slope indicating later median dates in modern years; NS refers to a nonsignificant slope (not different from zero). Bonferroni's correction was used to adjust the individual p-values to produce an experiment-wise critical value of $p = 0.05$.

high outlier data. Using this approach, we found no evidence of changes in phenology, with none of the ten species analyzed showing a significant relationship between time and median flight date.

A more nuanced analysis of the Acadian region's flight data suggests subtle but significant impacts of global warming on butterfly phenology. The extreme portions of a data distribution are often informative for detecting emerging trends, including butterfly phenology (Polgar et al. 2013). As such, we extracted the earliest 10% (the first decile) of flight records for the ten most common univoltine butterfly species and performed regressions of median yearly flight data versus time. Similarly, we extracted the latest 10% (the last decile). Only one species, the Northern Pearly-Eye, showed significant regressions for each region–decile combination (Figure 36). For the first decile, the slope is negative, indicating earlier contemporary flight. The regressions for the last decile show a positive slope, suggesting a later contemporary flight as well. For the remaining species (Table 6), eight of the ten species analyzed (three

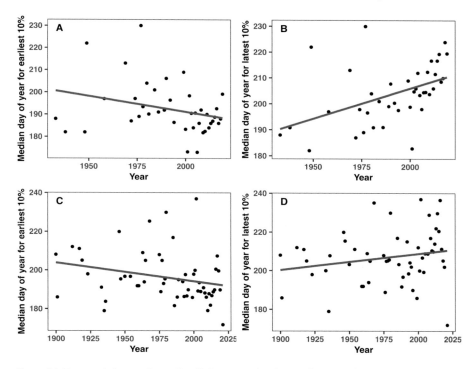

Figure 36. Temporal changes in median flight season for the Northern Pearly-Eye in the Acadian region using linear regression. (A) Earliest 10% of records for the Maritimes, (B) Latest 10% of records for the Maritimes, (C) Earliest 10% of records for Maine, (D) Latest 10% of records for Maine. All regressions are statistically significant ($p < 0.05$).

of ten from Maine; seven of ten from the Maritimes) showed a negative slope for the first decile, evidence of accelerated emergence in recent years. Similarly, for the last decile, eight of the ten species in Maine and the Maritimes are being recorded later in current years. While this latter pattern needs further investigation, it may be a result of extended flowering times of important nectar plants or amelioration of the late-season climate. Differences in collecting methodology are also likely a significant factor. Lepidopterists building personal or museum collections tend to target recently emerged adults as they make superior specimens. Because the historical data set is made up largely of specimens from such collections, it may be biased toward the beginning of flight periods.

If the behavior of all members of the population follows the pattern of these extreme portions of the flight distribution, we can expect earlier first appearances of these univoltine species and later persistence. The effect would thus be a lengthening of the flight season. Ironically, if these opposite effects are nearly equal in magnitude, the mean emergence date would not change despite overall effects on butterfly phenology related to climate change.

To summarize, when examining the full flight record for our region's ten most common univoltine butterfly species, we detected no significant changes in the median flight dates. However, phenological effects were noted when analyzing the first and last decile (10%) of flight records (Table 6). These results suggest that climate change is potentially affecting the biology of numerous butterflies in the Acadian region (ten of ten analyzed here), a pattern consistent with phenological changes reported for butterfly communities elsewhere, including in Massachusetts (Polgar et al. 2013), California (Forister and Shapiro 2003), and the United Kingdom (Roy and Sparks 2000).

Changes in Distribution—Examination of historical and modern era species maps reveals compelling cases of recent distribution changes in some species. For example, the Wild Indigo Duskywing and Broadwinged Skipper have clearly expanded their ranges in response to the continuing spread of exotic larval food plants (Webster and deMaynadier 2005; Gobeil and Gobeil 2016). Northward expansions of the Black Dash, Mulberry Wing, Silver-spotted Skipper, Delaware Skipper, Bronze Copper, and White M Hairstreak are also apparent. While some of these expansions may be related to a warming climate, others may be in response to changes in land management. For example, recent increases in the abundance and distribution of the Bronze Copper in the Maritimes may partly be explained by the increase in available habitat associated with waterfowl impoundments, such as those established by Ducks Unlimited. There are, however, fewer convincing examples of southward distribution expansions, such as that possibly being experienced by the Salt Marsh Copper and Western Tailed-Blue.

In some instances, increased search efforts may be partly responsible for the outward appearance of range expansions. For example, targeted surveys of swamps and salt marshes in recent years may be responsible for the apparent expansion of the Appalachian Brown and Salt Marsh Copper, respectively. In others, search effort is almost certainly responsible for perceived changes in range. The apparent increase in the modern distributions of the Bog Fritillary and Clayton's Copper is undoubtedly attributable to increased survey efforts focused on the specialized peatlands preferred by these species of regional interest.

In addition to expansions, there are some notable northward range contractions, including that of the Hoary Comma, Arctic Fritillary, and (in Maine) the Hoary Elfin. Such changes may be associated with climate warming and habitat loss but require further study. The significant contraction in the range of the Greenish Blue is less easily explained. This species presumably colonized the Maritimes in the early twentieth century, was well established in Maine by about 1950, and was expected to continue its expansion southward (Klots 1951). In a stunning reversal, it virtually disappeared from the region within a few decades, and it has been recorded at only two locations recently, one in New Brunswick and one in extreme northern Maine.

The recent establishment of the European Peacock in Nova Scotia probably represents an introduction rather than a natural range expansion. Given the localized nature of species records and their isolation from other populations, it seems likely that this species was transported into the Halifax, Nova Scotia, area. Halifax is a busy international port, so the importation of stowaway butterflies on a European cargo ship is a real possibility. It could also have been accidentally carried in from Montreal, Quebec, where the European Peacock has been established since the 1990s. It remains to be seen whether this exotic butterfly becomes a permanent resident of our region's fauna.

Stray and Historical Species

We define stray (or transient) butterflies as those that occur rarely or infrequently in our region, but they are not known to reproduce. The number of stray species reported from Maine and the various Maritimes Provinces are ten and thirteen, respectively (Figure 37). The number of new transients reported per decade (since the first available record in 1880) is relatively consistent, except for an upturn in the early 2000s, which is likely explained by the greater amount of survey effort during the recent Acadian atlas projects (Figure 28). Interestingly, while the total number of documented transient species has increased over time, the historical data set contains a greater proportion of stray species records than does the modern era data set. In Maine, 0.39% of historical records

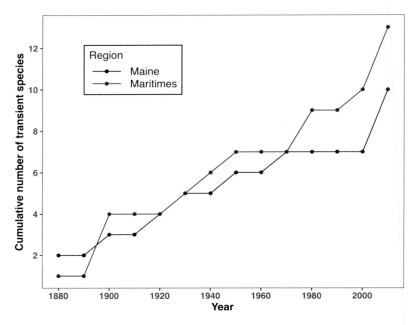

Figure 37. Decadal accumulation curves for transient butterfly species recorded for Maine and the Maritimes.

are of strays, compared with 0.06% of modern records, while in the Maritimes, 0.14% of historical records are of transients, compared with 0.11% of modern records. This pattern is likely attributable to differences in collecting techniques. Historically, a greater percentage of records were documented by lepidopterists who placed more scientific (and sometimes monetary) value on uncommon or rare species, including strays. Also, the investment of time and space needed to mount and preserve specimens meant that common species were less frequently collected. In contrast, the composition of the modern era data set was heavily influenced by the Maine Survey and Maritimes Atlas projects, for which participants sought specimens or photos of *all* species present at a survey site, thereby reducing the proportion of transients among their records.

Stray species with modern records in Maine, and their year of first discovery, include the Little Yellow (1880), Pipevine Swallowtail (ca. 1900), Cloudless Sulphur (1922), American Snout (1974), Long-tailed Skipper (2012), Fiery Skipper (2012), and Ocola Skipper (2016). Maritimes transients with modern records include the Fiery Skipper (1947), Common Buckeye (1954), American Snout (2004), and Ocola Skipper

(2014). Perhaps as expected, the records of most nonbreeding, one-way migrants (e.g., American Snout and Fiery Skipper) are irregularly dispersed throughout the region, with the highest density along the coast, including islands (e.g., Williams 2019). This contrasts with the geographically clustered records characteristic of species showing apparent range expansions into our region (e.g., Black Dash and Mulberry Wing). The causes for increases in the ranges or abundance of stray species in our region and elsewhere are species specific and likely linked to such factors as a warming climate and changes in land-use patterns that affect regional habitat availability.

The Common Sootywing (two records), Orange-barred Sulphur (two records), and Checkered White (one record) were only historical rare transients in the Acadian region and have not been recorded in modern times. The Common Sootywing and Orange-barred Sulphur are uncommon at the periphery of their ranges, while the Checkered White has declined precipitously throughout the Northeast. Historically, the Early Hairstreak was probably an uncommon resident in Maine and the Maritimes, but this species has been documented from only a few localities in New Brunswick since 1996, leaving uncertainty about its current status and distribution in the Acadian region. While only a rare stray in New Brunswick, the Regal Fritillary was historically a relatively widespread resident in southern Maine. This distinctive butterfly was last documented in Maine in 1941 and is now considered extirpated in our region and virtually throughout its entire eastern range. The Regal Fritillary occurred at the periphery of its range in Maine, where ecological succession and changes in land use likely contributed to its disappearance prior to its decline elsewhere in the East during the late twentieth century.

Comparisons of historical and modern species records are often complicated by differences in sampling effort and methods. Both the Maine Survey and Maritimes Atlas projects benefited from the recruitment and training of large numbers of volunteers, geographically targeted surveys encouraged by digital communications, improved road access to remote regions, and better information about the habitat requirements of many species provided by recent butterfly field guides. As a result, direct comparisons between historical and modern differences in species distribution and abundance must be interpreted cautiously to ensure that perceived changes are not simply the products of more intensive, modern survey efforts and enhanced recognition of cryptic species by trained volunteers. Genuine butterfly population fluctuations are usually driven by changes in critical habitat elements associated with natural disturbance and succession, such as larval food plant abundance, nectar availability, microclimate, and predator or parasite populations. Increasingly in the modern era, human settlement patterns and land-use practices have become primary forces for more permanent habitat change resulting from sprawling development, intensification of agriculture, widespread herbicide and pesticide use, and global climate change. These and other anthropogenic factors affecting regional butterfly populations are further considered in the next chapter, Butterfly Conservation in the Acadian Region.

CLOSING THOUGHTS

After more than 150 years of study, our understanding of the status, distribution, and (to a lesser extent) biology of butterflies in the Maine and Maritimes region is now relatively comprehensive, especially following the intensive surveys by trained community scientists. It is unlikely that the Acadian region currently hosts many breeding populations of undocumented butterfly species, with a few notable exceptions of those in specialized habitats (see Butterflies of Possible Occurrence). Rather, future additions to our region's butterfly fauna are more likely to be occasional strays (e.g., Sleepy Orange, Southern Dogface), introduced species (e.g., European Common Blue), or more southerly species undergoing range expansion (e.g., Zabulon Skipper, Red-banded Hairstreak). The latter phenomenon is especially likely given current climate warming projections for northeastern North America. Sustained vigilance for new species discoveries is worthwhile, but there is also much more that can be learned about the distribution and biology of our known fauna. The Acadian landscape is vast, and there are remote areas that remain poorly sampled, especially in northwestern and eastern Maine, northern New Brunswick, and southwestern Nova Scotia. We hope this book serves not only as a summary of existing knowledge but also as a tool for inspiring generations of further butterfly field study.

BUTTERFLY CONSERVATION IN THE ACADIAN REGION

If all mankind were to disappear, the world would regenerate back to the rich state of equilibrium that existed ten thousand years ago. If insects were to vanish, the environment would collapse into chaos.

E.O. Wilson

GLOBAL AND REGIONAL PATTERNS OF INSECT AND BUTTERFLY DECLINE

Recent scientific studies have shone a spotlight on the issue of global insect declines. Manifested as contractions in species range, reduced abundance, and even localized extirpation, insect declines have now been documented across a wide range of aquatic and terrestrial groups and from nearly every continent (reviews by van Klink et al. 2020; Wagner 2020; Sánchez-Bayo and Wyckhuys 2019, 2021). Insects serve a fundamental role in anchoring a wide range of ecosystem services, including nutrient cycling, biological pest control, food for other wildlife, and pollination of more than 80% of the world's flowering plants (Losey and Vaughan 2006; Dirzo et al. 2014; Tallamy 2019). As a result, their decline has attracted the attention of not only entomologists and ecologists but also environmentalists, politicians, and the global media. As a dominant form of animal life, measured by biomass or diversity, it is not an exaggeration to consider the health of global insect populations as a bellwether of the condition of our planet's biological diversity writ large, no less the fate of humankind. For this reason, the eminent Harvard biologist E. O. Wilson (1987) bluntly cautions that "we need invertebrates, but they don't need us."

Despite their ecological importance, there is little question that our understanding of the diversity, let alone the status and trends, of most insect groups lags far behind that of birds, mammals, and other vertebrates. Fortunately for our purposes, one of the best-studied insect orders worldwide is the Lepidoptera (butterflies and moths), which translates into a greater number of long-term population monitoring efforts. Several studies, conducted mostly in Western Europe and North America, indicate that Lepidoptera are one of several insect groups signaling strong population declines. For example, in a recent meta-analysis of twenty-nine separate long-term monitoring studies (median period = 33 years), Sánchez-Bayo and Wyckhuys (2021) reported that 50.6% of the butterfly and moth species analyzed exhibited significant population declines, a value exceeding that of any of the other ten insect orders considered, as well as that for global vertebrates (Dirzo et al. 2014). This calculation is alarming, suggesting that as many as half of the world's Lepidoptera—one of the most species-rich groups of living animals (Wagner 2000)—are declining in the modern era. However, we must be cautious in extending the results of taxon-specific declines from one geographic area to another. To better understand what this troubling pattern might mean for our Acadian regional butterfly fauna, it is worth taking a closer look at studies of long-term dynamics in butterfly communities closer to home, both geographically and taxonomically.

Although we do not yet have an analysis of butterfly population trends specific to Maine or the Maritimes, two of North America's most rigorous long-term monitoring studies were conducted relatively close to our borders. Analyzing population trends of 100 butterfly species observed in Massachusetts by volunteers of the Massachusetts Butterfly Club between 1992 and 2010, Breed et al. (2013) reported significant declines in thirty-nine species that have geographic ranges that include the Acadian region. This equates to 1.6 times the number of shared species reported to be experiencing population increases in Massachusetts. Slightly farther afield, one of the most extensive and systematic butterfly monitoring efforts in North America was organized by the Ohio Lepidopterists. Using data gathered by trained volunteers assigned to monitor transects distributed across Ohio during the period 1996 to 2016, Wepprich et al. (2019) reported an annual rate of butterfly population decline of 2%, equivalent to a total cumulative decline in overall abundance of 33% for the 21-year period. As with the Massachusetts study, these results are sobering, especially when one considers the significant overlap in the composition of the Ohio butterflies analyzed (~80%) with that of our own region. Among the Ohio butterflies exhibiting statistically significant declines are twenty-seven species in common with the Acadian region (4.5 times the six shared species reported to be experiencing population increases). The annual 2% rate of decline in Ohio is consistent with annualized trends calculated by rigorous long-term butterfly monitoring programs in Spain (Melero et al. 2016), the United Kingdom, the Netherlands, and Belgium (Warren et al. 2021), signaling that multiple localities across North America and Europe are witnessing declines in one of their most popular and closely monitored insect groups.

To the extent that the declines documented above are, to some degree, also affecting our Acadian butterfly populations should be unsettling to anyone concerned about potential cascading effects on all manner of ecosystem function. Furthermore, it is disconcerting that North American and European studies have documented significant declines across species from all major butterfly families, encompassing a wide breadth of habitats and food plant associations. These declines have been detected not only within rare, range-restricted butterflies but also among previously common, more wide-ranging species. For example, among the species experiencing the most precipitous declines in abundance in Massachusetts and Ohio are numerous widespread generalists common to the Acadian region, such as the American Copper, European Skipper, Dreamy Duskywing, Meadow Fritillary, and Common Wood-Nymph. A crash in populations of such formerly abundant species can be expected to impact valuable ecosystem services, including pollination, herbivory, and food-web-wide wildlife nutrition.

There is perhaps no better example of a decline of a formerly abundant, widely distributed insect than the Monarch butterfly, a conspicuous species familiar to even our region's most casual naturalists. The Monarch is a regular breeding colonist of Maine and the Maritimes, migrating over 4000 km (~2500 mi) from our region in late summer and early fall to the Transvolcanic Mountains of central Mexico, where nearly the entire eastern North American population overwinters across a dozen montane roost sites (Slayback et al. 2007). By annually measuring the area of habitat occupied by roosting butterflies since 1995, biologists now estimate that the population of Monarchs breeding east of the Rocky Mountains has declined by over 80% during the last two decades (Semmens et al. 2016). Such declines have not yet been apparent in the Acadian region, but they are nonetheless concerning. If the overwintering roosts are lost, most of the eastern North American population could disappear. Monarch declines have precipitated continental concern, with the Committee on the Status of Endangered Wildlife in Canada (COSEWIC) assessing the species as nationally Endangered in 2016, and the U.S. Fish and Wildlife Service identifying the Monarch as a candidate for Endangered or Threatened listing (USFWS 2020). It is impossible to predict the ecological ripple effects of removing millions of Monarch caterpillars from the food chain of our northeastern grasslands and farmlands, but most ecologists would agree that the impact is likely significant. Now consider that this is just a single, unusually well-studied species of declining butterfly in our region.

ACADIAN SPECIES OF CONSERVATION CONCERN

As a mostly static assessment of butterflies, the Acadian regional atlas is not suited for quantifying long-term trends in our region's butterfly populations. Yet it serves as an important baseline from which to track future changes. We trust that future colleagues will do just that by adding data from Maine and the Maritimes to our global knowledge on regional butterfly population trends. Presently, our atlas permits a detailed assessment of the current conservation status of individual species based on criteria that are commonly used to characterize rarity and vulnerability, including number of populations, extent and continuity of range, endemism, and degree of habitat specialization. One of the greatest initial obstacles to invertebrate conservation worldwide is a lack of basic knowledge about the distribution and biology of regional faunas. An accurate, periodic assessment of species conservation status, across all taxa, is critical for identifying those species at risk of decline and extirpation. Thus informed, natural resource agencies and their partners can direct limited conservation resources where they are needed most.

In Maine, conservation status reviews of nearly all vertebrates and better-studied nonmarine invertebrates (including butterflies) is conducted by biologists at the Maine Department of Inland Fisheries and Wildlife (MDIFW). These reviews can lead to a recommended status of Endangered, Threatened, or Special Concern depending on the degree of species rarity and the risk of state extirpation. Species listed as state

Figure 38. Insect drawer from Maine's L. C. Bates Museum containing Regal Fritillary specimens collected in the late 1800s at Manchester, ME by Mattie Wadsworth (Bryan Pfeiffer).

Endangered or Threatened are eligible for protections under the Maine Endangered Species Act. In addition to leveraging increased resources for population monitoring, life history research, habitat restoration, and public outreach, an Endangered or Threatened listing status permits MDIFW to provide limited regulatory protections from the effects of habitat loss due to development. For insect species with only a few documented populations statewide (e.g., Hessel's Hairstreak and Edwards' Hairstreak), often on private land, the habitat protections afforded by the Maine Endangered Species Act can potentially mean the difference between local population persistence and state extirpation. Consider that our region's last documented Regal Fritillary butterfly graced Maine's meadows and grasslands in the early 1940s, well before Maine's legislature enacted the state's first Endangered Species Act in 1975 and before most northeastern state wildlife agencies began developing robust invertebrate conservation programs (McCollough 1997). The reasons for the disappearance of the Regal Fritillary in the East are still not well understood, but we wonder if closer attention to endangered insects at the time in Maine, and throughout the Northeast, might have helped preserve this colorful member of our natural heritage (Figure 38).

The conservation status of species in the Maritime Provinces are reviewed every 5 years as part of the General Status of Species in Canada Program (Canadian Endangered Species Conservation Council 2016). These assessments produce province-specific S-ranks based on

NatureServe guidelines (see Appendix B for rank definitions). The Atlantic Canada Conservation Data Centre maintains a database of occurrence data for nearly all wildlife species within the Maritimes. The occurrences of species of conservation concern (i.e., those with ranks ranging from S1 to S3S4, or those that are provincially or federally listed) may be taken into consideration in environmental assessments and are used for the identification of high-value habitats and other conservation applications. In the Maritimes, species can also be afforded legal protection under the federal Species at Risk Act, the New Brunswick Species at Risk Act, and the Nova Scotia Endangered Species Act. Under all these acts, habitat critical to the survival of the species can be designated and protected, even on private property. Such habitat has been delineated for the Maritime Ringlet in New Brunswick (NBMRRT 2005; Environment Canada 2012).

Informed by the compilation of data underlying this atlas effort, officials in Maine and the Maritimes recently updated the Acadian region's butterfly species legal status and Natural Heritage (NatureServe) conservation status ranks. These ranks are reported in the individual species accounts and in our regional checklist (Appendix A). In total, twenty-six butterfly species and subspecies in the Acadian region (over 20% of our fauna) are taxa of conservation concern at either the global, federal, state, or provincial level (Table 7). A brief review of the list's composition reveals some potentially interesting patterns of risk associated with phylogeny, natural history, and geography.

Maine has assigned a status of Endangered, Threatened, or Special Concern to twenty-one butterfly taxa, while two species occurring in the Maritimes are listed under provincial and federal acts. The discrepancy in the number of species listed in the two regions is explained partly by need (there are more species at risk of extirpation in Maine) and partly by policy (the Maritime Provinces rely more on NatureServe S-ranks to accomplish conservation goals). The region's species of conservation concern are diverse at the family level, but a disproportionate (30%) number of lycaenids (mainly hairstreaks, elfins, and blues) are considered at elevated risk, likely as a result of the group's frequent association with specialized habitats and larval food plants. Indeed, a review of species–habitat relationships for vulnerable butterflies of the Acadian region reveals that most are tied to local patch scale (versus larger, matrix forming) habitats, conditioned by specific soils, drainage, or

Table 7. Butterfly species and subspecies of conservation concern in Maine and the Maritimes (twenty-six taxa), including status, habitat, and potential threats

Taxon	Legal status	NatureServe G-rank	Primary habitat	Potential threats
Skippers				
Sleepy Duskywing (*Erynnis brizo*)	ME-T	G5	Dry woodlands, barrens	Development, succession, Spongy Moth spraying
Leonard's Skipper (*Hesperia leonardus*)	ME-SC	G4	Xeric grasslands, dry open woodlands, barrens	Development, succession, Spongy Moth spraying
Cobweb Skipper (*Hesperia metea*)	ME-SC	G4	Xeric grasslands, dry open woodlands, barrens	Development, succession, Spongy Moth spraying
Dusted Skipper (*Atrytonopsis hianna*)	ME-SC	G4G5	Xeric grasslands, dry open woodlands, barrens	Development, succession, Spongy Moth spraying
Swallowtails				
Short-tailed Swallowtail (*Papilio brevicauda gaspeensis*)	ME-SC	G5T4?	Moist meadows and stream margins in inland northern forests	Climate change, Spruce Budworm spraying
Short-tailed Swallowtail (*Papilio brevicauda bretonensis*)		G5T3	Coastal dunes and salt marshes	Climate change (sea level rise), development, pollution
Spicebush Swallowtail (*Pterourus troilus*)	ME-SC	G5	Rich seepage forest	Development, intensive forestry, invasive species
Gossamer-wings				
Clayton's Copper[a] (*Tharsalea dorcas claytoni*)	ME-T	G5T3T4	Rich calcareous fens, wet meadow	Succession, flooding, invasive species
Salt Marsh Copper (*Tharsalea dospassosi*)		G3	Salt marsh	Climate change (sea level rise), development, pollution
Juniper Hairstreak (*Callophrys gryneus*)	ME-E	G5	Rocky hilltops, rights-of-way	Succession, intensive rights-of-way management, development
Hessel's Hairstreak (*Callophrys hesseli*)	ME-E	G3	Swamp, peatland	Development, intensive forestry
Hoary Elfin (*Callophrys polios*)	ME-SC	G5	Dry open woodlands, barrens	Spruce Budworm and Spongy Moth spraying, development, succession, climate change

Table 7. *continued*

Taxon	Legal status	NatureServe G-rank	Primary habitat	Potential threats
Early Hairstreak (*Erora laeta*)	ME-SC	G2G3	Northern hardwood forest	Intensive forestry, Beech scale-fungus disease, Beech leaf-mining weevil, Spruce Budworm spraying
Coral Hairstreak (*Satyrium titus*)	ME-SC	G5	Dry shrubland, old fields	Development, succession, intensive agriculture, Spongy Moth spraying
Edwards' Hairstreak (*Satyrium edwardsii*)	ME-E	G4	Dry open woodlands, barrens	Development, succession, Spongy Moth spraying
Silvery Blue[a] (*Glaucopsyche lygdamus mildredae*) (possibly extinct)		G5TNR	Coastal shoreline	Potential competition and genetic swamping from *G. l. couperi*
Crowberry Blue (*Plebejus idas empetri*)	ME-SC	G5T5	Coastal peatland	Peat mining, development
Northern Blue (*Plebejus idas scudderii*)	ME-SC	G5T5	Boreal conifer woodland and edges	Climate change, Spruce Budworm spraying
Brushfoots				
Monarch (*Danaus plexippus*)	ME-SC USA-C CAN-SC NB-SC NS-E	G4	Old fields, gardens, rights-of-way	Mexico winter habitat loss, development, climate change, succession, herbicides, intensive agriculture, invasive species
Arctic Fritillary (*Boloria chariclea*)	ME-T	G5T5	Boreal conifer woodland	Intensive forestry, climate change, Spruce Budworm spraying
Frigga Fritillary (*Boloria frigga*)	ME-E	G5	Peatland	Climate change, Spruce Budworm spraying, peat mining
Regal Fritillary (*Argynnis idalia*) (extirpated)		G3?	Wet meadows, old fields	Unknown; likely multiple factors
Satyr Comma (*Polygonia satyrus*)	ME-SC	G5	Stream margins and openings in northern forest	Intensive forestry, Spruce Budworm spraying

Taxon	Legal status	NatureServe G-rank	Primary habitat	Potential threats
Silvery Checkerspot (*Chlosyne nycteis*)	ME-SC	G5	Wet meadows, open woodlands near streams	Development, succession, deer grazing, parasitism, pesticides
Maritime Ringlet (*Coenonympha nipisiquit*)	CAN-E NB-E	G1G2	Salt marsh	Climate change (sea level rise), development, pollution
Katahdin Arctic[a] (*Oeneis polixenes katahdin*)	ME-E	G5T1	Alpine tundra	Climate change, trampling, illegal collection

Notes: Official status is abbreviated as follows: E = Endangered, T = Threatened, SC = Special Concern, C = USA Candidate. Legal status and NatureServe G-ranks are defined in Appendix B. ME = Maine, NB = New Brunswick, NS = Nova Scotia, CAN = Canada.
[a] Subspecies is endemic to the Acadian region.

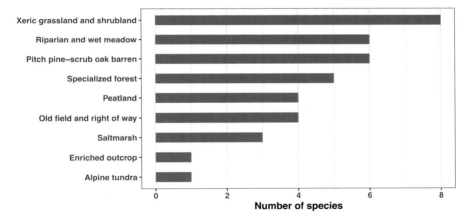

Figure 39. Habitat associations of butterflies of conservation concern (twenty-six taxa) in the Acadian region. Some species are assigned to more than one habitat.

elevation patterns (Figure 39). Organisms specialized for rare or patchily distributed site conditions are at increased vulnerability to localized habitat loss from development, pesticide use, climate change, ecological succession, and natural disturbance (e.g., wind, fire, flooding). This inherent vulnerability is confirmed by observations of butterfly communities elsewhere, which exhibit a pattern of sharper population declines among habitat specialists (Melero et al. 2016; Wagner 2020; Warren et al. 2021). Additionally, butterfly specialists are often more geographically restricted than are habitat generalists, which is more likely to lead to localized or regional extirpation. We may be witnessing this phenomenon in New England's formerly widespread population of the Hoary Elfin. A dry woodland and barren specialist in the southern part of its range, this species is now known only from historical records in southern Maine and is extirpated from many other parts of the eastern United States (NatureServe Explorer 2021).

Species of conservation concern are found throughout the Acadian region, but there is a distinct concentration in southwestern Maine (Figure 40), which serves as a regional hotspot for biological diversity, hosting an exceptional richness of butterflies (Figure 33), woody plants (McMahon 1990) and vertebrates (Krohn et al. 1999). This pattern exists largely because southwestern Maine is the northern limit of the Northeastern Coastal Zone (U.S. EPA Level III ecoregional scheme, described in Omernik and Griffith 2014), an ecoregion that hosts several unique habitats, such as pitch pine woodlands and sandplain barrens, as well as southern-affinity forest types, which are mostly absent elsewhere in the Acadian region. Closely tied to these unusual habitats are several species approaching the northern edge of their range in southern Maine, including Dusted Skipper, Cobweb Skipper, Juniper Hairstreak, Edwards' Hairstreak, Hessel's Hairstreak, Spicebush Swallowtail, and others. This phenomenon of biogeography presents an acute conservation challenge for the Acadian region's butterflies and biodiversity as this same area also hosts some of Maine's greatest human population densities, highest rates of development and land conversion, and least amount of protected conservation land (Schlawin et al. 2021).

While these relatively southern species are acutely threatened by habitat loss in the near term, several boreal-associated species may be lost from our region over longer periods because of global warming. For example, the Bog Fritillary, Arctic Fritillary, Northern Blue, Hoary Comma, Bog Elfin, and Jutta Arctic are at the southern limits of their ranges in the Acadian region. As their climate niche shifts northward, our region will likely be one of the first areas from which they disappear. With common names that evoke their northern origins, observations of any of these unique species are considered a prized encounter by many North American naturalists, particularly those from southern latitudes, who have been known to travel great distances to our region in the hopes of encountering these and other boreal specialties.

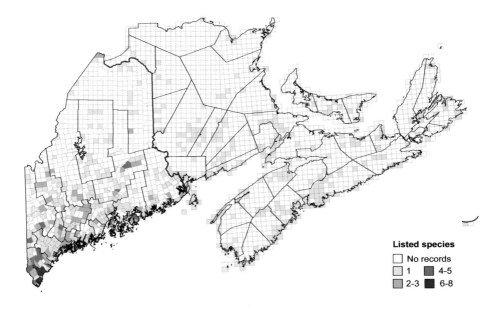

Figure 40. The distribution and richness of twenty-four butterfly taxa of conservation concern in the Acadian region. Historical records for the Regal Fritillary (extirpated) and the *mildredae* subspecies of the Silvery Blue (potentially extirpated) are omitted.

In terms of global butterfly diversity, the Acadian region's most important inhabitants include a suite of northeastern range-restricted species and endemic subspecies. The relatively narrow geographic range of these taxa is often reflected in lower global ranks (G1–G3) as assigned by NatureServe (Table 7). Specifically, the Acadian region is home to most of the global populations of the Salt Marsh Copper (G3) and the Maritime Ringlet (G1), the latter entirely restricted to the salt marshes of Chaleur Bay, the water body between Quebec's Gaspe Peninsula and New Brunswick. Similarly, three subspecies—Clayton's Copper (G5T3T4), the Katahdin Arctic (G5T1), and the *mildredae* subspecies of Silvery Blue (G5TNR)—are entirely restricted to the Acadian region. Two others, the Crowberry Blue (G5T5) and the *bretonensis* subspecies of the Short-tailed Swallowtail (G5T3), may be similarly restricted, though additional survey efforts are needed in adjacent Quebec.

THREATS TO THE BUTTERFLIES OF MAINE
AND THE MARITIMES

Consistent with patterns elsewhere (Black and Jepsen 2007; Hall 2009; Schweitzer et al. 2011; Warren et al. 2021), several stressors are affecting the health of butterfly populations in the Acadian region. Here we touch briefly on those that are believed to be most prevalent and impactful to our region's species of conservation concern (Figure 41), including habitat loss, ecological succession, pesticides, invasive species, and climate change. Readers are cautioned that assigning threats to individual species (as done in Table 7) is a partly speculative exercise given the dearth of documentation in our region linking individual species declines to specific stressors. Nonetheless, the scientific consensus is that all the threats reviewed below are of significant concern to butterflies and many other insects. Furthermore, multiple threats often magnify the effects of any one factor. For example, the impacts of climate change on some specialized butterflies are compounded by the destruction and fragmentation of habitat from development, which can reduce the likelihood for successful dispersal to climatically suitable areas (Pöyry et al. 2009).

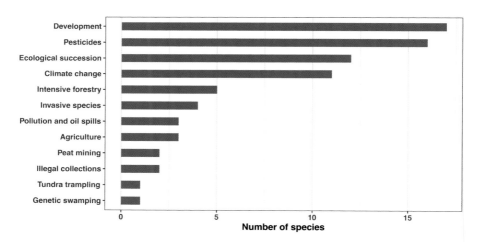

Figure 41. Primary threats to butterflies of conservation concern (twenty-six taxa) in the Acadian region. Multiple threats are assigned to some species.

Habitat Loss and Alteration

Butterflies worldwide are vulnerable to the same threats facing most other animals, with habitat loss near the top of the list of destructive forces (Black and Jepsen 2007; Sánchez-Bayo and Wyckhuys 2019, 2021). Indeed, two independent national assessments of the status and threats to biological diversity in the United States (Stein et al. 2000) and Canada (Venter et al. 2006) identified habitat loss and fragmentation as a leading contributor to the imperilment of butterflies in North America. Habitat loss is among the most important threats to butterfly species in the Acadian region (Figure 41), especially those with patchy or narrow distributions, as in the case of many of our habitat specialists (Table 7).

Development—Fortunately, the Acadian region, as with much of the Northeast, retains most of its forest and natural land cover. This is notable given a history, especially in southern and coastal areas, of intensive clearcutting, conversion to cropland and pasture in the 1800s, and agricultural abandonment and recovery during the twentieth century (Foster 1995; Barton et al. 2012). However, scientists recently reported that after more than 150 years, the tide has changed in New England (and Maine), where annual forest and natural land loss now exceeds the rate of recovery because of a second wave of disturbance: residential development (Drummond and Loveland 2010; Olofsson et al. 2016). While overall rates of land-use conversion in most of the Acadian region are lower than much of the northeastern United States seaboard, the pace of development remains high in parts of central and south-coastal Maine (Foster et al. 2010; Theobald 2010). The rate and geography of current and projected habitat loss and fragmentation in our region is especially alarming given the concentration of imperiled butterflies (Figure 40) and other at-risk species in southern Maine. The authors are personally familiar with several formerly rich butterfly sites in our region—meadows, barrens, and woodlands—that have been lost to the bulldozer. These areas will likely lack specialized insect communities, and other native biota, until long after the next glacier plows them under thousands of years hence.

Agriculture—Historically, one of the greatest sources of habitat loss for our region's inland and coastal biota was the conversion of forest and salt marsh to agriculture. This change in land use presented both opportunity and loss for many of our native butterflies. Far more limited in extent than forests, the degradation of salt marsh ecosystems has negatively affected a variety of northeastern coastal biota, from birds to butterflies, including some notably specialized species in our region, such as the Broad-winged Skipper, Salt Marsh Copper, Maritime Ringlet, and the coastal subspecies of the Short-tailed Swallowtail.

In a landscape previously dominated by mature woodland, the patchy conversion of forest to low intensity agriculture was arguably a boon for inland butterfly diversity, providing a pulse of habitat opportunity for grassland and shrubland species. However, since the late nineteenth century, much of the Acadian region's marginal farmland has been abandoned. Although large areas of transitional conditions remain that are at least temporarily favorable to many early successional species, the extent of anthropogenic-influenced grassland and wet meadow habitat in our region has shrunk dramatically in the last century. This has likely contributed to declines in many species of grassland butterfly, including some of our region's at-risk species, such as the Monarch, Leonard's Skipper, and the Regal Fritillary. Most agricultural lands in the Acadian region that remain viable today have done so through the adoption of modern methods, such as mechanical tilling, fertilizers, pesticides, grain-supplemented livestock, and monocultural crop rotations. These intensive practices, while compatible for many open land generalists (e.g., Common Ringlet, Cabbage White, Clouded Sulphur, Black Swallowtail, and several grass skippers) are not as favorable to specialized species that require a diversity of management-sensitive host plants, including violets, asters, milkweeds, and native sunflowers, among others.

Forestry—Forestry is historically, and currently, the greatest source of habitat alteration in the Acadian region. As a result, the composition of our region's woodlands has changed significantly. In the Maritimes, most major types of climax hardwood forest and nearly all types of old growth forest, have declined or nearly disappeared, while some types of mixed forest dominated by short-lived species (e.g., poplar, *Populus* spp.; Red Maple, *Acer rubrum*; Balsam Fir, *Abies balsamea*) have increased (Loo et al. 2010; Thompson et al. 2013). While less prevalent than in the past, clearcutting is widely used in contemporary commercial forestry throughout the region, especially in the Maritime Provinces. This practice replaces overstory forest conditions with early successional habitat, which results in a replacement of local butterfly species, shifting from forest-inhabiting species (such as commas and satyrs) to sun-loving skippers, crescents, greater fritillaries, and others. While it is tempting to argue that forestry practices are generally benign or beneficial for most butterflies, such an assumption is likely overly simplistic. Studies of the effects of timber management on forest moth communities in mixed hardwood forests of Indiana and Ohio (Summerville and Crist 2002; Summerville 2011, 2013) suggest that lepidopteran community structure (richness, abundance, dominance) is affected across a range of harvest intensities, including patch cuts, shelterwood cuts, and clearcuts. Dietary specialists were generally most affected because of reductions in the availability of larval food plants, with long-term effects predicted to depend on the speed at which stands recover to a precut floristic composition. Supporting this notion are several studies documenting that intensive forest management practices can have significant effects on the presence and abundance of northeastern forest understory plants (Meier et al. 1995; Moola and Vasseur 2008; Goltz 2014), including some species that are important larval food sources. There are undoubtedly both winners and losers associated with forest management, but the impact of decades of intensive silviculture in our region on overall butterfly and moth diversity and abundance is still not well understood.

Ecological Succession

A more subtle factor contributing to habitat loss and degradation is ecological succession. As is the case in southern New England (Wagner et al. 2003), many of our region's most vulnerable butterfly species are associated with grasslands, shrublands, and open barrens (Figure 39). While partly a natural process, the gradual succession of disturbance-maintained, open land communities to common, closed canopy forest types (often pine–oak or mixed hardwoods) is also partly a function of widespread development patterns in southern Maine. As the landscape becomes increasingly fragmented by roads, store fronts, and subdivisions, patches of remnant grasslands, shrublands, and barrens become smaller, more insular, and less capable of sustaining the frequency and severity of wildfires and windstorms required periodically to stem succession (Latham 2003). In short, while important in the near term, simply targeting these naturally dynamic habitats for protection from development is not sufficient. Active management and restoration practices, including prescribed fire and mechanical cutting, are often necessary to conserve the unique structure and composition of flora that supports diverse butterfly and moth communities within the increasingly fragmented landscapes of southern Maine.

In areas where natural disturbance no longer maintains large areas of early successional habitat, powerline and pipeline rights-of-way can provide a valuable refuge and corridor for shrubland and barren-associated butterflies and other pollinators (Wojcik and Buchmann 2012; Gobeil and Gobeil 2014). The value of rights-of-way in this regard is especially pertinent in parts of the Acadian region, where much of the landscape is dominated by closed canopy forest.

Pesticides

Insecticides—Butterflies are directly and indirectly affected by the chemicals commonly used by farmers, foresters, and homeowners to control populations of unwanted plants and pests. Many insecticides cause direct mortality to butterflies. Control programs for the Spongy Moth (*Lymantria d. dispar*) and the Spruce Budworm (*Choristoneura fumiferana*) are of particular concern in our region because of the potential scope of response by government authorities in charge of pesticide application policy. The Spongy Moth is an introduced species with periodic outbreaks resulting in the defoliation of trees, particularly hardwoods like oaks, poplars, and birches. Within our region, only southern and central Maine have been affected by severe outbreaks to date, with events in the early 1980s and early 1990s resulting in over 240,000 hectares (600,000 acres) of defoliation (Maine Forest Service, unpublished data). The Spruce Budworm is a native moth species that periodically reaches outbreak numbers, resulting in the defoliation of spruce and Balsam Fir, species of significant economic value in the Acadian regional forest. The last major outbreak in the 1970s and 1980s affected spruce and fir dominated forests across northeastern North America. The New Brunswick Spruce Budworm spraying program, which started in 1952 and lasted 40 years, was the largest and longest aerial insecticide spraying program in the world (Klassen and Locke 2010), with up to 38,000 km^2

(~15,000 mi²) of forest sprayed annually in the province (Armstrong and Cook 1993). A new outbreak, detected in 2006 in eastern Quebec, just on our region's border, has grown to more than 13 million hectares (~30 million acres) (Healthy Forest Partnership 2021), prompting renewed preparations for aerial spraying by concerned forestry officials in Maine and the Maritimes. Consultation with natural resource agency maps of spray-sensitive habitats, along with carefully planned spray buffers, can help prevent unintended impacts to rare butterfly populations.

The pesticides currently used for Spongy Moth and Spruce Budworm, such as the biological agent *Bacillus thuringiensis* var. *kurstaki* (*Btk*) and the insect growth regulator tebufenozide, affect a wide variety of larval Lepidoptera. Broad-scale applications have the potential to reduce butterfly abundance at both the population and landscape level. Since 2014, New Brunswick has been effectively containing the spread of a Spruce Budworm outbreak from Quebec into northern parts of the province with a strategy that treats budworm populations along the leading edge of the outbreak (Johns et al. 2019). While limiting its geographic scope, this strategy does not eliminate the threat of incidental spray impacts to butterflies. For example, 350 km² (135 mi²) were still sprayed in New Brunswick in 2020 (Healthy Forest Partnership 2021).

Insecticides also pose a threat to butterflies and other nontarget insects in agricultural settings. This is especially true when they are applied along field margins and hedgerows that support plants that are consumed by butterfly larvae or provide nectar sources for adults (Davis et al. 1991). Insecticides can drift into adjacent habitats or make their way into them via groundwater and runoff. These chemicals can persist in such environments for a long time. For example, the half-lives of neonicotinoids, a widely used class of insecticide, can exceed 1000 days in soils (Bonmatin et al. 2014). While it has been demonstrated that insecticides used in agriculture can have negative effects on butterflies, there has been relatively little research on the subject, especially in light of the billions of dollars spent on insecticide development, production, and use globally (Mulé et al. 2017).

Herbicides—The direct toxicity of herbicides to vertebrates (when applied at recommended field application levels) is generally reported to be low, but effects on invertebrates are not well studied. At least some research indicates that arthropods, including larval Lepidoptera, can suffer both lethal and sublethal effects upon ingestion (Russell and Schultz 2010; Stark et al. 2012).

Of less question is the indirect effect that herbicides can have on butterflies and moths by eliminating essential larval food and nectar plants (Schweitzer et al. 2011). This effect can be especially significant where management is used to create and perpetuate early successional habitat, such as along roadside verges and utility rights-of-way. While not exclusive to these areas, we know of several colonies of butterfly species of conservation concern residing in powerline rights-of-way in Maine and the Maritimes, including Leonard's Skipper (and other bluestem-feeding skippers), Juniper Hairstreak, Hoary Elfin, and Monarch. A lack of communication between natural resource agencies and utility companies about the needs of these populations risks unintended impacts to butterfly populations from periodic vegetation control measures. For example, in the 1990s a colony of Arctic Fritillary disappeared from a powerline right-of-way in Gloucester County, New Brunswick, after herbicide was applied to clear vegetation. Generally, and especially in areas of rights-of-way where at-risk species are present, the use of mechanical tools, such as mowers and brush saws, is a more butterfly friendly method of vegetation control than is broad-scale aerial herbicide applications (Russell et al. 2005). Nonetheless, a population of Juniper Hairstreak was apparently destroyed in York County, Maine, when its food plant, Red Cedar (*Juniperus virginiana*), was cut following new requirements that limit the presence of vegetation within a certain distance of high voltage powerlines.

The effects of herbicides on butterflies in forest-dominated landscapes is of particular interest in the Acadian region, where much of our woodlands are owned or managed by large corporations focused on commercial silviculture. Herbicides used for forest management purposes can be expected to have negative effects on habitat quality for some Lepidoptera by reducing or eliminating essential caterpillar food and nectar plants (Moola and Vasseur 2008; Goltz 2014), in turn affecting survivorship and fecundity of larvae and adults (Boggs and Freeman 2005; Russell and Schultz 2010). This risk can be significant where herbicides are used to specifically reduce understory competition and cover in intensive management prescriptions, such as clearcuts and plantations. Arguably, because the forests of Maine and the Maritimes are vast and the status of most of the region's Lepidoptera are thought to be secure, the overall effects of herbicides on most species are not likely to be significant. However, in specific localities where rare and imperiled butterflies are known or predicted to occupy areas targeted by herbicides, there is the potential for unintended impacts. Indeed, Maine's Wildlife Action Plan (MDIFW 2015) identifies aerial pesticide use as one of several primary threats to the state's Lepidoptera.

Nonnative Species

Fauna—Introduced insects affect butterflies in a variety of ways. Some, like the Spongy Moth and Brown-tail Moth (*Euproctis chrysorrhea*), compete with butterflies for food and thereby have the potential to affect food availability for native summer-feeding larvae (Schweitzer et al. 2011). Furthermore, these and other unwanted pests (e.g., mosquitoes, ticks) may invite the application of problematic nontarget insecticides. Other exotic insects can have more direct impacts on our native Lepidoptera. Introduced lady beetles, such as the Harlequin Lady Beetle (*Harmonia axyidis*), are voracious predators of soft-bodied insects, including caterpillars. Another introduced predator, the European Mantis (*Mantis religiosa*), captures and consumes adult butterflies that visit showy flowers, to the dismay of butterfly gardeners.

Nonnative insects can also serve as vectors for plant diseases, posing an indirect threat by killing or damaging larval food plants. The accidental introduction of the beech scale insect (*Cryptococcus fagisuga*) from Europe in the 1890s is an example that hits close to home. Following initial detection in the Halifax area of Nova Scotia, the scale insect is now established throughout northeastern and midwestern North America. Together with several exotic fungal species, the scale insect causes beech bark disease, stunting and killing American Beech (*Fagus grandifolia*), one of the few hardwood trees that produces nutritious hard mast for wildlife throughout Maine and the Maritimes. The decline of American Beech is of particular concern to one of our region's most elusive butterfly species, the Early Hairstreak, whose larvae feed primarily on beech nuts. The Emerald Ash Borer beetle (*Agrilus planipennis*), which causes nearly 100% mortality to ash trees (*Fraxinus* spp.), is our region's most recent example of an exotic insect introduction with the potential for devastating impacts to our native Lepidoptera. Native to Southeast Asia, the Emerald Ash Borer was first reported from Michigan in 2002 and has spread rapidly across eastern North America, reaching Maine and New Brunswick in 2018. Unless a practical means of controlling beetle populations is found soon, most of our region's ash trees will die with severe effects on native ash-feeding Lepidoptera in swamps and uplands where the trees are now abundant. Over forty species of Lepidoptera native to eastern North America feed on various species of ash (Wagner 2005), including the Canadian Tiger Swallowtail, Baltimore Checkerspot, and several spectacular species of giant silkworm moths (subfamily Saturniinae). Several moths are believed to be ash specialists, and thus particularly vulnerable to decline or possibly extinction (Wagner 2007).

Many exotic insects have been purposely introduced to control the populations of pests. When carefully researched, these biological control agents can be effective and ecologically benign compared with many chemical pesticides. However, they can also be highly problematic if they affect nontarget species. Before current, strict host-choice testing requirements were instituted by permitting agencies, several biocontrol insects were released that had undesired off-target impacts. One ruinous example is the tachinid fly *Compsilura concinnata*, first introduced in 1906 to control the Spongy Moth and the Brown-tail Moth. This parasitoid is known to attack at least 200 species of Lepidoptera, including some butterflies (O'Donnell et al. 2007) and is thought to have had a devastating impact on populations of giant silk moths (Saturniidae) and other large-bodied moth families.

Flora—Nonnative flora, particularly those species that become invasive in natural areas, can also cause harm to native butterfly communities. Some nonnative plants function as dead-end hosts by attracting oviposition, even though the plants are unfavorable to growth or even toxic to the larvae. This is the case for the Mustard White, which may oviposit on Garlic-mustard (*Alliaria petiolata*) and Garden Yellow-rocket (*Barbarea vulgaris*) in our region (Renwick 2002). A similar troubling example is the oviposition of Monarch butterflies on Black Swallowwort (*Cynanchum louiseae*), an extremely invasive vine that is not only lethal to Monarch caterpillars but is also capable of outcompeting and replacing native milkweed species in some habitats. Many additional nonnative plants are successful invaders of natural ecosystems in the Acadian region, where they have the potential to alter the structure and composition of entire plant communities. These changes can have cascading ecological impacts for native Lepidoptera and their insectivorous predators (Tallamy 2019). For example, Reed Canary Grass (*Phalaris arundinacea*) and Japanese Knotweed (*Fallopia japonica*) form dense, near-monoculture stands in areas of rich soils, particularly along open river shores, an ecotone that otherwise supports a diverse assemblage of native herbs, grasses, and shrubs of greater forage value to Lepidoptera and other herbivorous species. Glossy Buckthorn (*Frangula alnus*) and Common Reed (*Phragmites australis*) are becoming increasingly abundant in a variety of northeastern wetland habitats, displacing native food and nectar sources for a wide variety of wetland-associated entomofauna. Our region's uplands are also vulnerable to invasion by exotic flora where, among many others, Japanese Barberry (*Berberis thunbergii*) and shrubby honeysuckles (*Lonicera* spp.) are known to proliferate in both open field and forest settings, reducing growing space for important understory larval food plants. For a more detailed accounting by habitat of the most problematic invasive plants in the Acadian region, including recommendations for their control, readers are encouraged to consult Olmstead and Yurlina (2019).

Climate Change

Climate is a fundamental factor in determining where species live, as well as the size of their populations. Therefore, ecologists generally agree that current human-accelerated climate warming will have profound effects on the distribution and status of biota worldwide. We have no reason to doubt that this will similarly be the case for the Lepidoptera of Maine and the Maritimes. Undeniably one of our region's most pervasive potential threats, climate change is also one of the most complex and least understood stressors in terms of predicting specific impacts to individual species and habitats. Certainly, a shifting climate will have an influence on our butterfly community as it has elsewhere (Black 2018). Some species will likely thrive, while others will decline or disappear if they cannot adapt physiologically or find acceptable refugia (e.g., suitably high latitudes and elevations) within the Acadian region.

One predictable outcome of a warming climate is that many insects in our area will respond by changing their flight periods. Our preliminary analysis suggests this has already happened with several butterfly species in our region (see Methods and Biogeographical Findings), with most of the univoltine species examined showing evidence of accelerated emergence in the modern era in either Maine (five species), the Maritimes (eight species), or across the Acadian region (e.g., Northern Pearly-Eye, Northern Crescent, and Silvery Blue). These observations are consistent with reports from Europe (Roy and Sparks 2000), California (Forister and Shapiro 2003), and Massachusetts (Polgar et al. 2013), where many elfins and hairstreaks (Lycaenidae) are responding to a changing climate by flying significantly earlier in warmer years. One conservation implication of this dynamic is the potential for trophic mismatch, whereby ecologically associated species do not respond to warming or other climate-related changes at the same rate (Nakazawa and Doi 2012). For example, it is possible that the tightly linked life cycle of some specialized butterflies and their host plants will become misaligned, negatively affecting the survival of their caterpillars, whose growth is no longer timed for optimal nutrition, as during host plant senescence. Such trophic mismatches can cascade through the food web in complex ways and have implications for insectivores, such as many songbirds that rely on an abundant crop of moths and butterflies (both larvae and adults) to feed developing young.

Another likely response of our region's butterfly fauna to a changing climate is a shift in species ranges as they track optimal conditions across the landscape. Such climate-related distribution shifts have been well documented for several European butterflies, often with a pattern of contraction near the southern edge of their range (Parmesan et. al. 1999; Pöyry et al. 2009; Devictor et al. 2012). One of the few North American studies of this phenomenon in butterflies was conducted in Massachusetts. In analyzing nearly two decades of observational data gathered mostly by community scientists, Breed et al. (2013) reported distinct differences in long-term population trends for species depending on their overall range distributions. Specifically, population trajectories indicated increases for many warm-adapted butterflies near the northern edge of their range in Massachusetts (e.g., Fiery Skipper, Spicebush Swallowtail) and declines in nearly all of the cold-adapted butterflies approaching the southern edge of their range in the state (e.g., Atlantis Fritillary, Acadian Hairstreak). The investigators concluded that major climate-induced range shifts are already underway in North American insect communities and that climate warming may be more impactful than habitat loss as a leading cause of endangerment for some northeastern butterflies.

Qualitatively, a comparison of historical versus modern distributions in the Acadian region reveals some potentially notable range shifts among our own butterfly fauna. Several skippers, including the Black Dash and Mulberry Wing, appear to have expanded their northern ranges into our region only recently, while others with previously restricted distributions, like the Delaware Skipper and Silver-spotted Skipper, have greatly increased. This pattern may be the case for the White M Hairstreak as well. Historically absent from New England, this small, colorful butterfly was first recorded in Massachusetts in 1979, but it is now observed almost annually. It was a welcome surprise when it was found in Maine in 2018. On the other hand, there are some apparent northward range contractions in our region, as appears to be the case for the Arctic Fritillary and Hoary Comma. These partially boreal species approach the southern limits of their distribution here, possibly making them more susceptible to the effects of climate warming. In lieu of waiting for long-term quantitative data on the effects of climate change on potentially at-risk species, many natural resource agencies are consulting experts to predict the most vulnerable elements of our biodiversity based on life history traits and habitat exposure. Such vulnerability assessments can help flag species, guilds, and habitats likely to be most affected by the impacts of climate change (Glick et al. 2011). One such vulnerability assessment conducted in Maine, involving over 100 biologists (Whitman et al. 2013), identified alpine tundra, boreal and montane forest, peatlands, and coastal marshes as among the most vulnerable ecosystems to decline based on future climate projections for the state as provided by Jacobson et al. (2009). Unfortunately, these same ecosystems are known to host many of the Acadian region's most specialized and unique butterflies, including the Katahdin Arctic (alpine tundra), Arctic Fritillary and Northern Blue (boreal forest), Bog Elfin and Frigga Fritillary (northern bogs), and Maritime Ringlet and Salt Marsh Copper (northern salt marshes). While it is impossible to predict the exact ecological implications of losing one or more of these individual species, we worry that their decline portends the potential for widespread loss of some of our most iconic northern flora and fauna, including, the Moose (*Alces alces*), Common Loon (*Gavia immer*), Canada Jay (*Perisoreus canadensis*), Brook Trout (*Salvelinus fontinalis*), and Red Spruce (*Picea rubens*).

CONSERVATION OF HIGH VALUE HABITATS

We are heartened by the recent wave of popularity and conservation attention that pollinating insects are receiving. In response to this interest, much has been written about specific land management actions that landowners can take to attract and sustain robust populations of native insect pollinators. These recommendations are designed to enhance populations of butterflies, bees, flower flies, and other beneficial species in landscapes of largely anthropogenic origin, such as backyards and gardens, urban parks, grasslands, roadsides, and rights-of-way. When thoughtfully managed, these semidomestic settings can provide exceptional habitat for a wide variety of native insects and other biota, especially less specialized species that tolerate, or even thrive in, moderately disturbed landscapes. The pollinator-friendly practices necessary to improve ecological integrity to these near-home, mostly private, lands are increasingly well studied and encouraged by organizations such as the Xerces Society and North American Butterfly Association. Instead of describing these same management techniques here, we refer readers to other well-researched sources on this topic, such as provided by Mader et al (2011), Sardinas et al. (2018), and Tallamy (2019). In the remainder of this chapter, we focus on the importance of identifying and conserving select natural ecosystems that serve as valuable refuges for some of the Acadian regions most noteworthy native butterfly species.

Ecologists agree that one of the most enduring ways to conserve the full array of biotic diversity (whether mammals, mayflies, birds, or butterflies) is to ensure that a regionally representative network of natural habitats is protected and well managed. The premise, often referred to as a coarse-filter approach to conservation (Hunter 1991), is that by strategically selecting a diverse and representative array of ecosystems one can ensure that necessary habitat is available for the vast majority of species in a region: those known and those yet to be discovered. One reason the coarse-filter approach is most efficient is that there are simply too many native plants, animals, fungi, and other organisms to effectively permit a comprehensive, species-centric (or fine-filter) approach toward conservation. This problem is especially true for insects and other nonmarine invertebrates, a group conservatively estimated to include at least 15,000 species in Maine (Gawler et al. 1996) and 24,000 species in the Maritimes, or more than 97% of our region's animal diversity (McAlpine and Smith 2010). Supporting this staggering diversity is a complex evolutionary story of individual species life history requirements, a story more complicated than we currently understand and, in the case of invertebrates, more complex than we will likely ever fully know. To advance the conservation of most of our entomofauna, we recommend concentrating on habitat-based or coarse-filter conservation efforts across Maine and the Maritime Provinces. To this end, we highlight examples below of especially valuable habitats for Acadian Lepidoptera, areas that host both a variety of narrowly distributed habitat specialists as well as a robust cross section of our overall regional butterfly fauna. Of course, these are only a sample of the entire suite of regional habitat types required to support a rich and resilient community of Acadian Lepidoptera. Readers wishing to dig deeper into the full breadth of the natural community and habitat diversity of the Acadian region are encouraged to consult publications by Neily et al. (2010), Gawler and Cutko (2018), and Porter et al. (2020).

Dry Pine–Oak Woodlands and Barrens

Reaching their northeastern limit in southern Maine, dry pine–oak woodlands and barrens (Figure 42) are lightly forested to mostly open areas with well-drained, acidic, and often sandy soils. Pitch Pine (*Pinus rigida*), Scrub Oak (*Quercus ilicifolia*), Gray Birch (*Betula populifolia*), Common Lowbush Blueberry (*Vaccinium angustifolium*), Sheep-laurel (*Kalmia angustifolia*), and Bracken Fern (*Pteridium aquilinum*) are characteristic of the vegetation in these distinctive natural communities. Once viewed as unproductive wastelands, Maine's few remaining intact pitch pine woodlands and barrens are now recognized by biologists as providing valuable habitat for a variety of unique plants and insects, including many butterflies that are highly specific in their choice of larval food plants. Examples are the **Sleepy Duskywing, Cobweb Skipper, Dusted Skipper, Leonard's Skipper, Sassacus Skipper, Edwards' Hairstreak**, and **Hoary Elfin**. Total Lepidopteran richness can also be high in this ecosystem. In the Waterboro Barrens, one of the Northeast's best examples of a boreal variant of the pitch pine–scrub oak barrens community, at least 387 species of moths (mostly) and butterflies have been recorded (Sferra and Patton 1996; Schuerman and Puryear 2001).

Xeric woodlands and barrens were historically maintained by fires,

Figure 42. Pitch Pine–Scrub Oak woodland, York County, ME (Phillip deMaynadier).

which prevented succession to a more common pine–oak forest type. However, natural wildfire is now mostly short-circuited by active suppression and habitat fragmentation (Copenheaver et al. 2000). It has been conservatively estimated that at least a quarter of Maine's original pine barren acreage has been lost to residential development, agriculture, and gravel mining (Widoff 1987). Much of what remains is now in state and private conservation ownership (e.g., Kennebunk Plains Wildlife Management Area (WMA), Vernon Walker WMA, Wells Barrens Preserve, and Waterboro Barrens Preserve), where managers often employ prescribed fire and mechanical cutting for vegetation control. Given the unique biota associated with Maine's barren ecosystems, and their extreme vulnerability to development, a concerted effort should be made to conserve as much of the remaining unprotected acreage as possible.

Freshwater Wetlands

The Acadian region is home to an exceptional abundance and variety of freshwater wetland types, including bogs, fens, swamps, marshes, and wet meadows (Figure 43). Freshwater wetland, including forested

Figure 43. Marshy pond surrounded by diverse herbaceous plants, Shelburne County, NS (Sean Basquill, Nova Scotia Natural Resources and Renewables).

wetland, covers upward of 20% of Maine (MDEP 2003) and the Maritimes. Most seasoned naturalists are aware that wetlands and their riparian margins are bastions of biodiversity, yet novice observers are often surprised to learn that some aquatic ecosystems host a diverse and specialized suite of butterflies in our region. Because these habitats tend to be overlooked, there are still valuable discoveries to be made for those willing to get their feet wet. Among the most diverse settings for butterflies are open, herbaceous wetlands, including marshes and wet meadows dominated by grasses, sedges, and emergent nectar "magnets" such as Swamp Milkweed (*Asclepias incarnata*), Blue Iris (*Iris versicolor*), and Pickerelweed (*Pontederia cordata*). These are the preferred habitats of the **Two-spotted Skipper, Mulberry Wing, Black Dash, Least Skipper, Harvester, Bronze Copper, Acadian Hairstreak, Silver-bordered Fritillary, Meadow Fritillary, Eyed Brown, Baltimore Checkerspot**, and **Harris' Checkerspot**, among others.

Fortunately, the Acadian region retains much of its original wetland footprint relative to other areas of the eastern United States and southern Canada. Nonetheless, Maine is estimated to have lost up to 20% of its wetland area since the 1780s (Dahl 1990), and incremental losses continue to take a toll through wetland conversion and general degradation from invasive plants, illegal dumping, excessive nutrient runoff, and fragmentation of the riparian fringe, especially in rapidly developing areas of southern Maine. Laws protecting freshwater wetlands are relatively strong in Maine compared with many jurisdictions, but increased capacity is needed by the state's natural resource agencies to improve public awareness, compliance, and enforcement.

Peatlands

Partly owing to historical glaciation, northeastern North America hosts an exceptional number and diversity of bogs and fens ("peatlands"; Figure 44), which represent a special class of freshwater wetland. In Maine alone, seven ecosystems and sixteen natural community types are recognized by the Maine Natural Areas Program (Gawler and Cutko 2018) to describe different peat-dominated habitats. Cold, acidic, and saturated, such peatlands provide a harsh environment that is not generally conducive to high biotic richness. However, the plants and animals that are found in these settings are often quite specialized and thus of scientific and conservation interest. Some of the butterflies most closely associated with peatlands in the Acadian region include the **Pink-edged Sulphur, Bog Elfin, Bog Copper, Clayton's Copper, Crowberry Blue, Bog Fritillary, Frigga Fritillary,** and the often pursued but rarely captured **Jutta Arctic.**

Commercial peat extraction poses a direct threat to our region's peatlands. While this activity was previously more frequent in Maine, it is now conducted by only a small handful of operators (M. Stebbins, Land Division Director, Maine Department of Environmental Protection, pers. comm. to P. deMaynadier, October 2021). The practice is more significant in the Maritimes, particularly in New Brunswick. Canada is the world's largest producer of horticultural peat, and despite its relatively small size, New Brunswick accounts for 38% of this production, more than any other province (Peatland Ecology Research Group 2021). New Brunswick has approximately 1400 km^2 (540 mi^2) of peatland, approximately 100 km^2 (38 mi^2) of which has been disturbed, mostly from peat extraction (E. Prystupa, Peat Resource Specialist, New Brunswick, pers. comm. to J. Klymko, March 2021). Nearly all peat mines in New Brunswick are on government land (crown land). Under the province's peat mining policy, all such land that is leased for peat

Figure 44. The senior author scans for the Bog Fritillary and Jutta Arctic in a peatland, Northeast Carry Twp, ME (Bryan Pfeiffer).

production must be restored to a "natural wetland state" once mining is concluded (New Brunswick Department of Energy and Mines 2014). It is unclear if peatland-associated butterflies are capable of repopulating restored sites. Fortunately, New Brunswick has already protected approximately 350 km^2 (135 mi^2) of peatland habitat, and most of Maine's peatlands are protected from direct disturbance by the state's Natural Resources Protection Act. Climate change is another important threat to our region's peatlands given their potential heightened sensitivity to warming temperatures and changing precipitation patterns (Smith and Hayden, 2009). Given the fragility and irreplaceability of peatlands, and the unique biota found there, natural resource agencies should enact strict environmental review standards prior to allowing further development or commercial activity that could degrade remaining peatlands in our region.

Salt Marsh and Coastal

Salt marshes, dunes, and islands of the Atlantic Coast are home to a suite of unique plants and animals that are adapted to a saline environment (Figure 45). The Acadian region has three species and two subspecies of butterflies that are largely restricted to coastal habitats: **Maritime Ringlet**, **Salt Marsh Copper**, **Crowberry Blue**, the *mildredae* subspecies of **Silvery Blue** (possibly extinct), and the *bretonensis* subspecies of **Short-tailed Swallowtail**. Notably, four of these taxa are endemic to

Figure 45. Coastal salt marsh, Halifax County, NS (Sean Basquill, Nova Scotia Natural Resources and Renewables)

the southern Gulf of St. Lawrence region and therefore of global conservation significance. Thousands of small marine islands provide an extension of coastal habitat that remains largely unexplored in our region. Many support breeding residents and serve as off-shore refugia for migrants and strays. For example, Williams (2019; K. Lindquist, poet and naturalist, pers. comm. to P. deMaynadier, January 2022) have recorded forty-nine butterfly species (~41% of the Acadian fauna) on Monhegan Island, an isolated land area of about 2.2 km^2 (~1 mi^2) located 22 km (~14 mi) off midcoast Maine. In addition, there is hope that the *mildredae* subspecies of Silvery Blue persists on the very isolated St. Paul Island, located 24 km (15 mi) off the northern tip of Cape Breton, Nova Scotia.

Large amounts of salt marsh habitat has been lost in our region, including up to 65% from Canada's Atlantic Coast (Cox 1993), and nearly all coastal habitats are currently threatened by development, particularly in southern Maine and northern New Brunswick. Cottage and residential development, and the associated pollution, are one of the primary threats to the Maritime Ringlet in salt marshes in the Bathurst area of New Brunswick, which supports the largest population of the species in the world (COSEWIC 2009). Future sea-level rise is a significant threat to coastal biota throughout the Acadian region, especially to the Maritime Ringlet and Salt Marsh Copper. Between 2000 and 2100, the projected sea-level rise in the southern Gulf of St. Lawrence is 50–59 cm (± 35 cm) (19–23 in [± 14 in]), which is more than double the rate of the previous century (Forbes et al. 2006). It is unclear if salt marsh habitat will retreat inland fast enough to keep up with this accelerated rise. Given the looming threat of marine inundation, the future of our region's salt marshes and adjacent upland parcels require immediate planning and protection.

Sub-Boreal Woodlands

Several boreal plants (e.g., White Spruce, *Picea glauca*; Jack Pine, *Pinus banksiana*; Common Labrador-tea, *Rhododendron groenlandicum*) and animals (e.g., Canada Lynx, *Lynx canadensis*; Boreal Chickadee, *Poecile hudsonicus*; and historically Woodland Caribou, *Rangifer tarandus caribou*), including butterflies, approach the southern edge of their range in the Acadian region. Some of these species are known from relatively few, scattered records in our current atlas (e.g., **Arctic Fritillary** and **Northern Blue**), as may be expected for insects that have a preferred climate niche and habitat conditions that are predominantly north of our region. However, the apparent rarity of other butterflies in this group, for example **Western Pine Elfin, Satyr Comma,** and **Hoary Comma**, is more likely due to reduced survey efforts by naturalists in the remote wildlands of northern Maine and the Maritimes (Methods and Biogeographical Findings, Figures 29 and 30). Other more common species that are characteristic of our northern mixed conifer forests and openings (natural glades, outcrops, logging roads, and openings) include the **Arctic Skipper, Common Branded Skipper, Mustard White, Green Comma,** and **Atlantis Fritillary.**

Much of the habitat of these species in our region is managed by industrial forest companies whose land holdings often include entire watersheds across multiple townships, amounting to tens of thousands of hectares (hundreds of thousands of acres). In these remote settings, the risk of habitat loss to development is generally low, but intensive silvicultural practices (e.g., clearcutting, stand-type conversion, herbicide application, and short rotation management) can degrade the flora of the forest floor, leading to local extirpation for some specialized Lepidoptera. A warming climate is likely a more permanent threat to our north woods

Figure 46. Black Spruce woodland heath, Yarmouth County, NS (Sean Basquill, Nova Scotia Natural Resources and Renewables)

butterflies and other boreally affiliated biota. Conducting comprehensive biotic inventories of exemplary patches of sub-boreal woodland (Figure 46) in Nova Scotia, New Brunswick, and northern Maine is critical if we hope to document impending climate-related shifts in the distributions of some of our region's most vulnerable forest species.

Mixed Hardwood Forests

Sugar Maple (*Acer saccharum*), Yellow Birch (*Betula alleghaniensis*), White Ash (*Fraxinus americana*), American Beech (*Fagus grandifolia*), Northern Red Oak (*Quercus rubra*), White Oak (*Quercus alba*), American Basswood (*Tilia americana*), and Hop-hornbeam (*Ostrya virginiana*) are some of the notable trees constituting the mid- to late-successional hardwood forests of the Acadian region (Figure 47). Beneath their canopy, particularly in mature stands underlain by richer soils, a verdant layer of woodland sedges, herbs, and shrubs is often found, offering a diversity of larval food opportunities for butterflies. Depending on geography and understory composition, mixed hardwood forests in our region may be inhabited by the **Pepper and Salt Skipper, Hobomok Skipper, Canadian Tiger Swallowtail, Spicebush Swallowtail, Banded Hairstreak,**

Figure 47. Northern hardwood forest, Stow, ME (Donald Cameron)

Early Hairstreak, Gray Comma, White Admiral, Little Wood Satyr, Appalachian Brown and **Northern Pearly-Eye.**

Many of the trees dominating the climax hardwood forest of the Acadian region are of significant economic value. As a result, they are harvested long before their natural age of mortality (>150–200 years for many species) for products such as saw timber, pulpwood, building veneer, and firewood. The effects of forest harvesting on most invertebrates, including Lepidoptera, is not well studied in the Northeast. Based on our observations, northern hardwood forest silviculture is compatible with the conservation of most native butterflies when conducted in a manner that preserves native understory vegetation and buffers impacts to the margins of streams, swamps, and other depressions. More intensive harvesting of this forest type risks opening the canopy to an extent that favors weedy invasives, such as exotic shrubby honeysuckles (*Lonicera* spp.), barberry (*Berberis* spp.), Asiatic Bittersweet (*Celastrus orbiculatus*), and Glossy Buckthorn (*Frangula alnus*), all of which may outcompete a formerly diverse understory of native wildflowers, ferns, sedges, and shrubs. Land managers and loggers can minimize the potential for invasive species impacts in managed forests by checking and cleaning one of the most important vectors: contaminated harvesting equipment, most notably truck and skidder tires and feller buncher tracks.

Closing Conservation Thoughts

We wish to emphasize a final point regarding the use of habitat protection as a tool for conserving butterfly communities of the Acadian region and beyond. Biologists and planners frequently focus their attention on conserving large, intact landscapes that are expansive enough to host viable populations of large-bodied vertebrates (e.g., wolves, bears, and eagles) and remote enough to be mostly unaffected by the fragmenting effects of roads, crop fields, and subdivisions. While there are many reasons to value large, ecologically intact natural areas, smaller patches of unique habitat (e.g., sandplain grasslands, pocket swamps, enriched outcrops, riparian meadows) should not be neglected. In fact, most of our region's biological diversity (>99%) is composed of smaller, less space-demanding species, such as salamanders, shrews, wildflowers, mosses, mushrooms, and myriad invertebrates. Of course, among the latter are insects, including our beloved butterflies. And as readers will learn in the species accounts that follow, many butterflies have highly specialized habitat affinities. Their very presence is determined by specific soil type, topography, and understory flora; all elements that may be defined in just fractions of a hectare. Even small, discrete islands of habitat can serve an out-sized role in furnishing the ecological heterogeneity required to support a region's full complement of native flora and fauna (Volenec and Dobson 2019; Wintle et al. 2019). This is especially true throughout much of Maine and the Maritimes, where smaller patches of high-value habitats are often well-buffered and connected by a matrix of wildlands and working forests. Ecologists have long recognized that habitat diversity begets species diversity, a tenet that is as true for butterflies as it is for other biota. A systematic approach to protecting representatives of all our natural ecosystems, small and large, is called upon if we hope to preserve the Acadian region's full array of biological diversity.

SPECIES ACCOUNTS

FORMAT AND USE OF THE SPECIES ACCOUNTS

The species accounts that follow provide regionally specific information for the 121 butterfly species that have been documented in Maine and the Maritime Provinces. For ease of reference and to facilitate comparison between species, each account is arranged using a standardized format.

Popular and scientific names are provided. With few exceptions, popular names follow the North American Butterfly Association's *Checklist of North American Butterflies Occurring North of Mexico* (NABA 2018). Popular names can change over time to reflect species relationships or evolving cultural perceptions, such as replacing the long-held name Indian Skipper with the more suitable Sassacus Skipper as done here. Scientific nomenclature is generally in accordance with the online *A Catalogue of the Butterflies of the United States and Canada* (Pelham 2022), which is regularly updated to reflect current research. Ongoing taxonomic studies, including emerging genomics techniques, have resulted in suggested name changes, but disagreement will always exist over accepted usage. Rare deviations from Pelham (2022) in this book are based on our own observations within the region. The scientific name is composed of the genus, species epithet, and author. The author is the person who originally described the species. Because, J. C. Fabricius originally described the Black Swallowtail, the full scientific name of this species is *Papilio polyxenes* Fabricius. Subsequent taxonomists sometimes move a species name from the original genus to another. For example, the Spicebush Swallowtail was originally described as *Papilio troilus*, but we now use the name *Pterourus troilus*. When this occurs, the author name is placed in parentheses. As a result, the full scientific name of the Spicebush Swallowtail is *Pterourus troilus* (Linnaeus) because the author (Carl Linnaeus) originally described the species in the genus *Papilio*. Following scientific convention, the scientific name is always given in italics.

The **subspecies** name (if applicable) is also given. Subspecies represent groups of populations that are believed to be genetically and geographically distinguishable from other populations within the species' range. A nominotypical subspecies is the race first described; thus it defines the original concept of the species. For such subspecies, the species name is duplicated. For example, populations in our region of the Northern Cloudywing are considered to represent the nominotypical

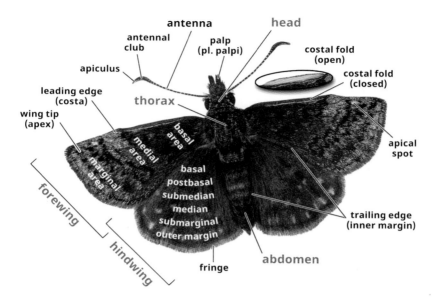

Dorsal aspect of adult Dreamy Duskywing (Hesperiidae) showing external morphology. Named wing regions apply to both the forewing and hindwing.

subspecies and are identified as *Thorybes pylades pylades*, or *T. p. pylades* for short.

The **distinguishing characters** section summarizes key morphological features that are useful in distinguishing each species. Relative sizes provide perspective. For example, our smallest species, the Bog Elfin with an average wingspan of less than 1 in (2.5 cm), is described as very small. On the other hand, the Eastern Giant Swallowtail, our largest species with a wingspan that can exceed 5 in (13 cm), is considered very large. The use of technical terms is kept to a minimum, and frequently used anatomical features are labeled in the figures shown. Species with similar-looking adults in the Acadian region are listed under **similar species.**

The **status and distribution** section is specific to Maine and the Maritime Provinces, with standardized details on residency status (Resident, Frequent colonist, Rare colonist, Stray, Extirpated), relative abundance (Rare, Uncommon, Common, Abundant), regional distribution, and conservation status (both legal status and NatureServe S-ranks, as defined in Appendix B).

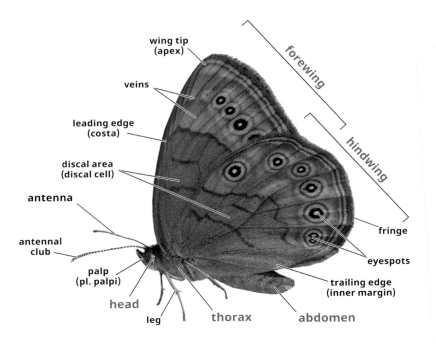

Ventral aspect of adult Eyed Brown (Nymphalidae) showing external morphology.

The **distribution map** illustrates records at the township level in Maine and at the 10 km × 10 km Universal Transverse Mercator grid square in the Maritime Provinces. Modern records (those from 1996 and later) are based on specimens and photographs. Historical records (those from 1995 and earlier) are based on specimens, photographs, and published records.

The **biology** section identifies the number of generations per year, flight period, larval food, and overwintering stage. Further specifics on flight dates can be gleaned from the **flight phenogram**, which depicts the number of adult butterfly records per 6-day block, as well the earliest and latest dates that adults have been recorded. For species that occur across the region, separate phenograms are given for Maine and the Maritime Provinces. Only records of wild adults are included (i.e., laboratory rearings are omitted).

The **adult behavior** section summarizes various behaviors associated with each species, including flight style and primary nectar and other adult food sources within the Acadian region.

The **comments** section offers regionally specific information and commentary on various topics, including, for example, unusual phenotypes, interesting behaviors and adaptations, historical occurrence in the region, and relevant taxonomic details.

Where appropriate, acute conservation **threats** that affect the profiled species are identified. More general threats that affect multiple species are discussed in the chapter Butterfly Conservation in the Acadian Region.

FAMILY HESPERIIDAE

A large family of 300 species in the United States and Canada, skippers are mostly small or medium-sized, drab-colored butterflies. They have a rapid, darting flight, which long ago earned them the name "skippers." They differ from true butterflies and were previously segregated into their own superfamily, the Hesperioidea. Skippers are now generally grouped together with other butterflies in the superfamily Papilionoidea. Because skippers look and behave differently from other butterflies, and can be difficult to identify, they are often ignored. Skippers tend to have more robust, moth-like bodies, and most bear a distinctive hook (apiculus) on the antennal club, which other butterflies lack. Males of some species possess a costal fold along the leading edge of each forewing that contains scent scales (androconia) that are used during courtship. Skippers are found in a variety of habitats, from coastal salt marshes to mountaintops. Some of the rarest and most interesting species are found in at-risk habitats, such as barrens and wetlands. In our region, skippers are grouped into four subfamilies that can often be distinguished by their resting posture. Dicot skippers (Eudaminae) tend to hold their wings partially opened, while spread-winged skippers (Pyrginae) usually hold their wings outstretched. Our single species of skipperling (Arctic Skipper; Heteropterinae) also perches with open wings. Grass skippers (Hesperiinae) do both, resting with partially opened forewings and outstretched hindwings; this posture has been described as resembling a miniature jet fighter. Food plants of skipper larvae vary, but many species feed on grasses and sedges.

Thirty-four species of Hesperiidae have been recorded in Maine and the Maritimes. At least thirty are breeding residents, most of which produce a single generation per year. Four species are state listed in Maine as Threatened (Sleepy Duskywing) or Special Concern (Leonard's Skipper, Cobweb Skipper, and Dusted Skipper).

Recent genomics work by Cong et al. (2019) and Li et al. (2019) suggests additional taxonomic changes that would affect the names of our skippers (such as placing most of our duskywings in the genus *Gesta*), but we follow Pelham (2022) pending additional research.

European Skippers—Williamsburg, MA (Frank S. Model). Accidentally introduced into North America (Ontario) around 1910, this skipper is now one of the most abundant butterflies in northeastern North America.

SOUTHERN CLOUDYWING

Thorybes bathyllus (J. E. Smith)

Subspecies: None.

Distinguishing characters:
   Size: Small to medium.
   Upperside: Dark brown; forewing with three small glassy spots near apex and several larger glassy spots in loose row across center of wing; male lacks costal fold (present in Northern Cloudywing).

Acton, MA, 13 July 2013 (Erik Nielsen)

   Underside: Mottled brown with light gray scaling near wing margins; forewing with glassy spots like upperside; hindwing with two irregular dark bands.
   Additional: "Face" (palpi) pale.

Similar species: Northern Cloudywing and Juvenal's Duskywing.

Status and distribution in Maine: Resident or rare colonist. Rare; limited to York County. S1

Status and distribution in Maritimes: Not documented.

Habitat: Various dry sites including woodland openings and edges, barrens, meadows, and rights-of-way.

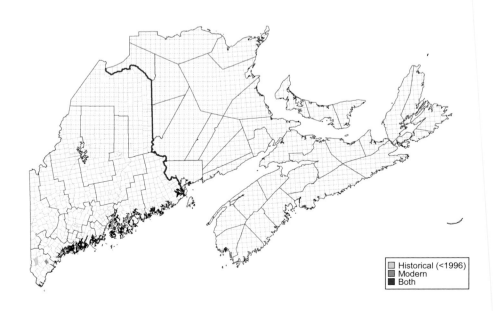

Historical (<1996)
Modern
Both

Biology: One generation per year. Maine records are from 1 June and 14 June. In Massachusetts, it flies from early June to mid-July (Stichter 2015). Larvae feed on legumes, such as bush-clovers (*Lespedeza* spp.) and tick-trefoils (*Desmodium* spp.). Overwinters as a fully grown larva.

Adult behavior: The flight is low and rapid. Males await females from low perches and occasionally visit damp soil. Both sexes frequently nectar, but the only documented flower visited in our region is dewberry (*Rubus* sp.).

Comments: The Southern Cloudywing is known in our region from just two specimens collected in York County many years apart: Shapleigh barrens in 1981 (Dana MacDonald and Dale Schweitzer), and Union Falls Dam in Dayton in 2012 (Pamela and Ronald Moore). Although twenty individuals were reported at the Dayton locality, none were found there by John Calhoun during a visit in early June 2019, and it remains unclear whether the species is a permanent resident or occasional colonist. Brower (1974) identified an old female specimen from Norway (Oxford County) as this species, but it is actually a Northern Cloudywing.

Threats: The dry, open habitats where this species is most likely to be resident in southern Maine are threatened from development, gravel mining, and ecological succession.

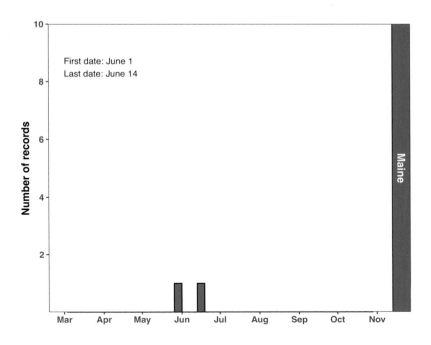

First date: June 1
Last date: June 14

NORTHERN CLOUDYWING

Thorybes pylades (Scudder)

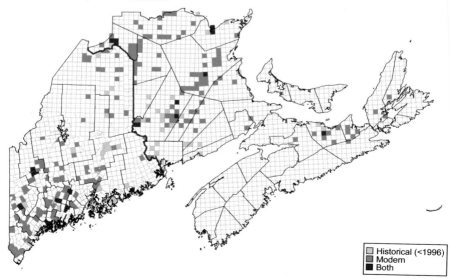

Subspecies: Only the nominotypical subspecies (*T. p. pylades*) occurs in eastern North America.

Distinguishing characters:
  Size: Small to medium.
  Upperside: Dark brown; forewing with scattered small white angular spots (often reduced or lacking in male), and male with costal fold (lacking in Southern Cloudywing).

Swan Island, ME, 21 June 2014 (Rose Marie Gobeil)

  Underside: Mottled brown with gray scaling near wing margins; forewing with small white spots like upperside; hindwing with two irregular dark bands.
  Additional: "Face" (palpi) dark.

Similar species: Southern Cloudywing and Juvenal's Duskywing.

Status and distribution in Maine: Resident. Common; widespread. S5

Status and distribution in Maritimes: Resident. Common in northern and western New Brunswick, rare in southeast; uncommon in northern Nova Scotia, including Cape Breton Island. NB-S5, NS-S2S3

Habitat: Various open (often dry) sites including forest clearings, utility corridors, old fields, and roadsides.

☐ Historical (<1996)
▨ Modern
■ Both

Biology: One generation per year. Adults are active from late May to late July, with a peak abundance in mid-June. Larvae feed on various legumes, including American Hog-peanut (*Amphicarpaea bracteata*), vetches (*Vicia* spp.), tick-trefoils (*Desmodium* spp.), bush-clovers (*Lespedeza* spp.), and clovers (*Trifolium* spp.). Overwinters as a fully grown larva.

Adult behavior: This skipper has a rapid, erratic flight. Males vigorously defend territories, flying out from low perches to investigate passing insects in search of females. Males also frequently imbibe nutrients from damp soil and animal dung. Both sexes visit flowers, including Cow Vetch (*Vicia cracca*), hawkweeds (*Hieracium* spp.), and Red Clover (*Trifolium pratense*).

Comments: The Northern Cloudywing occurs in an array of habitats across its broad North American range, from dry montane forest in Mexico to aspen parkland in British Columbia. Assuming all such populations are represented by the same species (and not a group of unrecognized sibling species), this apparent adaptability makes its patchy distribution in the Maritimes difficult to explain. At a distance, cloudywings look very much like duskywings. However, cloudywings tend to fly higher and faster and rest with their wings partially closed, while duskywings often rest with their wings fully outstretched.

Threats: None documented in our region.

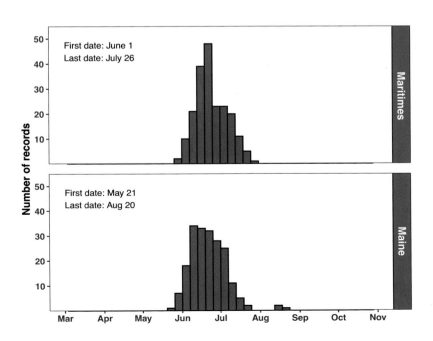

LONG-TAILED SKIPPER

Urbanus proteus (Linnaeus)

Subspecies: Only the nomino-
typical subspecies (*U. p. proteus*) occurs in eastern North
America.

Distinguishing characters:
  Size: Medium.
  Upside: Brown, with
    iridescent green or
    bluish-green wing bases
    and body; forewing with
    scattered translucent
    spots (larger in female).

Orange County, FL, 30 September 2017 (Edward Perry IV)

  Underside: Forewing like upperside; hindwing with dark bands and spots.
  Additional: Hindwing with broad, black tails.

Similar species: Individuals that have lost their tails resemble Southern
Cloudywing.

Status and distribution in Maine: Stray or possibly rare colonist. Rare;
recorded only from Kennebec and Knox Counties. SNA

Status and distribution in Maritimes: Not documented.

Habitat: Migrants could be found in virtually any sunny, open area, including
city parks and gardens.

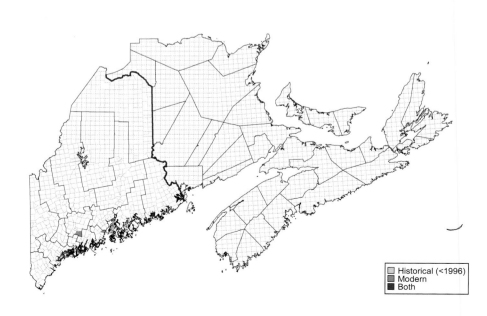

Historical (<1996)
Modern
Both

Biology: Multiple generations annually; adults are active all year where resident in the Deep South. Our few records are from late July and late August. Larvae feed on various legumes (Fabaceae), including tick-trefoils (*Desmodium* spp.) and beans (*Phaseolus* spp.). This species is known as the Bean Leafroller in the South, where it can damage cultivated legumes, such as soybeans (*Glycine max*). Overwinters as an adult in reproductive diapause but not in our region.

Adult behavior: This skipper has a rapid, erratic flight. Adults pause often to nectar and in our region have been observed visiting Orange-eye Butterfly-bush (*Buddleja davidii*), bee-balms (*Monarda* spp.), Purple-topped Vervain (*Verbena bonariensis*), and Cow Vetch (*Vicia cracca*).

Comments: The four individuals of this species recorded in Maine were photographed in 2012 and 2020. In Massachusetts, many Long-tailed Skippers were documented from August to October 2012 (Stichter 2015), which was a good year for southern migrants in northeastern North America. Two of the adults found in Maine were in fresh condition, suggesting that this species can establish temporary populations in our region (though it has not yet been confirmed to breed in Massachusetts). Climate change may expand this species' range northward, possibly resulting in a greater number of future records, especially during eruption years. Observers in the Acadian region should watch for this striking skipper, particularly in coastal areas, while keeping in mind that worn individuals are less conspicuous when missing their distinctive tails.

Threats: None documented in our region.

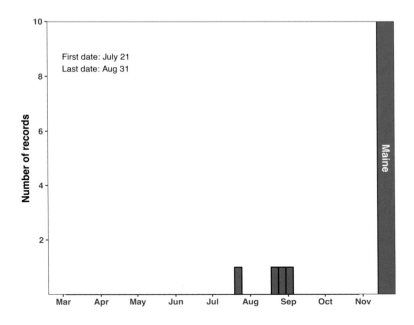

First date: July 21
Last date: Aug 31

Number of records

Maine

SILVER-SPOTTED SKIPPER

Epargyreus clarus (Cramer)

Camden, ME, 12 June 2012 (Roger Rittmaster)

Subspecies: Only the nomino-typical subspecies (*E. c. clarus*) occurs in eastern North America.

Distinguishing characters:
    Size: Medium (our largest resident skipper).
    Upperside: Dark brown; forewing with row of large orange spots.
    Underside: Dark brown; forewing with row of large orange spots, hindwing with large, elongate white patch.
    Additional: Forewings especially long and pointed.

Similar species: None.

Status and distribution in Maine: Resident. Common; limited to southern half of state. S4S5

Status and distribution in Maritimes: Resident. Rare; known only from southern half of New Brunswick. NB-S3

Habitat: Forest openings and edges, old fields, roadsides, gardens, and even urban green spaces.

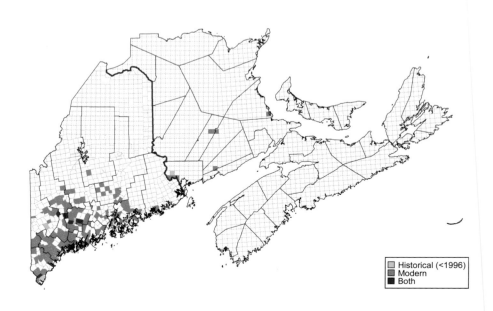

Historical (<1996)
Modern
Both

Biology: One protracted generation per year, with a partial second brood in some years. Adults are active from mid-May until late August, with a peak abundance in early July. Larvae feed on legumes (Fabaceae), including Black Locust (*Robinia pseudoacacia*), tick-trefoils (*Desmodium* spp.), Common Ground-nut (*Apios americana*), and bush-clovers (*Lespedeza* spp.). Over-winters as a pupa.

Adult behavior: This skipper has a powerful, direct flight. Males defend territories and fly out to investigate passing objects in search of females, which tend to remain near food plants and nectar sources. Both sexes visit a variety of flowers, such as Common Milkweed (*Asclepias syriaca*), Orange-eye Butterfly-bush (*Buddleja davidii*), Cow Vetch (*Vicia cracca*), Spreading Dogbane (*Apocynum androsaemifolium*), and Red Clover (*Trifolium pratense*).

Comments: The Silver-spotted Skipper is a successful opportunist that is spreading north and east within our region. Scudder (1888–1889) listed only two collection localities in Maine, in Oxford and Kennebec Counties. It was not recorded in Penobscot County until 2004, despite this county having been well collected historically. In the Maritimes, it was long known only from a single specimen collected in 1982 at Moores Mills, Charlotte County, New Brunswick, and its breeding status in that province was uncertain. However, since 2017 colonies have been discovered in Charlotte, Sunbury, and Queens Counties, and individual adults have been found in Kings and Kent Counties. This expansion is probably the result of the widespread planting of Black Locust as an ornamental and for soil stabilization and possibly of a warming climate.

Threats: None documented in our region.

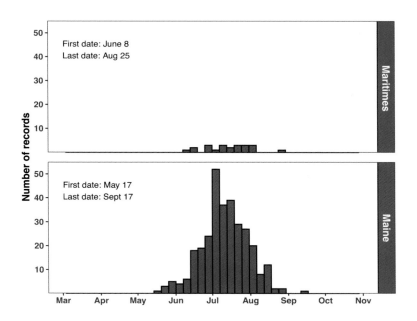

COMMON SOOTYWING

Pholisora catullus (Fabricius)

Subspecies: None.

Distinguishing characters:
   Size: Small to medium.
   Upperside: Sooty black or
      blackish brown, becom-
      ing all brown when worn;
      outer forewing with nu-
      merous small white spots
      (more pronounced in
      female); male with costal
      fold on forewing.

Northampton, MA, 21 August 2006 (Frank S. Model)

   Underside: Similar to
      upperside but browner.
   Additional: Distinctive coat of white scales on underside of head and
      thorax.

Similar species: None.

Status and distribution in Maine: Stray or rare colonist. Rare; recorded only
from the southern counties of York and Oxford. SNA

Status and distribution in Maritimes: Not documented.

Habitat: Strays can be found in virtually any open, disturbed area; colonies
are typically found in areas with food plants and patches of low vegetation,
such as pastures, gardens, vacant lots, and edges of agricultural fields.

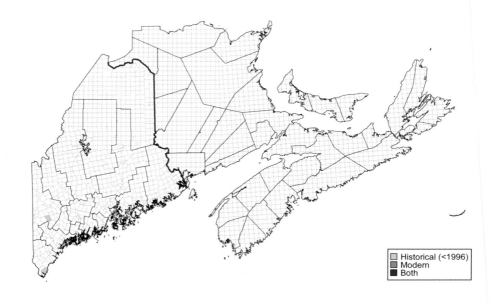

	Historical (<1996)
	Modern
	Both

Biology: Two to three generations where resident; Maine's only detailed record is from 18 August. In Massachusetts, at the northern edge of its resident range, the Common Sootywing flies from late April to early September (Stichter 2015). Larvae are reported to feed on naturalized Lambsquarters (*Chenopodium album*) and other members of the amaranth family (Amaranthaceae). Overwinters as a fully grown larva but not in our region.

Adult behavior: This wary species has a low, erratic flight and usually rests with its wings held outstretched or partially open. Males patrol for females low to the ground, usually at the edges of taller vegetation, and sometimes visit moist soil. Both sexes visit flowers, but there are no nectaring observations in our region.

Comments: The Common Sootywing has been recorded in the Acadian region only two times, over a century apart. It was reportedly collected at Norway (Oxford County) by Sidney Smith (Scudder 1888–1889), possibly during the 1860s. More recently, a specimen was collected by Richard Folsom at York (York County) on 18 August 1988. While this butterfly's original habitat likely included exposed river shores, sand plains, and possibly coastal salt marshes (Opler and Krizek 1984; Cech and Tudor 2005), it now occupies a wide variety of weedy, disturbed sites. Despite its name, the Common Sootywing is generally uncommon. It is widespread as far north as Massachusetts, where it is known to overwinter regularly and produce at least two generations (Stichter 2015). More study is needed to determine if this skipper can establish temporary (nonoverwintering) colonies in southern Maine.

Threats: None documented in our region.

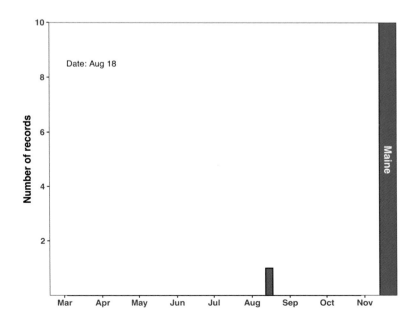

DREAMY DUSKYWING

Erynnis icelus (Scudder & Burgess)

Subspecies: None.

Distinguishing characters:
   Size: Small to medium.
   Upperside: Dark brown; forewing with dark-edged, gray, chain-like pattern on outer half and usually a single, small, white spot on leading-edge (missing in Sleepy Duskywing); hindwing with dark brown indistinct spotting on outer half.

Female—Magalloway Plantation, ME, 9 June 2018 (Josh Lincoln)

   Underside: Brown; hindwing with rows of indistinct pale spots.
   Additional: Wingspan lesser and labial palps longer than in Sleepy Duskywing.

Similar species: Sleepy Duskywing.

Status and distribution in Maine: Resident. Abundant; widespread. S5

Status and distribution in Maritimes: Resident. Abundant; widespread. NB-S5, NS-S5, PE-S5

Habitat: A variety of open areas, including forest edges, meadows, stream-sides, and gravel roads; often in moist areas.

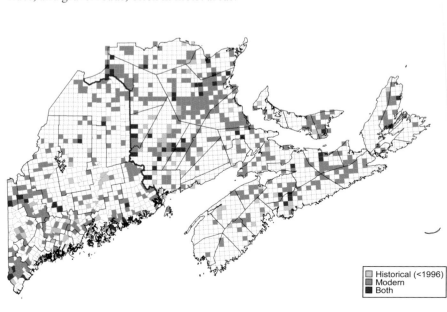

Historical (<1996)
Modern
Both

Biology: One generation per year. Adults are active from early May to mid-July, with peak abundance in mid-June. Larvae feed on aspens (*Populus* spp.), willows (*Salix* spp.), and birches (*Betula* spp.). Overwinters as a fully grown larva.

Adult behavior: The flight is low and rapid. Males perch or course back and forth within small territories in search of females and occasionally congregate on damp soil and dung. Adults often rest on open ground, with wings outstretched. They nectar at a variety of flowers in our region, particularly hawkweeds (*Hieracium* spp.), Common Lowbush Blueberry (*Vaccinium angustifolium*), Common Dandelion (*Taraxacum officinale*), Red Clover (*Trifolium pratense*), and Cow Vetch (*Vicia cracca*).

Comments: Duskywings (*Erynnis* spp.) are notoriously difficult to identify, but with only four confirmed species from our region (compared with eight to ten species in southern New England and Ontario), even novice observers can learn to differentiate them. Geography, habitat, and flight period are helpful clues. Both the Dreamy Duskywing and the larger, very similar Sleepy Duskywing are named for the lack of conspicuous eyespots on their wings, which looked "closed" (hence sleeping) to early researchers. Confusion over the identification of these two species resulted in erroneous historical reports of the Sleepy Duskywing in our region. For example, Scudder (1888–1889) described the Sleepy Duskywing as "very common" in Nova Scotia, where only the Dreamy Duskywing is known to occur.

Threats: None documented in our region.

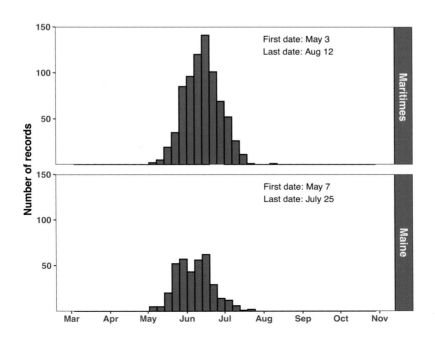

SLEEPY DUSKYWING

Erynnis brizo (Boisduval & Le Conte)

Subspecies: Only the nomino-typical subspecies (*E. b. brizo*) is found in eastern North America.

Distinguishing characters:
  Size: Small to medium.
  Upperside: Mottled brown with dark-edged, gray, chain-like pattern across the outer forewing; little to no grayish overscaling; hindwing dark brown with indistinct spotting on outer half.
  Underside: Brown; hindwing with rows of indistinct pale spots.
  Additional: Wingspan greater and labial palps shorter than in Dreamy Duskywing.

Female—Hollis, ME, 19 May 2016 (Bryan Pfeiffer)

Similar species: Dreamy Duskywing.

Status and distribution in Maine: Resident. Rare; restricted to York and southern Oxford County. State Threatened. S2

Status and distribution in Maritimes: Not documented.

Habitat: Dry oak–pine woodlands, barrens, and rocky outcrops.

Biology: One generation per year. Adults are active from early May to early June, with peak abundance in late May. In our region, Scrub Oak (*Quercus*

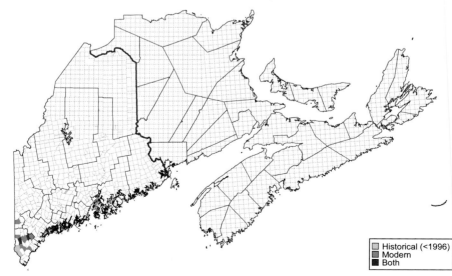

Historical (<1996)
Modern
Both

ilicifolia) is the only known larval food plant. Oaks are generally fed upon throughout this species' range, though Shapiro (1966) reported American Chestnut (*Castanea dentata*) in Pennsylvania, where females were observed ovipositing on regenerating sprouts (A. M. Shapiro, University of California, pers. comm. to J. Calhoun). Overwinters as a fully grown larva.

Adult behavior: This skipper has a low and rapid flight. Males patrol most of the day in search of females and visit damp soil along dirt roads and trails. Adults in Maine have been recorded nectaring at Common Low-bush Blueberry (*Vaccinium angustifolium*) and Highbush Blueberry (*V. corymbosum*).

Comments: The Sleepy Duskywing is difficult to distinguish from its congener, the Dreamy Duskywing, and both are active during spring in southern Maine. Habitat can be a helpful clue, with the Sleepy restricted to dry oak–pine woodlands and the Dreamy more often associated with moister forest edges and streamside openings. Nonetheless, they are found together in some barrens where Scrub Oak co-occurs with patches of aspens (*Populus* spp.) or birches (*Betula* spp.), which are larval food plants of the Dreamy Duskywing. A disjunct northern report of the Sleepy Duskywing from Waterville in central Maine (Scudder 1888–1889) is undoubtedly based on a misidentification of the Dreamy Duskywing. Owing to their similarity, duskywings are often misidentified in older literature.

Threats: The dry woodlands and barrens occupied by this butterfly (and other rare biota) are limited in distribution to southern Maine, where they are increasingly degraded by development and ecological succession resulting from fire prevention. In addition, aerial pesticide spraying for introduced Spongy Moths (*Lymantria dispar*), a pest that periodically defoliates large areas of oak, poses a potential risk to this and other barrens-inhabiting Lepidoptera.

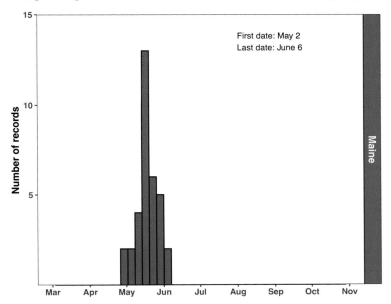

JUVENAL'S DUSKYWING
Erynnis juvenalis (Fabricius)

Subspecies: Only the nomino-typical subspecies (*E. j. juvenalis*) occurs in eastern North America.

Distinguishing characters:
    Size: Medium.
    Upperside: Dark brown (male) to mottled light and dark brown (female); forewing with row of small glassy spots.
    Underside: Forewing brown, with row of small glassy spots; hindwing darker brown, usually with two small white spots near leading edge.

Male—Sanford, ME, 5 June 2013 (W. Herbert Wilson Jr.)

Similar species: Wild Indigo Duskywing.

Status and distribution in Maine: Resident. Common in York, Cumberland, and southern Oxford Counties, uncommon elsewhere in southern third of state. S4

Status and distribution in Maritimes: Resident. Common in southern Nova Scotia, only two records from New Brunswick. NB-SU, NS-S4

Habitat: Various open areas, such as forest clearings, edges, trails, and barrens.

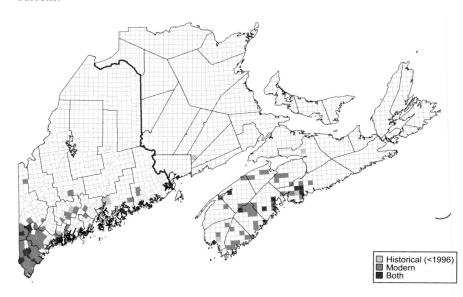

Historical (<1996)
Modern
Both

Biology: One generation per year. Adults are active from late April to late June (rarely July), with peak abundance in late May and early June. Larvae feed on oaks (*Quercus* spp.), including Northern Red Oak (*Q. rubra*), Eastern White Oak (*Q. alba*), and Scrub Oak (*Q. ilicifolia*). Overwinters as a fully grown larva.

Adult behavior: This skipper has a low, rapid flight, making it difficult to approach. Males patrol small territories in search of females. Like other dusky-wings, adults perch with wings outstretched, usually on open ground or low vegetation. Both sexes visit damp soil (females rarely) and flowers, especially blueberries (*Vaccinium* spp.), blackberries (*Rubus* spp.), and Common Dandelion (*Taraxacum officinale*).

Comments: No other butterfly in our region has such a strikingly disjunct distribution, being common in southern Maine and southern Nova Scotia yet virtually absent elsewhere. The Juvenal's Duskywing does not occur wherever oaks are abundant, suggesting additional habitat requirements are involved. With only two records in New Brunswick, nearly 40 years apart (1983 and 2021), it is unclear if the species has established a few small populations in the province (Red Oak is common in the vicinity of each record) or it is present only as a temporary colonist or vagrant. Sometimes found in the company of Dreamy and Sleepy duskywings, the more robust, energetic Juvenal's Duskywing is easily recognized.

Threats: None documented in our region.

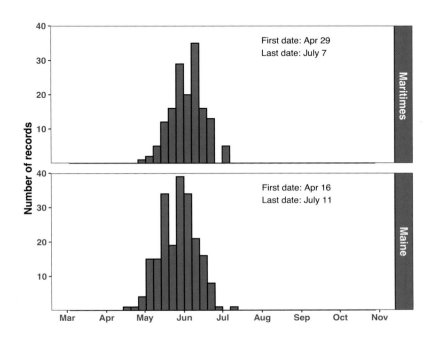

WILD INDIGO DUSKYWING
Erynnis baptisiae (W. Forbes)

Subspecies: None.

Distinguishing characters:
  Size: Small to medium.
  Upperside: Forewing brown, darker basally, with a paler patch near center of wing and several small, glassy spots (spots and pattern more pronounced in female); hindwing dark brown with indistinct paler spotting on outer half.

Male—Eliot, ME, 27 July 2018 (Bryan Pfeiffer)

  Underside: Brown, with several small, glassy spots on forewing and two rows of indistinct paler spots on hindwing.

Similar species: Juvenal's Duskywing.

Status and distribution in Maine: Resident. Locally common in southern third of state, with outlying recent records from Somerset and Washington Counties; range rapidly expanding. S4

Status and distribution in Maritimes: Not documented.

Habitat: Various (mostly disturbed) open areas, including roadsides, old fields, sparse woodlands, and barrens.

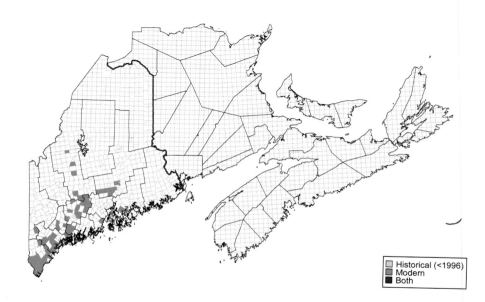

Historical (<1996)
Modern
Both

Biology: Two generations per year. Adults fly from early May to late September, with peak abundances in early June and late July/early August. Larvae feed on introduced Purple Crown-Vetch (*Securigera varia*). Overwinters as a fully grown larva.

Adult behavior: Like other duskywings, adults have a low, rapid flight and often perch with wings outstretched, usually on open ground or low vegetation. Males visit damp soil and dung, and both sexes visit flowers, including Purple Crown-vetch, clovers (*Trifolium* spp.), burdocks (*Arctium* spp.), coneflowers (*Rudbeckia* spp.), and ornamental plantings.

Comments: The Wild Indigo Duskywing originally was limited to the range of its native larval food plants, most notably Yellow Wild Indigo (*Baptisia tinctoria*). Although there are historical records of Yellow Wild Indigo in York County, Maine, the butterfly was not known to occur in the state. During the past few decades, Purple Crown-Vetch has widely been planted to stabilize slopes and prevent erosion along roads and highways, and it has spread into various other disturbed and natural habitats. The butterfly has adapted to feeding on this exotic plant, allowing it to colonize many new areas across much of eastern North America. It was first documented in Maine in 2007 in York County (Steve Walker and Phillip deMaynadier), though it was likely present before that (Gobeil and Gobeil 2016). By 2010 it had been recorded in three additional southern counties, and it had arrived in central Maine by 2011. The butterfly continues to expand its range, apparently via road corridors. It is just a matter of time before this invasion reaches New Brunswick.

Threats: None documented in our region.

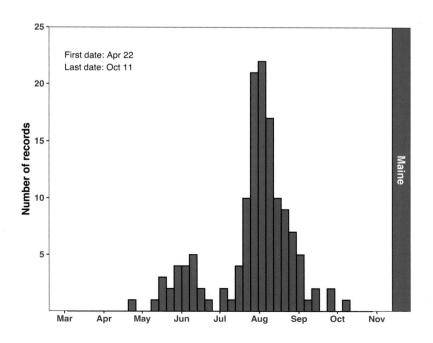

ARCTIC SKIPPER

Carterocephalus mandan
(W. H. Edwards)

Dartmouth, NS, 21 June 2011 (Jim Edsall)

Subspecies: None (see comments).

Distinguishing characters:
    Size: Small.
    Upperside: Dark brown, checkered with angular orange spots.
    Underside: Hindwing orange, with large, dark margined white spots.

Similar species: None.

Status and distribution in Maine: Resident. Common; widespread. S5

Status and distribution in Maritimes: Resident. Common; widespread. NB-S5, NS-S5, PE-S5

Habitat: Moist open grassy areas, including forest margins, streamsides, and wetland edges.

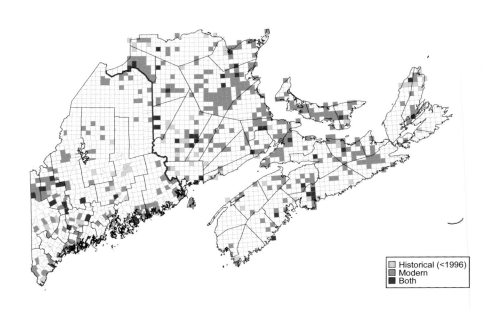

Historical (<1996)
Modern
Both

Biology: One generation per year. Adults are active from late May until mid-July, with peak abundance during mid-June. Larvae feed on various grasses (Poaceae), including bromes (*Bromus* spp.), bluejoints (*Calamagrostis* spp.), and blue grasses (*Poa* spp.). Overwinters as a fully grown larva.

Adult behavior: This skipper has a low, weak flight, and males perch on vegetation awaiting passing females. They are frequently observed basking with wings held partially or fully open. It regularly visits flowers, including Cow Vetch (*Vicia cracca*), clovers (*Trifolium* spp.), Common Dandelion (*Taraxacum officinale*), Common Blackberry (*Rubus allegheniensis*), strawberries (*Fragaria* spp.), and buttercups (*Ranunculus* spp.). Adults also frequent damp soil.

Comments: As its name suggests, the Arctic Skipper is a denizen of predominantly northern latitudes, albeit more boreal than Arctic in distribution. This small, attractive skipper is the only member of the subfamily Heteropterinae (known as skipperlings) in North America, where it was long considered to be conspecific with the widespread Eurasian species *C. palaemon*. Genomics research by Zhang et al. (2020) supports the conclusion of Pohl et al. (2010) that New World populations represent a discrete species named *C. mandan*. Although Pelham (2022) recognizes the subspecies *C. m. mesapano* (Scudder)—originally described as a species based on specimens from Norway (Oxford County), Maine—we agree with Pohl et al. (2018) that the subtle characters that define *mesapano* are insufficient to separate it from other populations of *C. mandan*.

Threats: None documented in our region.

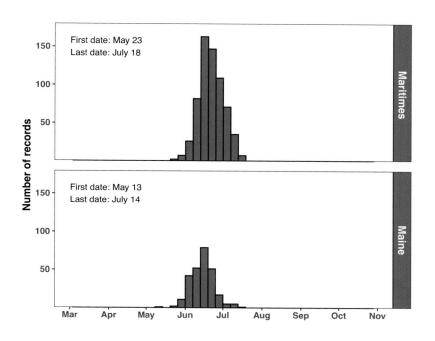

BLACK DASH

Euphyes conspicua
(W. H. Edwards)

Eliot, ME, 21 July 2020 (Bryan Pfeiffer)

Subspecies: Only the subspecies *E. c. orono* (Scudder) occurs east of the Great Lakes region.

Distinguishing characters:
  Size: Small.
  Upperside: Dark brown; male forewing with thick black stigma ("black dash") surrounded by orange; female forewing with diagonal row of prominent pale dots, the central-most nearly white.
  Underside: Brownish orange; hindwing with curved row of pale spots, two of which are elongated.

Similar species: Long Dash.

Status and distribution in Maine: Resident. Rare; localized, recorded only in Oxford, Cumberland, and York Counties. S3

Status and distribution in Maritimes: Not documented.

Habitat: Sedge-dominated marshes and wet meadows, straying to nearby uplands for nectar.

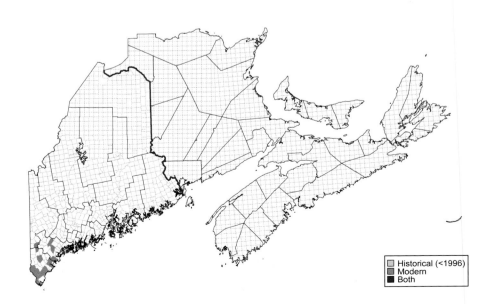

Historical (<1996)
Modern
Both

Biology: One generation per year. Adults fly from early July to mid-August, with a peak abundance in late July. Larvae feed on Tussock Sedge (*Carex stricta*) and potentially other sedge species. Overwinters as a young larva.

Adult behavior: Males usually perch on sedges to await passing females, which tend to stay low within marsh vegetation. Both sexes are strongly attracted to flowers, including Boneset Thoroughwort (*Eupatorium perfoliatum*), Common Buttonbush (*Cephalanthus occidentalis*), Joe-Pye weeds (*Eutrochium* spp.), and milkweeds (*Asclepias* spp.).

Comments: The weakly differentiated eastern subspecies, *E. c. orono*, is named after Chief Joseph Orono of the Penobscot Nation, not the city in Maine. In fact, this skipper was not discovered in Maine until 20 July 2009, when it was found by John Calhoun at five localities in three townships of York County. Many additional colonies were subsequently discovered, and the Black Dash is currently known from fifteen townships in three counties. Such a sudden burst of records suggests that this species only recently arrived in the state. This parallels the 2008 discovery in Maine of the Mulberry Wing, with which the Black Dash is often found (the North American ranges of these two skippers are strikingly similar). Although most Maine records of these species are the result of targeted wetland surveys, adults frequently stray to upland nectar sources, where casual observers may encounter them. Had these skippers historically been residents of southern Maine, it seems likely that they would have been recorded long ago.

Threats: Smaller marshes and wet meadows are at risk of filling, draining, and colonization by invasive plants, especially in rapidly developing southern Maine where this skipper is restricted.

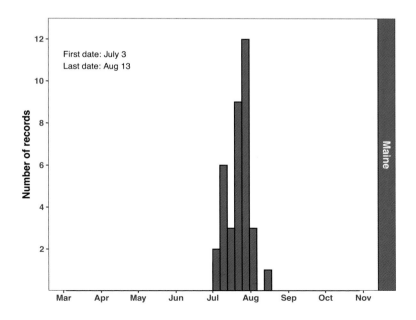

First date: July 3
Last date: Aug 13

Number of records

Mar Apr May Jun Jul Aug Sep Oct Nov

Maine

TWO-SPOTTED SKIPPER

Euphyes bimacula (Grote & Robinson)

Subspecies: Only the nomino-typical subspecies (*E. b. bimacula*) occurs in northeastern North America.

Distinguishing characters:
  Size: Small.
  Upperside: Dark brown; male forewing with long black stigma surrounded by yellowish orange; female forewing with a

Mating pair—Charlotte, NB, 12 July 2017 (John Klymko)

row of small, pale spots near the apex and two or three pale spots in the center of the wing; both sexes with white wing fringe when fresh.
  Underside: Orange; pale spots on forewing; paler scaling along hindwing veins; thick white fringe along inner edge of hindwing (adjacent to body).

Similar species: None.

Status and distribution in Maine: Resident. Uncommon; widespread but localized, just recently recorded in extreme north. S3

Status and distribution in Maritimes: Resident. Uncommon; widespread in New Brunswick but not reported from the northwest; otherwise known only from two locations in northern mainland Nova Scotia. NB-S3, NS-S1S2

Habitat: Sedge-dominated wet meadows, marshes, and fens.

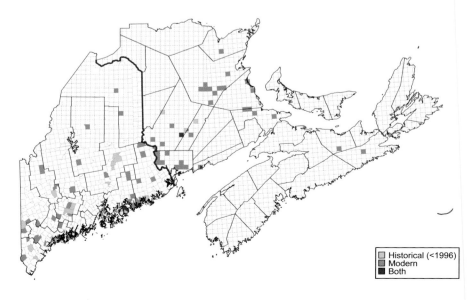

Historical (<1996)
Modern
Both

Biology: One generation per year. Adults fly from mid-June to late July, with a peak abundance in early July. Larvae feed on sedges, including Tussock Sedge (*Carex stricta*). Overwinters as a partially grown larva.

Adult behavior: The flight is rapid and adults are very wary. Males await females from perches in marsh vegetation or on the tops of sedge plants. Females are less active and mostly stay low within the sedges. Males occasionally visit damp earth, and both sexes frequent flowers in and around wetland habitats, including Blue Iris (*Iris versicolor*), Swamp Milkweed (*Asclepias incarnata*), Spreading Dogbane (*Apocynum androsaemifolium*), and Cow Vetch (*Vicia cracca*).

Comments: Generally uncommon or rare throughout its range, the Two-spotted Skipper is typically found in small numbers. However, its abundance can fluctuate at a given locality, being frequent one year while nearly absent the next. As with many of our wetland-associated butterflies, the increased survey efforts during the Maine and Maritimes atlas projects reveal just how little was previously known about the distribution of this species in our region. This is especially evident in the Maritimes, where it was first recorded in Nova Scotia in 2013 (Pictou County) by Ken McKenna.

Threats: The cumulative loss of wetland habitat to development is a localized threat, especially for smaller wet meadows and marshes that are not adequately protected by existing regulations.

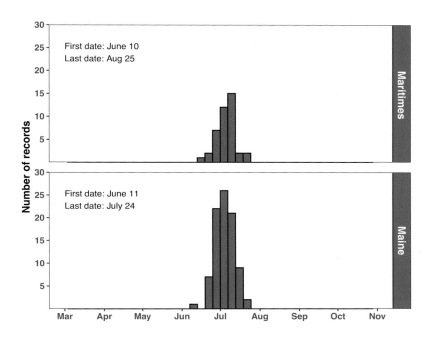

DUN SKIPPER
Euphyes vestris (Boisduval)

Subspecies: Only the subspecies *E. v. metacomet* (T. Harris) occurs in eastern North America.

Distinguishing characters:
Size: Small.
Upperside: Dark brown with bronze-colored head and thorax; male forewing with long black stigma (difficult to see); female with several small, pale spots across middle of forewing.

Waterford, ME, 18 July 2020 (Josh Lincoln)

Underside: Dark brown with purplish sheen when fresh; hindwing often with inconspicuous row of small, pale dots (especially in female).

Similar species: Little Glassywing and Northern Broken-Dash.

Status and distribution in Maine: Resident. Abundant; widespread. S5

Status and distribution in Maritimes: Resident. Abundant; widespread. NB-S5, NS-S5, PE-S4

Habitat: A wide variety of open, grassy areas, including roadsides, open woodlands, forest edges, old fields, and gardens.

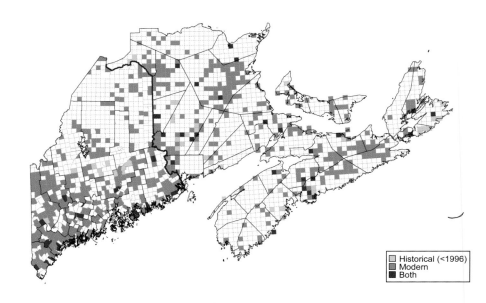

☐ Historical (<1996)
▨ Modern
■ Both

Biology: One generation per year. Adults fly from late June to late August, with a peak abundance in mid-July. Larvae feed on various sedges (Cyperaceae), including true sedges (*Carex* spp.), bulrushes (*Scirpus* spp.), and flatsedges (*Cyperus* spp.). Overwinters as a partially grown larva.

Adult behavior: Typical of small grass skippers, this species has a low, rapid flight. Males search for females from perches in low vegetation and frequent damp soil. Both sexes regularly visit flowers, such as Common Milkweed (*Asclepias syriaca*), Heal-all (*Prunella vulgaris*), Cow Vetch (*Vicia cracca*), and knapweeds (*Centaurea* spp.).

Comments: The common Dun Skipper can occur in virtually any grassy habitat in our region during early summer. Unlike other ubiquitous skippers that feed on grasses, Dun Skipper larvae eat sedges; a food preference shared with our other *Euphyes* skippers, the Two-spotted Skipper and Black Dash. Because the Dun Skipper lacks any conspicuous markings, it is sometimes difficult to differentiate in the field from the similar Little Glassywing and Northern Broken-Dash. Often flying together, the females of these three unassuming grass skippers are often collectively called the "three witches." Helpful guidance for distinguishing the females of these and other similar species in our region can be found in Iftner et al. (1992) and Monroe and Wright (2017).

Threats: None documented in our region.

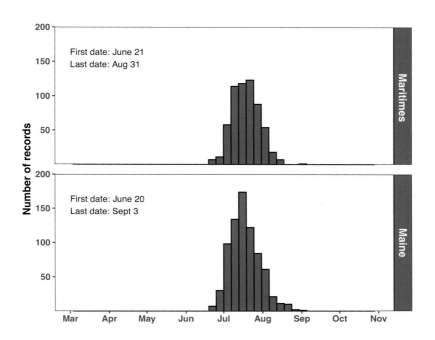

DELAWARE SKIPPER

Anatrytone logan
(W. H. Edwards)

Waterford, ME, 18 July 2020 (Bryan Pfeiffer)

Subspecies: Only the nomino-typical subspecies (*A. l. logan*) occurs in eastern North America.

Distinguishing characters:
  Size: Small.
  Upperside: Male bright orange without stigma, with dark wing margins and black scaling along veins; female similar but with wider dark wing margins, more extensive dark scaling, and dark wing bases.
  Underside: Uniformly bright orange yellow.

Similar species: European Skipper and Least Skipper (both much smaller).

Status and distribution in Maine: Resident. Uncommon; widespread in southern half of state. S4S5

Status and distribution in Maritimes: Not documented.

Habitat: Usually damp, open settings, such as marshes, riparian meadows, and roadside ditches, but also in dryer areas, including barrens, old fields, and gardens.

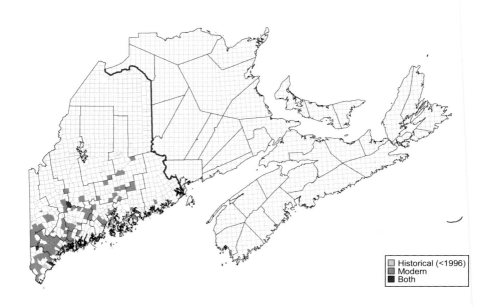

☐ Historical (<1996)
◼ Modern
■ Both

Biology: One generation per year. Adults fly from late June until early August, with a peak abundance in mid-July. Larvae feed on grasses (Poaceae), including Little Bluestem (*Schizachyrium scoparium*) and Switch Panicgrass (*Panicum virgatum*). Overwinters as a partially grown larva.

Adult behavior: This species has a rapid flight but usually higher off the ground than smaller grass skippers. Males watch for passing females from perches on or near the ground, and they frequent damp soil. Both sexes visit a variety of flowers, including Common Milkweed (*Asclepias syriaca*), knapweeds (*Centaurea* spp.), thistles (*Cirsium* spp.), Spreading Dogbane (*Apocynum androsaemifolium*), and Pickerel Weed (*Pontederia cordata*).

Comments: Although the Delaware Skipper sometimes occurs in dryer areas, it is most often found in or around wetlands. This species appears to be expanding in range northeastward. Scudder (1888–1889) and Farquhar (1934) were aware of few records from New England, none north of Massachusetts. Brower (1974) knew of only a single 1961 record from Biddeford, York County. An earlier specimen was located during the Maine Butterfly Survey, which was collected in 1950 at Kennebunk, York County. There are only five known records from before 1996, the most northeasterly from New Sharon, Franklin County (1989). It has since been found in eighty additional Maine townships, within every county except Aroostook. It is probably just a matter of time before this skipper is found in the Maritimes.

Threats: None documented in our region.

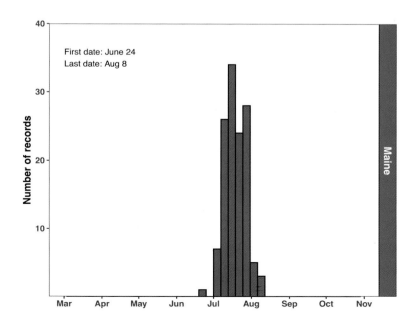

First date: June 24
Last date: Aug 8

FIERY SKIPPER

Hylephila phyleus (Drury)

Male—Halifax County, VA, 25 September 2013 (Allen Belden)

Subspecies: Only the nomino-typical subspecies (*H. p. phyleus*) occurs in North America.

Distinguishing characters:

Size: Small.

Upperside: Male yellowish orange with jagged black margin and a black stigma on forewing; female brown, with pattern of orange or yellowish-orange spots.

Underside: Male yellowish orange, with scattered black spots and black stripe on inner edge of hindwing; female brownish orange, with indistinct crescent of paler, rectangular spots on hindwing bordered on each end by black.

Additional: Antennae very short.

Similar species: Females of Sassacus Skipper and Long Dash.

Status and distribution in Maine: Stray. Rare; four modern records, from the southern coastal counties of York and Cumberland, and Piscataquis County in the north. SNA

Status and distribution in Maritimes: Stray. Rare; only one modern record from coastal Saint John County, New Brunswick, and one historical record from Queens County, Prince Edward Island. NB-SNA, PE-SNA

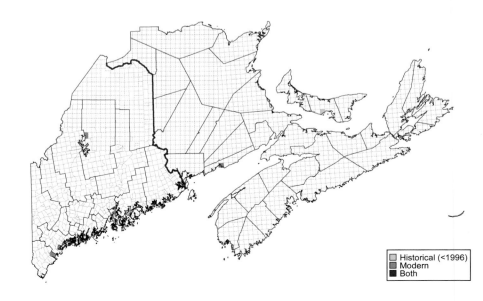

Historical (<1996)
Modern
Both

Habitat: This migratory species occurs in virtually any sunny, open area, including city parks, gardens, and lawns.

Biology: Multiple generations annually; adults are active most of the year in the South. Our records are from late July to early October; most are from August. Larvae feed on various weedy grasses (Poaceae), including bentgrasses (*Agrostis* spp.) and crabgrasses (*Digitaria* spp.). Overwinters as partly grown larva but not in our region.

Adult behavior: This skipper has a low, rapid flight. Adults are very wary, but they frequently nectar at flowers. In our region, they have been observed visiting asters (*Symphyotrichum* spp.), goldenrods (*Solidago* spp.), and ornamental Orange-eye Butterfly-bush (*Buddleja davidii*).

Comments: Named for the bright coloration of the male, the Fiery Skipper is a resident of the South that regularly migrates northward, reaching southern Canada and New England. It may occasionally produce one brood as far north as Massachusetts but does not survive the winter (Stichter 2015). The first record in our region was on 29 July 1947, when the entomologist R. Hazen Wigmore collected a single individual at Charlottetown, Queens County, Prince Edward Island (specimen in Canadian National Collection). The next records (including the first from Maine and New Brunswick) were not until 2012, which was an exceptional year for southern migrants throughout the Northeast. The Fiery Skipper was subsequently recorded in Maine in 2013 and 2019, suggesting that it may become a more frequent visitor, possibly in response to climate change (Gobeil and Gobeil 2020). Searching flowery coastal areas in late summer may reward observers with a glimpse of this sprightly wanderer.

Threats: None documented in our region.

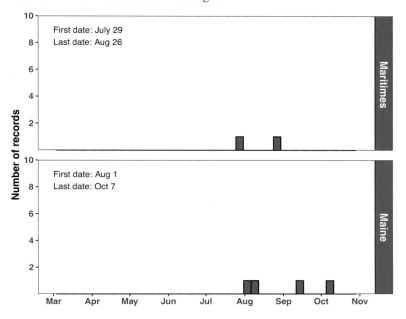

CROSSLINE SKIPPER

Polites origenes (Fabricius)

Subspecies: Only the nomino-typical subspecies (*P. o. origenes*) occurs in eastern North America.

Female—Eliot, ME, 28 July 2019 (Bryan Pfeiffer)

Distinguishing characters:
    Size: Small.
    Upperside: Dark brown; male forewing with yellowish-orange patch on leading edge, black stigma, and two or three small orange spots adjacent to the stigma tip (versus a single orange spot in Tawny-edged Skipper); female forewing dark brown with two or three large, light yellow angular spots; hindwing usually with row of indistinct pale spots.
    Underside: Forewing spots similar to upperside; hindwing brown, orange brown, or buff, usually with row of inconspicuous pale spots.

Similar species: Tawny-edged Skipper.

Status and distribution in Maine: Resident. Uncommon; restricted to southern half of state. S4

Status and distribution in Maritimes: Resident. Rare; known only from the Saint John River Valley in New Brunswick. NB-S1?

Habitat: Various open areas, including old fields, utility corridors, and roadsides.

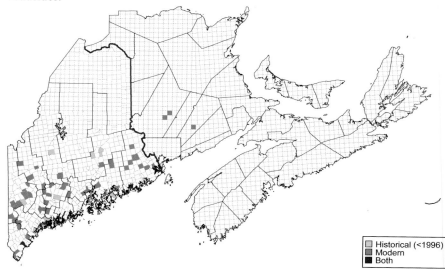

☐	Historical (<1996)
▨	Modern
■	Both

Biology: One generation per year. Adults are active from late June to early August, with a peak in abundance in early July. Larvae feed on grasses, including Little Bluestem (*Schizachyrium scoparium*). Overwinters as a fully or nearly fully grown larva.

Adult behavior: Like most of our small grass skippers, this species has a low, rapid flight. Males perch in low vegetation to await passing females and often visit damp earth. Both sexes regularly nectar at flowers, including Common Milkweed (*Asclepias syriaca*), Cow Vetch (*Vicia cracca*), and Garden Bird's-foot-trefoil (*Lotus corniculatus*).

Comments: The Crossline Skipper is a small, drab, uncommon butterfly that is named for the more diagonal position of the male stigma, which appears straighter and broader than that of the similar Tawny-edged Skipper. The female forewing typically lacks a tawny leading edge, which gives the Tawny-edged Skipper its name. Although these species are sometimes found together, the Crossline Skipper prefers dryer habitats and is much more localized than its slightly smaller look-alike. The similarity of these species may explain why the Crossline Skipper was overlooked in the Maritimes until 2012, when Scott Makepeace discovered a colony in the Saint John River Valley in Queens County, New Brunswick. Since then, this species has been found at two other Saint John River Valley locations. Additional colonies probably occur in the southern portion of the province.

Threats: None documented in our region.

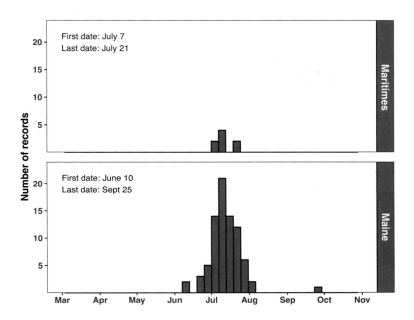

LONG DASH

Polites mystic (W. H. Edwards)

Subspecies: Only the nomino-typical subspecies (*P. m. mystic*) occurs in eastern North America.

Distinguishing characters:
  Size: Small.
  Upperside: Orange with broad brown border (male) or dark brown with yellowish-orange spots (female); forewing with orange spots aligned

Colchester County, NS, 27 July 2014 (Phil Schappert)

  along outside edge (compare with Sassacus Skipper); male with broad, elongate, black stigma ("long dash").
  Underside: Light orange (male) or reddish brown (female); forewing spotted like upperside; hindwing with curving row of pale square spots and a single central spot.

Similar species: Sassacus Skipper.

Status and distribution in Maine: Resident. Abundant; widespread. S5

Status and distribution in Maritimes: Resident. Abundant; widespread. NB-S5, NS-S5, PE-S5

Habitat: Various open, grassy areas, including meadows, streamsides, and roadsides.

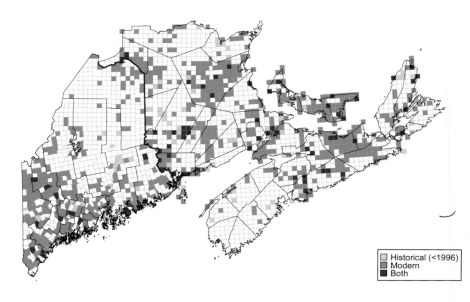

Historical (<1996)
Modern
Both

Biology: One generation per year. Adults are active from early June until early August, with a peak abundance in early July. The larvae feed on various grasses (Poaceae), including bluegrasses (*Poa* spp.) and presumably others. Overwinters as a partially grown larva.

Adult behavior: This skipper flies rapidly and tends to stay near the ground. Males perch on low vegetation to await passing females. Adults regularly visit a variety of flowers, including Red Clover (*Trifolium pratense*), Cow Vetch (*Vicia cracca*), Common Milkweed (*Asclepias syriaca*), Spreading Dogbane (*Apocynum androsaemifolium*), hawkweeds (*Hieracium* spp.), Ox-eye Daisy (*Leucanthemum vulgare*), and knapweeds (*Centaurea* spp.).

Comments: The Long Dash is named for the conspicuous stigma on the male forewing. Found on the forewings of most other small, orange and brown skippers in our region, a stigma is a specialized patch of scales that emits pheromones used during courtship. In the Long Dash, the stigma is surrounded by black scales, making it appear larger than it really is. This species usually occurs in moister areas than the very similar Sassacus Skipper, though they can sometimes be found visiting the same patches of flowers.

Threats: None documented in our region.

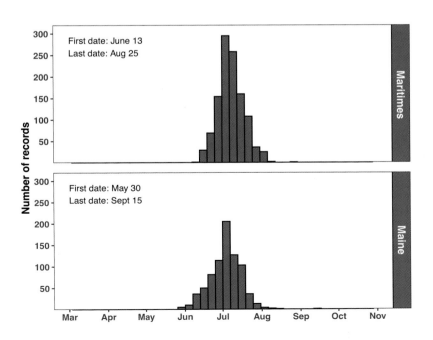

TAWNY-EDGED SKIPPER

Polites themistocles (Latreille)

Subspecies: Only the nomino-typical subspecies (*P. t. themistocles*) occurs in eastern North America.

Female—Eliot, ME, 6 August 2018 (Josh Lincoln)

Distinguishing characters:
  Size: Small.
  Upperside: Dark brown; forewing with central row of orange-yellow spots (larger in female) and orange patch along the leading edge (more extensive in male); three-part, uneven black stigma in male (compare with Crossline Skipper); hindwing unmarked.
  Underside: Grayish to orange brown; forewing with orange patch along the leading edge (more extensive in male); hindwing usually unmarked but sometimes with a row of small, indistinct pale spots or (more rarely) larger, blurry-looking spots.

Similar species: Crossline Skipper.

Status and distribution in Maine: Resident. Common; widespread. S5

Status and distribution in Maritimes: Resident. Common; widespread. NB-S5, NS-S5, PE-S5

Habitat: Various open, grassy areas including meadows, roadsides, and old fields.

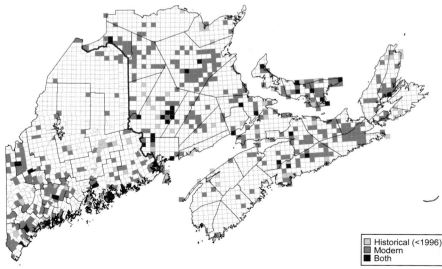

Historical (<1996)
Modern
Both

Biology: One generation per year, with a partial second brood in southern Maine. Adults appear in early June and fly until early to mid-August (late September southward), with a peak in abundance in early July. The larvae feed on various grasses (Poaceae), including Kentucky Bluegrass (*Poa pratensis*), Little Bluestem (*Schizachyrium scoparium*), and crabgrasses (*Digitaria* spp.). Overwinters as a pupa.

Adult behavior: As with most other small skippers, this species flies rapidly and remains near the ground. Males visit damp soil and perch on low vegetation to await passing females. Both sexes visit a variety of flowers in our region, including Cow Vetch (*Vicia cracca*), Spreading Dogbane (*Apocynum androsaemifolium*), hawkweeds (*Hieracium* spp.), milkweeds (*Asclepius* spp.), and Red Clover (*Trifolium pratense*).

Comments: Like other grass skippers (subfamily Hesperiinae), Tawny-edged Skippers bask with their hindwings outstretched, while their forewings are held more upright. This configuration has been cleverly described as a "jet fighter" posture. When perched with their wings closed, Tawny-edged adults are quite drab and can even be mistaken for the Dun Skipper. The namesake "tawny edge" refers to orange scales along the forewing (above and below) that help to differentiate this species from others, particularly females of the similar Crossline Skipper. Although it has been suggested that habitat loss and widespread chemical application have negatively affected the abundance of the Tawny-edged Skipper in the Northeast (Cech and Tudor 2005), this species remains a widespread and familiar member of our fauna.

Threats: None documented in our region.

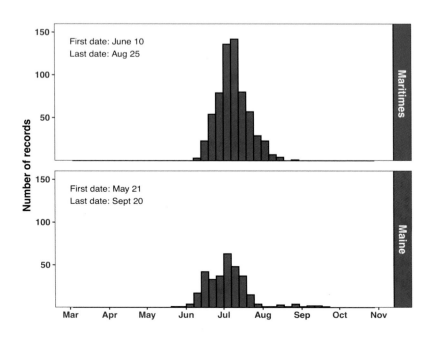

PECK'S SKIPPER

Polites peckius (W. Kirby)

Buxton, ME, 27 August 2019 (Bryan Pfeiffer)

Subspecies: Only the nomino-typical subspecies (*P. p. peckius*) occurs in eastern North America.

Distinguishing characters:
  Size: Small.
  Upperside: Dark brown with orange or yellowish-orange spots; reddish-orange stripe across center of wing (absent in some females), bordered below by large black stigma in male.
  Underside: Brownish orange; hindwing with large central and smaller basal yellow patches (often connected) covering most of the wing.

Similar species: None.

Status and distribution in Maine: Resident. Abundant; widespread. S5

Status and distribution in Maritimes: Resident. Abundant; widespread. NBS5, NS-S5, PE-S4

Habitat: Virtually any open, grassy areas, including meadows, roadsides, old fields, and lawns.

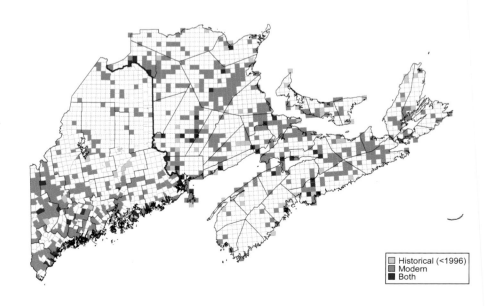

Historical (<1996)
Modern
Both

Biology: One generation per year, with a partial second brood. Adults fly from late May to mid-October, with a peak in early July. The larvae feed on various grasses (Poaceae), including Kentucky Bluegrass (*Poa pratensis*), Little Bluestem (*Schizachyrium scoparium*), and Rice Cut Grass (*Leersia oryzoides*). Overwinters as a partly grown larva.

Adult behavior: Like many small skippers, this species has such a rapid flight that adults seemingly disappear when disturbed. Males perch low in vegetation to await females and occasionally visit damp soil. Both sexes regularly visit flowers, particularly Red Clover (*Trifolium pratense*), Cow Vetch (*Vicia cracca*), Common Milkweed (*Asclepias syriaca*), Heal-all (*Prunella vulgaris*), and knapweeds (*Centaurea* spp.).

Comments: The Peck's Skipper is part of a suite of native skippers, including the Long Dash, Tawny-edged, and Dun, that can be expected during early summer in just about any open, grassy area. They may even be seen in urban lots and inner-city gardens, where few other butterflies frequent. A July stroll through a field of Red Clover in our region usually rewards the observer with dozens of Peck's Skippers. Adults found in September and October are presumably from a partial second brood. In the historical era, there were only two records of this species after August (1.3% of records). In the recent era, there are eighty-five such records (5.8% of records). South of our region, the Peck's Skipper regularly has multiple generations; the increase of second-generation individuals here may be the result of climate change. Research suggests that this species should be known as *P. coras*, a name used during the late twentieth century.

Threats: None documented in our region.

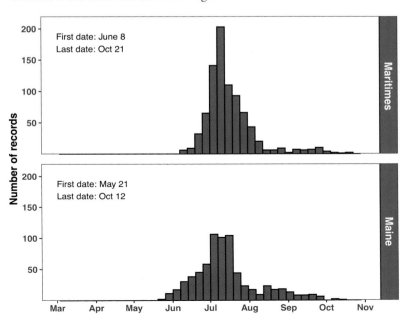

First date: June 8
Last date: Oct 21

Maritimes

First date: May 21
Last date: Oct 12

Maine

NORTHERN BROKEN-DASH

Polites egeremet (Scudder)

Harrison, ME, 18 July 2020 (Bryan Pfeiffer)

Subspecies: None.

Distinguishing characters:
   Size: Small.
   Upperside: Dark brown; male forewing with two-part, dark stigma, several pale spots, and tawny scaling along leading edge; female forewing with less tawny scaling and additional pale spots.
   Underside: Hindwing brown (often with purplish sheen), with indistinct crescent of paler spots or smudges that faintly resembles the number 3 in shape.

Similar species: Dun Skipper and Little Glassywing (especially females).

Status and distribution in Maine: Resident. Common to uncommon in southern third of state, rare or absent elsewhere. S4

Status and distribution in Maritimes: Not documented.

Habitat: Open grassy and brushy areas, including old fields, streamsides, and roadsides; usually near woodlands.

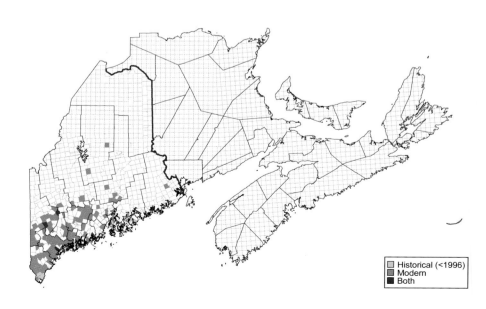

Historical (<1996)
Modern
Both

Biology: One generation per year. Adults fly from late June to mid-August, with a peak abundance in mid-July. Larvae feed on various grasses (Poaceae), including Deer-tongue Rosette-panicgrass (*Dichanthelium clandestinum*), and crabgrasses (*Digitaria* spp.). Overwinters as a partially grown larva.

Adult behavior: Like most grass skippers, adults fly rapidly and tend to stay near the ground. Males perch on low vegetation to await passing females, and both sexes visit flowers, including Common Milkweed (*Asclepias syriaca*), Spreading Dogbane (*Apocynum androsaemifolium*), Red Clover (*Trifolium pratense*), and Cow Vetch (*Vicia cracca*).

Comments: This unassuming species is typically found with its more common look-alike, the Dun Skipper, which probably leads to their being overlooked and underreported. In fact, females of the Northern Broken-Dash, Dun Skipper, and Little Glassywing (collectively known as the "three witches") often fly together, making field identifications more challenging. Helpful characters for distinguishing the females of these and other similar species that occur in our region can be found in Iftner et al. (1992) and Monroe and Wright (2017). The common name of this skipper is derived from the two-part, or "broken," dash-like stigma on the forewing of the male. Previously in the genus *Wallengrenia*, Pelham (2022) includes this species in the genus *Polites* based on genomics research of Zhang et al. (2019a).

Threats: None documented in our region.

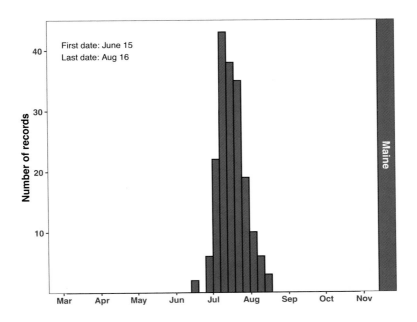

First date: June 15
Last date: Aug 16

Maine

LITTLE GLASSYWING

Vernia verna (W. H. Edwards)

Subspecies: None.

Distinguishing characters:

  Size: Small.

  Upperside: Dark brown; male forewing with long black stigma and row of translucent spots; female forewing similar but without stigma and largest spot is square shaped.

Whitefield, ME, 19 July 2020 (Josh Lincoln)

  Underside: Brown with subtle purplish sheen when fresh; forewing spotting like upperside; hindwing with curving row of faint pale spots.

  Additional: White scaling just below the club of each antenna.

Similar species: Dun Skipper and Northern Broken-Dash.

Status and distribution in Maine: Resident. Uncommon to common in southern third of state, absent elsewhere. S4

Status and distribution in Maritimes: Not documented.

Habitat: Damp grassy areas, such as meadows, forest edges, utility corridors, and roadsides.

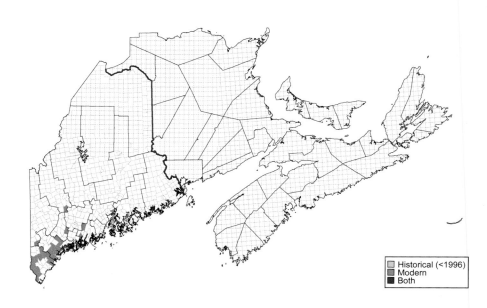

Historical (<1996)
Modern
Both

Biology: One generation per year. Adults fly from mid-June to early August, with a peak abundance in mid-July. Larvae feed on grasses (Poaceae), including panicgrasses (*Panicum* spp.), but it is unknown what species are used locally. Purpletop Tridens (*Tridens flavus*) is reported as the primary food plant elsewhere but is not known to occur in Maine. Overwinters as a larva.

Adult behavior: Like most small skippers, the Little Glassywing flies rapidly and tends to stay near the ground. Males perch on low vegetation to await passing females and often visit damp soil. Both sexes are strongly attracted to nectar sources, including Common Milkweed (*Asclepias syriaca*), Red Clover (*Trifolium pratense*), and Cow Vetch (*Vicia cracca*).

Comments: Usually found in low numbers, the Little Glassywing is often mistaken for the much more common and widespread Dun Skipper. This possibly contributed to the Little Glassywing's perceived absence in Maine prior to 2000, when the first confirmed record was documented by Gail Everett at Portland, Cumberland County. Dozens of new observations have since been recorded in southern counties, suggesting that this species was either previously overlooked or is actively expanding its range. As with many grass skippers, the question of what Little Glassywing larvae eat in our region is ripe for investigation. For many decades, this species was included in the genus *Pompeius*, but genomics studies by Cong et al. (2019) determined that it belonged in a different genus, which they proposed to call *Vernia* in honor of its type species, *verna*.

Threats: None documented in our region.

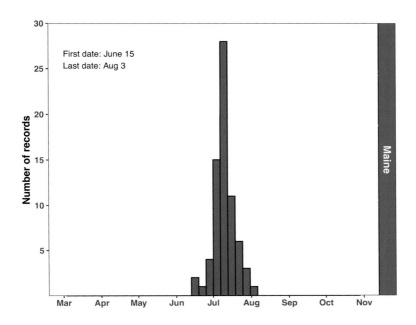

First date: June 15
Last date: Aug 3

COMMON BRANDED SKIPPER (LAURENTIAN SKIPPER)

Hesperia colorado (Scudder)

Colchester County, NS, 8 August 2012 (Jim Edsall)

Subspecies: Only the subspecies *H. c. laurentina* (Lyman) occurs in the eastern United States and southeastern Canada.

Distinguishing characters:
> **Size:** Small.
> **Upperside:** Orange brown, darker on outer half of wings, with orange spotting; male forewing with elongate black stigma.
> **Underside:** Orange brown with olive cast; hindwing with distinct white spots, including two basally (versus one in Leonard's Skipper).

Similar species: Leonard's Skipper.

Status and distribution in Maine: Resident. Uncommon; limited to north and east. S3S4

Status and distribution in Maritimes: Resident. Common in New Brunswick, uncommon to common in Nova Scotia and Prince Edward Island. NB-S5, NS-S4, PE-S4

Habitat: Coniferous and mixed forest openings and edges, including meadows, woodland roads, and margins of salt marshes.

Biology: One generation per year. Adults are active from late July until mid-September, with peak abundance in mid-August. Larvae feed on various grasses

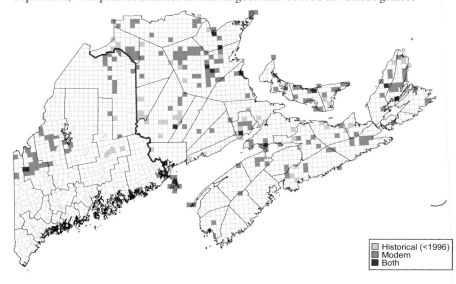

☐ Historical (<1996)
☐ Modern
■ Both

(Poaceae). The Common Branded Skipper has been observed ovipositing on Poverty Grass (*Danthonia spicata*) in Maine. Overwinters as a larva.

Adult behavior: Extremely wary with a rapid flight, this skipper is difficult to approach. Males perch on the ground or on low vegetation to await females and often take nutrients from damp soil. Males also hilltop. Both sexes visit flowers, including Common Grass-leaved Goldenrod (*Euthamia graminifolia*), Pearly Everlasting (*Gnaphalium margaritaceum*), Purple Crown-vetch (*Securigera varia*), Red Clover (*Trifolium pratense*), Spotted Joe-Pye Weed (*Eutrochium maculatum*), hawkweeds (*Hieracium* spp.), and American-asters (*Symphyotrichum* spp.).

Comments: The Common Branded Skipper and Leonard's Skipper are the latest skippers to take flight in our region each season. This may contribute to the Common Branded Skipper being overlooked, especially in northern Maine where it is undoubtedly more widespread than records indicate. It also tends to occupy habitats that are generally poor for butterflies, with patches of rocky, open ground, and sparse vegetation. This skipper appears to have expanded its range southward during the early twentieth century and was first recorded in Maine in 1931. There has been much debate about the scientific name of the Common Branded Skipper in North America. It was long treated as *H. comma*, which was believed to be Holarctic in distribution, occurring across the boreal regions of Eurasia and North America. However, recent genomics research by Cong et al. (2019) revealed that *H. comma* is an Old World species and that virtually all North American populations, including those in our region, should be treated as *H. colorado*. The subspecies *laurentina*, which ranges west to the Great Lakes, was long considered to be distinct from the Common Branded Skipper. As a result, the name Laurentian Skipper is still widely used in our region.

Threats: As one of the few species in our region that occurs at the southeastern edge of its range, this butterfly should be monitored for potential effects of climate change.

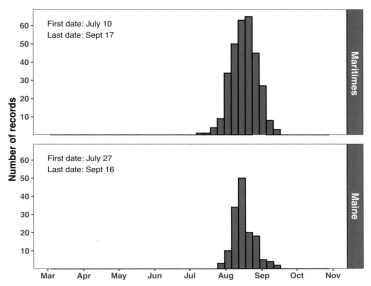

LEONARD'S SKIPPER

Hesperia leonardus T. Harris

Subspecies: Only the nomino-typical subspecies (*H. l. leonardus*) occurs in eastern North America.

Distinguishing characters:
  Size: Small to medium.
  Upperside: Brownish orange; wide dark brown borders, with orange spotting (more extensive in female); male forewing with elongate black stigma.

Kennebunk, ME, 20 August 2020 (Bryan Pfeiffer)

  Underside: Reddish brown to chestnut colored; hindwing with chevron pattern of white to yellow spots.

Similar species: Common Branded (Laurentian) Skipper.

Status and distribution in Maine: Resident. Rare; limited to southern half of state. State Special Concern. S3

Status and distribution in Maritimes: Not documented (see comments).

Habitat: Dry shrubby barrens, grasslands, rights-of-way, oak–pine forest clearings, and coastal dunes.

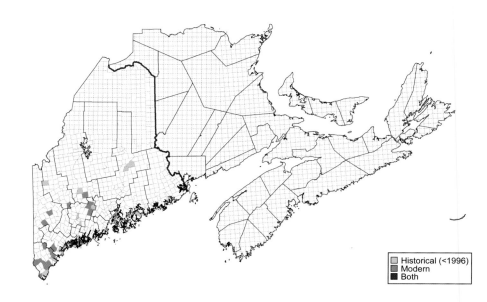

Historical (<1996)
Modern
Both

Biology: One generation per year. Adults are active from early to mid-August until mid-September, with peak abundance in late August. Larvae feed on various grasses (Poaceae), particularly Little Bluestem (*Schizachyrium scoparium*). Overwinters as a partly grown larva.

Adult behavior: This skipper has a very rapid flight. Males perch low in vegetation or on the ground to await passing females and occasionally visit moist soil. Both sexes are especially attracted to purple flowers, with observations in our region on Northern Blazing Star (*Liatris novae-angliae*), Swamp Milkweed (*Asclepias incarnata*), thistle (*Cirsium* sp.), and Spotted Joe-Pye Weed (*Eutrochium maculatum*).

Comments: The Leonard's Skipper is extremely wary and observing them at close range can sometimes be challenging. This species, and the Common Branded Skipper, are the latest native skippers to appear in our region each season. There is a dubious historical record of the Leonard's Skipper from Digby, Nova Scotia, which was first reported by Ferguson (1954). It is based on two specimens, a male and female, dated 20 June 1905, in the Canadian National Collection of Insects, Arachnids, and Nematodes (Ottawa, Ontario). Not only is the collector not indicated, the June date is far too early for this late-flying species; Perrin and Russell (1912) did not include it in their catalogue of Nova Scotia Lepidoptera. In addition, Digby is far removed from the nearest records in Maine, casting further doubt on the validity of this record.

Threats: The dry, barrens habitat where this species is most common in southern Maine is threatened from development, gravel mining, and ecological succession from fire prevention.

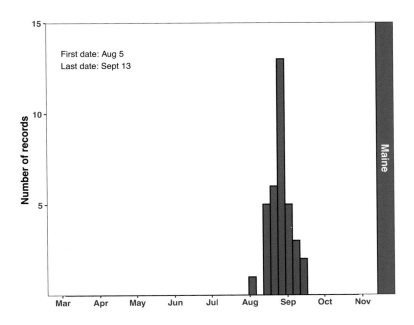

COBWEB SKIPPER

Hesperia metea Scudder

Subspecies: Only the nomino-typical subspecies (*H. m. metea*) occurs in northeastern North America.

Distinguishing characters:

    Size: Small.

    Upside: Male tawny brown, forewing with narrow black stigma bordered by patch of lighter orange, and small light spots near forewing margin; female darker brown with more defined forewing spots; hindwing of both sexes with indistinct, pale V-shaped pattern.

Male—Montague, MA, 22 May 2020 (Michael Newton)

    Underside: Brown with distinctive white chevron on hindwing; male with white scales along veins of both wings producing a "cobweb" appearance; female darker brown.

    Additional: Both sexes with distinctive white eye ring.

Similar species: Common Branded (Laurentian) Skipper.

Status and distribution in Maine: Resident. Rare; restricted to southern counties. State Special Concern. S2

Status and distribution in Maritimes: Not documented.

Habitat: Shrubby barrens, grasslands, rights-of-way, and other disturbed sites with dry soils.

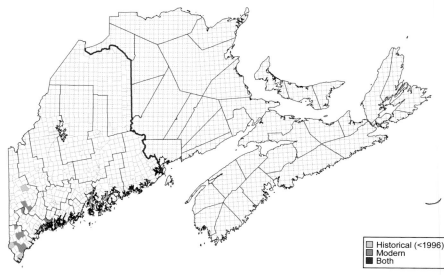

    ☐ Historical (<1996)
    ▨ Modern
    ■ Both

Biology: One generation per year. This skipper has an early and brief flight, with adults active only from mid-May to early June; it is generally most common during the last 2 weeks of May. Larvae feed on Little Bluestem (*Schizachyrium scoparium*) and possibly Big Bluestem (*Andropogon gerardii*), though the latter is rare or absent at most sites occupied by the Cobweb Skipper in Maine. Overwinters as either a fully grown larva or pupa.

Adult behavior: This skipper has a rapid flight, usually just above the ground. Often perched on bare patches of soil or on the ground near the base of the food plant, adults blend in with their surroundings and are easily overlooked. They are usually seen only when flushed from patches of bluestems, often returning to the same spot a few minutes later. The only nectaring observations in Maine are from Common Lowbush Blueberry (*Vaccinium angustifolium*) and blackberries (*Rubus* spp.), though this skipper likely visits other low, early blooming species, such as wild strawberries (*Fragaria* spp.).

Comments: The Cobweb Skipper is among a group of specialized "bluestem feeders." In Maine, it shares this distinction with the Dusted Skipper and Leonard's Skipper, and all three species may inhabit the same locality. Although this species reportedly occurs in Massachusetts in larger, more established stands of Little Bluestem (Stichter 2015), colonies in Maine are just as apt to be found in disturbed areas with a dense growth of the food plant. They are extremely localized and usually occur in low numbers within a relatively small portion of the available habitat.

Threats: Sandplain barrens and grasslands that provide habitat for this butterfly (and other rare biota) are limited in distribution to southern Maine, where they are threatened by development, gravel mining, and ecological succession. One significant population in Oxford County was recently lost to the installation of a commercial solar farm.

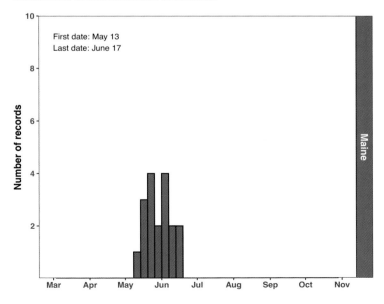

First date: May 13
Last date: June 17

SASSACUS SKIPPER

Hesperia sassacus T. Harris

Subspecies: Only the nomino-typical subspecies (*H. s. sassacus*) occurs in northeastern North America.

New Maryland, NB, 16 June 2010 (Jim Edsall)

Distinguishing characters:
  Size: Small.
  Upperside: Orange, with broad brown border; forewing with two isolated, square, orange spots near the tip (compare with Long Dash); male with narrow, elongate, black stigma surrounded by orange.
  Underside: Light orange; forewing spotted like upperside; hindwing with chevron pattern of indistinct, pale spots.

Similar species: Long Dash.

Status and distribution in Maine: Resident. Uncommon; recorded only in southern half of state. S4

Status and distribution in Maritimes: Resident. Uncommon; widespread in New Brunswick; rare on Prince Edward Island; not documented in Nova Scotia. NB-S3, PE-SU

Habitat: Various open, grassy areas including meadows, barrens, roadsides, and utility corridors.

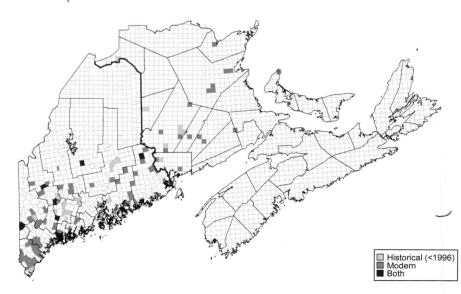

☐ Historical (<1996)
▨ Modern
■ Both

Biology: One generation per year. Adults are active from late May to mid-July, with a peak flight in early June. The larvae feed on various grasses (Poaceae), including Kentucky Bluegrass (*Poa pratensis*), Little Bluestem (*Schizachyrium scoparium*), and Poverty Oatgrass (*Danthonia spicata*). Overwinters as a larva.

Adult behavior: This skipper has a rapid flight and is extremely wary. Males perch on the ground and on low vegetation to await passing females. Adults regularly visit flowers, particularly Red Clover (*Trifolium pratense*), blackberries (*Rubus* spp.), and hawkweeds (*Hieracium* spp.).

Comments: The Sassacus Skipper is rarely seen in significant numbers, though it can be locally common. It is easy to overlook, as it often occurs with the similar, and much more common, Long Dash Skipper. However, Sassacus Skippers typically fly earlier, with only a few worn females still present by the time the Long Dash reaches its peak abundance. The Sassacus Skipper is the most widespread, and least habitat specific, of our four species of *Hesperia*. It is sometimes found in Maine with the much rarer Cobweb Skipper, which is the only other member of this group in the region that flies early in the season. The single Prince Edward Island record of the Sassacus Skipper is from Tidnish, where it was found on 25 June 2012 by Rosemary Curley. Instead of the often-used name Indian Skipper, we prefer to use Sassacus Skipper, which is consistent with the species' scientific name.

Threats: None documented in our region.

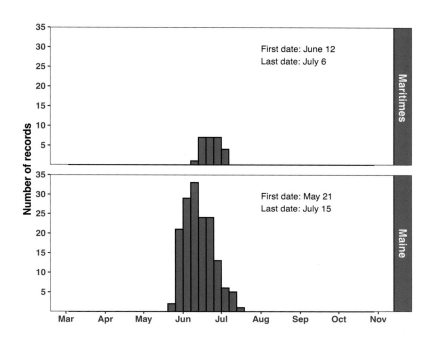

MULBERRY WING

Poanes massasoit (Scudder)

Subspecies: Only the nomino-typical subspecies (*P. m. massa-soit*) occurs in our region.

Distinguishing characters:
  Size: Small.
  Upperside: Blackish brown with subtle purplish luster (similar to the color of mulberries); male unmarked or with small spots on forewing and/or hindwing; female with series of small, pale spots on forewing and a C-shaped row of pale spots on hindwing.

Eliot, ME, 21 July 2020 (Bryan Pfeiffer)

  Underside: Forewing like upperside; hindwing orange brown, with large, yellow, arrow-shaped marking that resembles an airplane.

Similar species: Broad-winged Skipper and Hobomok Skipper.

Status and distribution in Maine: Resident. Rare; localized, recorded only in York and Cumberland counties. S3

Status and distribution in Maritimes: Not documented.

Habitat: Sedge-dominated marshes and wet meadows; occasionally nearby weedy fields.

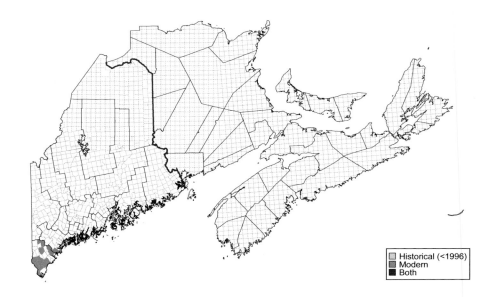

Historical (<1996)
Modern
Both

Biology: One generation per year. Adults fly from early July to early August, with a peak abundance in late July. Larvae feed on Tussock Sedge (*Carex stricta*), and potentially other *Carex* species. Overwinters as a partly grown larva.

Adult behavior: Males patrol low through sedges or along paths through marsh vegetation with a slow, weak, bouncing flight. Adults tend to remain within their marshy habitats, but they will stray to adjacent areas for nectar. In our region, they have been observed visiting Swamp Milkweed (*Asclepias incarnata*), Common Buttonbush (*Cephalanthus occidentalis*), Joe-Pye weed (*Eutrochium* sp.), and Fuller's Teasel (*Dipsacus fullonum*).

Comments: There are no historical records of the Mulberry Wing in our region; it was first recorded in Maine on 5 August 2008, in Waterboro, York County, by Phillip deMaynadier. Since then, targeted surveys have revealed many additional populations. It can inhabit very small patches of sedge, often in disturbed areas along roadsides or at the edges of pastures. Like elsewhere in its range, the Mulberry Wing often co-occurs with the Black Dash in southern Maine. That these two species remained unrecorded in Maine for so long, despite the number of recently documented localities, may indicate that their ranges are expanding northward, as suggested by Stichter (2015). This expansion is apparently continuing, as implied by the presence of the Mulberry Wing in a marsh in northern York County in 2021, where only the Black Dash was found in 2011. When searching for these rare skippers, prepare to get your feet wet and focus attention on flowering marsh plants.

Threats: Smaller marshes and wet meadows are at risk of filling, draining, and degradation by invasive plants, especially in rapidly developing southern Maine where this skipper is restricted.

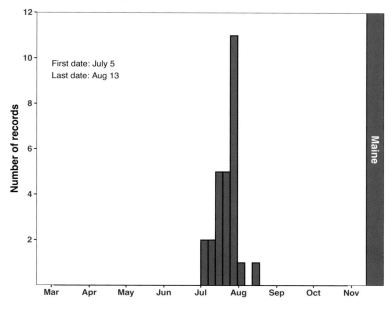

First date: July 5
Last date: Aug 13

Number of records

Maine

Mar   Apr   May   Jun   Jul   Aug   Sep   Oct   Nov

BROAD-WINGED SKIPPER

Poanes viator (W. H. Edwards)

Subspecies: Only the subspecies *P. v. zizaniae* Shapiro
occurs in our region.

Distinguishing characters:
Size: Medium
Upperside: Dark brown,
with large orange (male)
or creamy (female) spots
on the forewing and
extensive orange on the
hindwing; male lacks
stigma

Falmouth, ME, 6 August 2018 (Bryan Pfeiffer)

Underside: Forewing much like upperside; hindwing orange brown with
curving row of pale spots and long, tawny central streak through center
(markings sometimes indistinct)

Similar species: Mulberry Wing.

Status and distribution in Maine: Resident. Uncommon; localized, restricted
to southern third of state. S3

Status and distribution in Maritimes: Not documented.

Habitat: Freshwater and brackish marshes and ditches.

Biology: One generation per year. Adults fly from mid-July to late August,
with a peak abundance in early August. Larvae feed on Common Reed
(*Phragmites australis*). Overwinters as a larva.

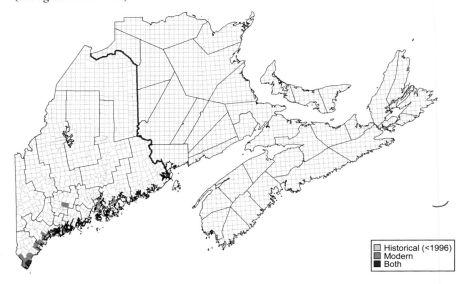

□ Historical (<1996)
▨ Modern
■ Both

Adult behavior: Males search for females through stands of Common Reed with a slow, weak, bouncing flight. Both sexes visit nearby flowers and in Maine have been recorded on goldenrods (*Solidago* spp.), American-asters (*Symphyotrichum* spp.), Cow Vetch (*Vicia cracca*), Purple Loosestrife (*Lythrum salicaria*), Red Clover (*Trifolium pratense*), marjorams (*Origanum* spp.), and Fuller's Teasel (*Dipsacus fullonum*).

Comments: The larvae of the Atlantic coastal subspecies of the Broad-winged Skipper originally fed upon native marsh grasses, such as Southern Wild Rice (*Zizania aquatica*) (hence the subspecies name, *zizaniae*). Within the past 40 years, however, its primary food plant in the Northeast has become Common Reed, an extremely successful invasive plant that forms extensive monocultures in many coastal marshes. Once considered rare in New England, this skipper has become more widespread since successfully exploiting this food source, much as the Wild Indigo Duskywing has done in association with Purple Crown-vetch (*Securigera varia*). First documented in Maine in 1991 by Robert Godefroi in York County, the Broad-winged Skipper was not recorded again in the state until 2007, when it was photographed in Cumberland County. It has since been found at many additional localities in southern Maine and is likely expanding northward. While most Maine records are from coastal marshes, the discovery of a colony in 2016 in a small roadside stand of Common Reed at Augusta (Kennebec County) suggests that it may be more frequent in inland areas than records reflect. This species can be abundant where it occurs, with dozens of adults jostling one another at flowers of Purple Loosestrife, which often grows in the same habitats. Because the Broad-winged Skipper tends to remain in and around dense beds of Common Reed, it can easily be overlooked.

Threats: None documented in our region.

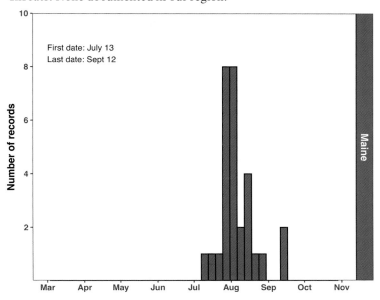

HOBOMOK SKIPPER

Lon hobomok (T. Harris)

Subspecies: Only the nominotypical subspecies (*L. h. hobomok*) occurs in northeastern North America.

Male—Northwest Piscataquis, ME, 15 June 2018 (Bryan Pfeiffer)

Distinguishing characters:
>   **Size:** Small.
>   **Upperside:** Male orange with broad brown borders and short, diagonal, black marking on forewing; female similar but with broader, more irregular brown borders and extensive basal darkening.
>   **Underside:** Hindwing brown, with large medial orange patch; fresh specimens have a frosting of purple scaling along wing borders.
>   **Additional:** Females of form "pocahontas" are mostly dark, with or without blurry orange markings, and with a series of small, white spots near tip of forewing. Intermediate forms often occur.

Similar species: Dusted Skipper (resembles "pocahontas" form).

Status and distribution in Maine: Resident. Abundant; widespread. S5

Status and distribution in Maritimes: Resident. Abundant; widespread. NB-5, NS-S5, PE-S5

Habitat: Grassy areas in or near woodlands, including forest edges, trails, and roadsides (strays to less wooded areas for nectar).

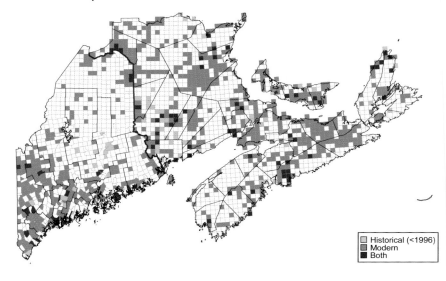

Historical (<1996)
Modern
Both

Biology: One generation per year. Adults are active from the latter half of May until early August, with a peak abundance in mid-June. Larvae feed on grasses, including rosette-panicgrasses (*Dichanthelium* spp.), bluegrasses (*Poa* spp.), and Orchard Grass (*Dactylis glomerata*). Overwinters as a larva.

Adult behavior: Males perch on sunlit foliage to await passing females, rapidly pursuing passing insects and often returning to the same perch. Males visit damp earth, and both sexes frequent a wide diversity of flowers in our region, especially buttercups (*Ranunculus* spp.), hawkweeds (*Hieracium* spp.), Red Clover (*Trifolium pratense*), milkweeds (*Asclepius* spp.), Blue Iris (*Iris versicolor*), Common Blackberry (*Rubus allegheniensis*), and Cow Vetch (*Vicia cracca*).

Comments: The Hobomok Skipper is a welcome sight in late May when it heralds the arrival of summer-like weather. It is one of our most familiar skippers, which is a testament to its abundance and approachability. Unlike most of our grass skippers, the Hobomok is often found in proximity to forested habitats. The dark female form "pocahontas" is rare in Maine, and it has not been reported from the Maritimes. The Hobomok Skipper is often confused with the closely related Zabulon Skipper, a more southern species that is expanding northward; it soon may turn up in southern Maine, if it is not already present (see Butterflies of Possible Occurrence).

Threats: None documented in our region.

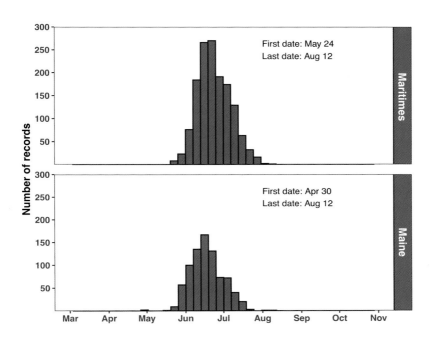

DUSTED SKIPPER

Atrytonopsis hianna (Scudder)

Sanford, ME, 5 June 2013 (W. Herbert Wilson Jr.)

Subspecies: Only the nomino-typical subspecies (*A. h. hianna*) occurs in northeastern North America.

Distinguishing characters:
    Size: Small.
    Upperside: Warm brown; forewing with three small white spots near apex and one or more pale spots near center of wing (larger in female).
    Underside: Dark brown with heavy gray scaling ("dusting") on outer portion of both wings; forewing spotted like upperside; hindwing often with one or two small white spots.
    Additional: White crescent above the eye.

Similar species: Northern Cloudywing, Southern Cloudywing, and dark form female of Hobomok Skipper.

Status and distribution in Maine: Resident. Rare; known only from York County. State Special Concern. S1S2

Status and distribution in Maritimes: Not documented.

Habitat: Dry, open areas, including shrubby barrens, sandplain grasslands, and utility corridors.

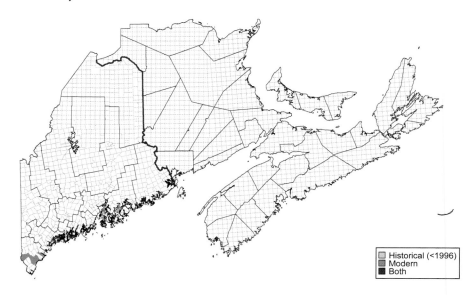

Historical (<1996)
Modern
Both

Biology: One generation per year. This skipper has an early flight, with adults active from late May to mid-June; most records are from the first week of June. Larvae feed on Little Bluestem (*Schizachyrium scoparium*) and possibly Big Bluestem (*Andropogon gerardii*), though the latter is mostly rare or absent at known sites in Maine. Overwinters as a fully grown larva.

Adult behavior: The flight of this skipper is low and rapid. They usually rest among grasses or on patches of bare ground, often going unnoticed unless flushed, when they fly a short distance before dropping into patches of the larval food plant. Adults have been observed in Maine nectaring on Common Lowbush Blueberry (*Vaccinium angustifolium*), blackberries (*Rubus* spp.), wild strawberries (*Fragaria* spp.), and Red Clover (*Trifolium pratense*).

Comments: Along with the Cobweb Skipper and Leonard's Skipper, the Dusted Skipper is a specialized "bluestem feeder." Although these skippers sometimes occupy the same locality (such as Maine's Wells Barrens and Kennebunk Plains), the Dusted Skipper has the most restricted distribution. Nonetheless, its range may be slowly expanding in the Northeast, with potential increases in abundance reported in Massachusetts (Stichter 2015) and Connecticut (O'Donnell et al. 2007). This intriguing skipper was only recently discovered in Maine in 2006 by Phillip deMaynadier. Future targeted surveys will likely reveal additional populations.

Threats: Sandplain barrens and dry grasslands that provide habitat for this butterfly are mostly limited to southern Maine, where they are threatened by development, gravel mining, and ecological succession. Aerial pesticide drift from adjacent oak–pine forests, targeted for Spongy Moth (*Lymantria dispar*) control, may be an additional threat.

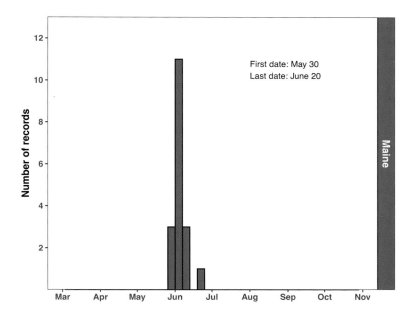

First date: May 30
Last date: June 20

PEPPER AND SALT SKIPPER

Amblyscirtes hegon (Scudder)

Alfred, ME, 29 May 2019 (Bryan Pfeiffer)

Subspecies: None

Distinguishing characters:
  Size: Very small.
  Upperside: Dark brown with checkered fringe and paler scaling basally; forewing with tight row of three small pale spots near tip and loose row of two or three larger pale spots across the center of the wing.
  Underside: Brown with grayish-green scaling ("pepper and salt" effect), becoming more dark brown with wear; numerous whitish spots on both wings.

Similar species: Common Roadside-Skipper.

Status and distribution in Maine: Resident. Uncommon; widespread. S4

Status and distribution in Maritimes: Resident. Uncommon in New Brunswick; uncommon to rare in Nova Scotia; not documented on Prince Edward Island. NB-S4, NS-S3S4

Habitat: Deciduous and mixed forest openings and edges, including roadsides, trails, and stream margins.

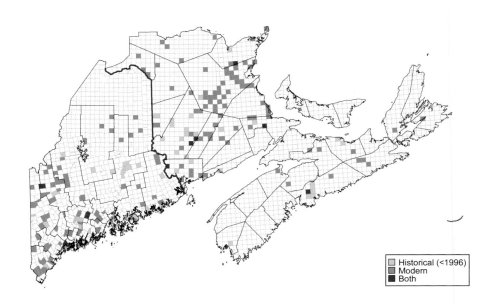

Historical (<1996)
Modern
Both

Biology: One generation per year. Adults are active from late May until mid-July, with a peak abundance in mid-June. Larvae feed on grasses (Poaceae), including introduced Kentucky Blue Grass (*Poa pratensis*). As with many skippers, the local suite of suitable food plants is unknown. Overwinters as a larva.

Adult behavior: This skipper has a rapid, darting flight near the ground. Their small size and dark coloration make them extremely hard to follow. Males perch on low vegetation to await passing females and frequently visit damp soil. Both sexes nectar at various low-growing flowers in our region, including Common Strawberry (*Fragaria virginiana*), Common Dandelion (*Taraxacum officinale*), Common Lowbush Blueberry (*Vaccinium angustifolium*), and Cow Vetch (*Vicia cracca*).

Comments: Rarely found in numbers, one or two Pepper and Salt Skippers can sometimes be glimpsed as they perch on sunlit leaves along trails or gravel roads through rich forest. Very wary, they usually dash off when approached, often to return to their original perch. Like the similar Common Roadside-Skipper, the Pepper and Salt Skipper is a small, inconspicuous butterfly that is probably far more frequent than records indicate. Future surveys should watch for this species in northern Maine, in southern Nova Scotia, and on Prince Edward Island.

Threats: None documented in our region.

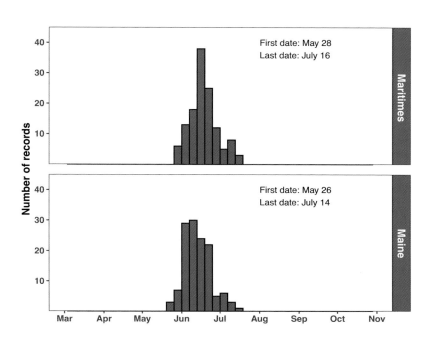

COMMON ROADSIDE-SKIPPER

Amblyscirtes vialis
(W. H. Edwards)

Cumberland County, NS, 2 June 2012 (John Klymko)

Subspecies: None.

Distinguishing characters:

    Size: Very small.

    Upperside: Dark brown with checkered fringe; forewing with tight row of two or three small white spots near anterior margin and sometimes with one or two inconspicuous white spots across center of wing.

    Underside: Dark brown, with gray scaling near tip; hindwing with gray scaling on outer half.

Similar species: Pepper and Salt Skipper.

Status and distribution in Maine: Resident. Uncommon; widespread. S4

Status and distribution in Maritimes: Resident. Common to uncommon in New Brunswick, uncommon in Nova Scotia, not documented on Prince Edward Island. NB-S5, NS-S4?

Habitat: Highly variable; includes woodland roadsides, dry forest openings, barrens, and bogs.

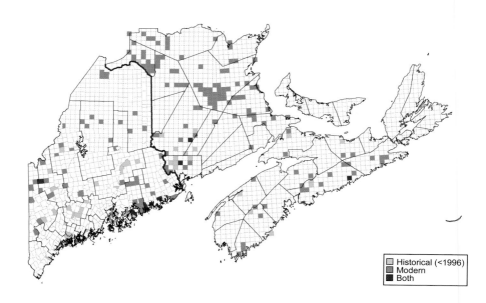

☐ Historical (<1996)
▨ Modern
■ Both

Biology: One generation per year. Adults are active from late May until mid-July, with a peak abundance in early to mid-June. Larvae feed on grasses, including Kentucky Blue Grass (*Poa pratensis*) and bentgrasses (*Agrostis* spp.). As with many grass skippers (subfamily Hesperiinae), the suite of local larval food plants is unknown. Overwinters as a fully grown larva.

Adult behavior: This skipper has a rapid, low flight. Its small size and drab coloration make it extremely hard to follow on the wing. Males perch on the ground or on low vegetation to await passing females. Adults are avid puddlers (often along roadsides) and occasionally nectar at flowers in our region, including hawkweeds (*Hieracium* spp.), Cow Vetch (*Vicia cracca*), Garden Bird's-foot-trefoil (*Lotus corniculatus*), blueberries (*Vaccinium* spp.), and Common Strawberry (*Fragaria virginiana*).

Comments: Although the Common Roadside-Skipper occurs along roadsides, it is rarely common anywhere within its extensive range. This is also true within our region, where only one or two are usually seen at a given locality per day. Possibly owing to habitat loss, this species is declining across much of the eastern United States (Cech and Tudor 2005), including Connecticut (O'Donnell et al. 2007) and Massachusetts (Stichter 2015). The lack of recent records from several well-surveyed counties in southern Maine may suggest a local decline, though adults are easy to overlook. Conversely, the abundance of recent records in northern Maine and northern New Brunswick could be the result of increased survey efforts, as well as the construction of numerous logging roads, which create easily accessible artificial habitats for this species.

Threats: The loss of natural barrens to development, gravel mining, and ecological succession is a potential threat to this skipper in southern Maine.

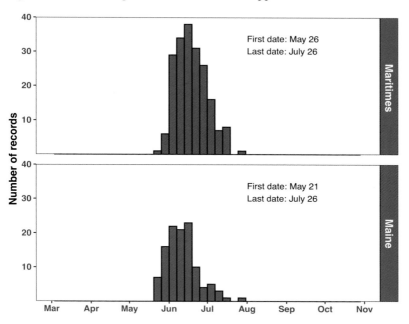

EUROPEAN SKIPPER

Thymelicus lineola
(Ochsenheimer)

Kent County, NB, 18 July 2011 (Denis Doucet)

Subspecies: Only the nomino-typical subspecies (*T. l. lineola*) is found in North America.

Distinguishing characters:
    Size: Very small.
    Upperside: Bright reddish orange, with narrow dark margin and darkened veins; hindwing extensively darkened basally; male forewing with narrow, gray stigma.
    Underside: Orange forewing with dark bases; hindwing dull, unmarked orange.
    Additional: In the form "pallida," the orange coloration is replaced with pale yellowish.

Similar species: Delaware Skipper and Least Skipper.

Status and distribution in Maine: Resident. Abundant; widespread. SNA

Status and distribution in Maritimes: Resident. Generally abundant; widespread but uncommon in southern Nova Scotia. NB-SNA, NS-SNA, PE-SNA

Habitat: Open grassy areas, such as old fields, roadsides, and waste areas.

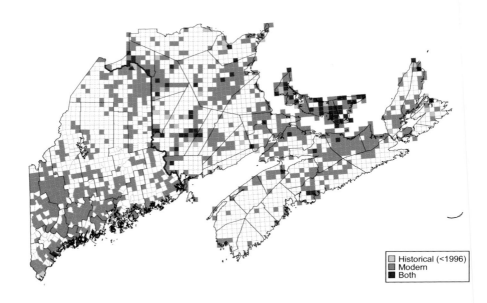

Historical (<1996)
Modern
Both

Biology: One generation per year. Adults are active from early June (southward) until early August, with peak abundance in early July. Larvae feed on various grasses (Poaceae), particularly introduced Common Timothy (*Phleum pratense*). Overwinters as an egg.

Adult behavior: European Skippers have a relatively weak, bobbing flight. Males typically patrol for females among their food plant grasses and visit damp ground, sometimes in large numbers. Adults frequent flowers, particularly Cow Vetch (*Vicia cracca*), clovers (*Trifolium* spp.), hawkweeds (*Hieracium* spp.), knapweeds (*Centaurea* spp.), and Common Milkweed (*Asclepias syriaca*).

Comments: The European Skipper was accidentally introduced to North America around 1910 near London, Ontario, with seeds of Common Timothy that were contaminated with eggs. Since that time, it has spread across southern Canada and the northern United States, both naturally and via transported hay or seed. The earliest record from the Acadian region is from 1966 at Halifax, Nova Scotia. It was first recorded in Maine in 1970, when it was reported as being "very common" at Madawaska, Aroostook County (Brower 1974). The European Skipper is now the most common butterfly in much of our region during July, sometimes reaching enormous numbers in damp hayfields. Adults of the rare, pale form "pallida" look old and faded, even when freshly emerged.

Threats: None documented in our region (introduced species).

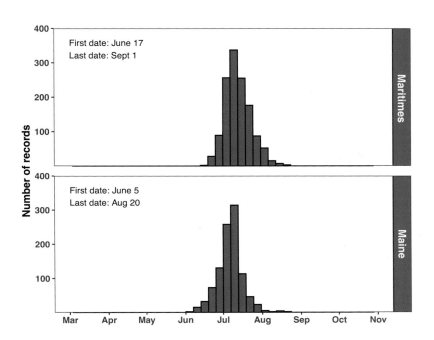

OCOLA SKIPPER

Panoquina ocola
(W. H. Edwards)

Subspecies: Only the nomino-typical subspecies (*P. o. ocola*) occurs in eastern North America.

Distinguishing characters:
  Size: Small to medium.
  Upperside: Brown; fore-wing with scattered translucent spots, the largest arrowhead shaped.

Ogunquit, ME, 25 September 2017 (Andrew Aldrich)

  Underside: Forewing like upperside; hindwing with slightly paler veins and row of indistinct pale spots (sometimes absent).
  Additional: Forewing noticeably elongated.

Similar species: Dun Skipper.

Status and distribution in Maine: Stray. Rare; recorded only from the coastal counties of Knox and York. SNA

Status and distribution in Maritimes: Stray. Rare; recorded only in New Brunswick, from coastal islands in Charlotte County. SNA

Habitat: Migrants could be found in virtually any sunny, open area, including city parks and gardens.

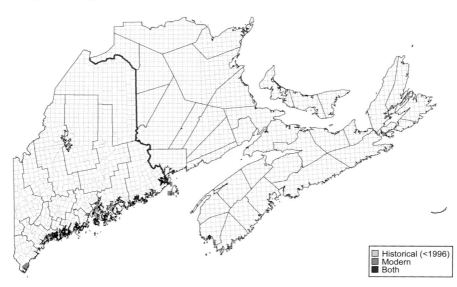

Historical (<1996)
Modern
Both

Biology: Multiple generations annually; adults are active all year where resident in the South. Our few records are from late August to late September. Food plants poorly known, but larvae are known to feed on various grasses (Poaceae), including cut grasses (*Leersia* spp.) and panicgrasses (*Panicum* spp.). Probably overwinters as adult in reproductive diapause but not in our region.

Adult behavior: This skipper has a rapid, low, erratic flight. Adults nectar often and in our region have been observed visiting Common Grass-leaved Goldenrod (*Euthamia graminifolia*) and ornamentals, such as Orange-eye Butterfly-bush (*Buddleja davidii*), Purple-topped Vervain (*Verbena bonariensis*), and zinnias (*Zinnia* spp.).

Comments: The Ocola Skipper is a common resident of the southeastern United States. It regularly migrates northward, sometimes reaching New England and extreme southern Canada between mid-August and mid-October. It was first recorded in our region by Marcy Wagner on 19 August 2014 on Campobello Island, Charlotte County, New Brunswick. This skipper was subsequently photographed at additional coastal localities in Maine and New Brunswick from 2016 to 2019. There are many records from Massachusetts, where it appears to be an increasingly regular visitor (Stichter 2015). Its northern abundance fluctuates from year to year, but reports from the Acadian region may become more frequent due to climate change. Although Ocola Skippers resemble several resident species (such as the Dun Skipper), they have a distinctive stretched-out appearance when perched, with their elongated forewings extending well beyond the hindwings. Any small, drab skipper encountered late in the season, especially in coastal areas, should be closely examined.

Threats: None documented in our region.

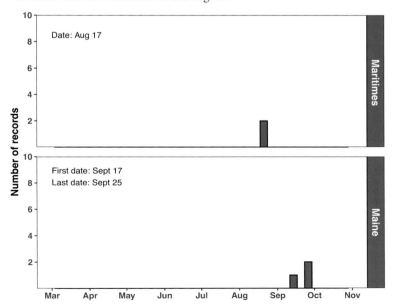

LEAST SKIPPER

Ancyloxypha numitor
(Fabricius)

Subspecies: None.

Distinguishing characters:
  Size: Very small.
  Upperside: Forewing dark brown to black, with variable amounts of orange; hindwing orange with broad dark margin.
  Underside: Forewing dark brown to black with broad orange margin; hindwing orange, with slightly paler veins.

Mating pair—Moncton, NB, 24 August 2008 (Jim Edsall)

Similar species: European Skipper.

Status and distribution in Maine: Resident. Common; widespread but localized. S4S5

Status and distribution in Maritimes: Resident. Common; widespread but localized. NB-S5, NS-S5, PE-S4

Habitat: Damp grassy areas, including ditches, streamsides, marshes, and wet meadows.

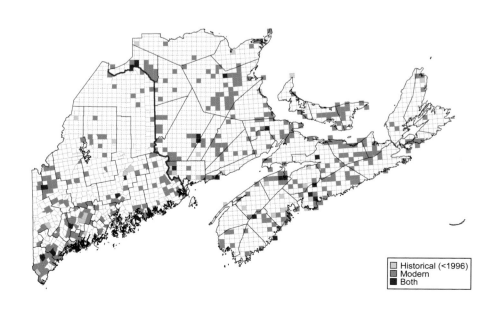

Historical (<1996)
Modern
Both

Biology: Two overlapping generations per year. Adults are active from late May (southward) until early to mid-September, with peak abundance in mid-July and late August. Larvae feed on various grasses (Poaceae), including blue grasses (*Poa* spp.), Reed Canary Grass (*Phalaris arundinacea*), and Rice Cut Grass (*Leersia oryzoides*). Overwinters as a late-stage larva.

Adult behavior: Least Skippers have a low, weak flight. Males are typically seen weaving their way through wetland vegetation, patrolling for females. While not frequently observed nectaring, they can be found at a variety of flowers, including Cow Vetch (*Vicia cracca*), clovers (*Trifolium* spp.), hawk-weeds (*Hieracium* spp.), Heal-all (*Prunella vulgaris*), and American Sea-rocket (*Cakile edentula*).

Comments: This is our only resident grass skipper with two generations per year. This species' diminutive size and habit of flying slowly through wetland vegetation add to its distinctiveness, though they increase the chances that it is overlooked. In the American South, it is occasionally a pest of cultivated rice. Although this is not a concern to agriculture in our region, a Least Skipper colony was documented in association with ornamental Reed Canary Grass in Milton, Nova Scotia, by Dorothy Poole in 2013.

Threats: None documented in our region.

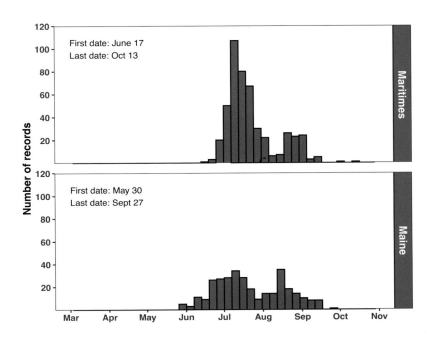

FAMILY PAPILIONIDAE

Named for the projections, or tails, on the hindwings of many species, the swallowtails include the largest butterflies in the world. Far from merely aesthetic, the tails are thought to improve aerodynamic performance and may draw the attention of predators away from the body. With about forty species recorded in the United States and Canada, these stunning, powerful flyers occupy a wide range of forested and open habitats. Because of their size and striking coloration, swallowtails are among those butterflies most often noticed by the general public. Swallowtails are eager flower visitors and may be approached while they are preoccupied with this activity. Some species nectar calmly with motionless wings, while others ceaselessly flutter from blossom to blossom. Groups of males gather at damp earth in "puddle clubs" to imbibe salts and minerals. Some swallowtails are distasteful to predators, and their appearance is mimicked by other butterflies, even fellow swallowtails. For example, some believe the wing pattern of the Pipevine Swallowtail, whose larvae feed on toxic plants of the Dutchman's Pipe family (Aristolochiaceae), is mimicked by several palatable butterflies, including the Spicebush Swallowtail. Larval food plants of the Papilionidae range from trees and shrubs to weedy herbs and vines.

Seven species of swallowtails are known to occur in Maine and the Maritimes, all in the subfamily Papilioninae (true swallowtails). Four are breeding residents that produce one or two generations per year. Three species are rare colonists or strays. One resident, the Short-tailed Swallowtail, is represented by two subtly differentiated subspecies (one mostly endemic to our region) that inhabit strikingly different habitats. The Spicebush Swallowtail is listed in Maine as a species of Special Concern.

Because of subtle morphological and genetic differences, controversy has existed for decades over the taxonomic status of certain groups of swallowtails. Specifically, some argue that the subgenera *Heraclides* and *Pterourus* in the genus *Papilio* should be elevated to genera. We follow Zhang et al. (2019b) and Pelham (2022), who recognize *Heraclides* for the Eastern Giant Swallowtail and *Pterourus* for the tiger swallowtails and the Spicebush Swallowtail. Ongoing studies of tiger swallowtails found in northeastern North America may reveal the presence of at least one additional species in our region.

Canadian Tiger Swallowtails—Halifax County, NS (Krista Melville). Striking in size and color, this butterfly is a conspicuous member of the Acadian region's fauna.

PIPEVINE SWALLOWTAIL
Battus philenor (Linnaeus)

Subspecies: Only the nomino-typical subspecies (*B. p. phile-nor*) occurs in eastern North America.

Distinguishing characters:
    Size: Large.
    Upperside: Black with several small, white spots near forewing margin (more pronounced in female); hindwing with

Paxton, MA, 29 August 2013 (Garry Kessler)

row of larger white spots and wash of iridescent blue or greenish blue on outer two-thirds (less brilliant in female).
    Underside: Black with white spots near forewing margin; hindwing with white marginal spots and seven large, black-rimmed orange spots (often with white) within field of iridescent blue or greenish blue.
    Additional: Hindwing with large spatulate tail.

Similar species: Black Swallowtail and Spicebush Swallowtail.

Status and distribution in Maine: Formerly stray, now rare colonist; few records with specific data, all from coastal areas. SNA

Status and distribution in Maritimes: Not documented.

Habitat: Open woodlands, fields, and gardens where resident; strays may be encountered virtually anywhere.

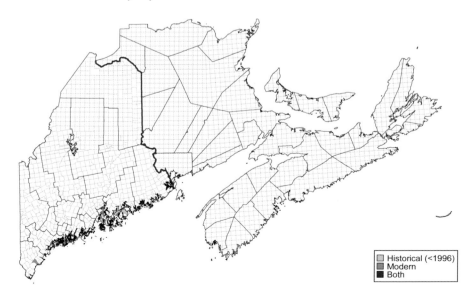

☐ Historical (<1996)
☐ Modern
■ Both

Biology: Two generations where resident, spring and summer. Adults have been recorded in Maine in mid to late September. Larvae feed on pipevines (birthworts), including Virginia Serpentaria (*Endodeca serpentaria*) and Large-leaved Dutchman's Pipe (*Isotrema macrophyllum*), a southeastern species that is sometimes cultivated in Maine. Recently, larvae have been found on cultivated Dutchman's Pipe in coastal Maine.

Adult behavior: A strong flyer, males erratically dash about with shallow wingbeats in search of females. Males frequently visit damp earth, and both sexes nectar at flowers, where they are easily observed. In Maine, adults have been documented visiting phlox (*Phlox* spp.) and Joe-Pye weed (*Eutrochium* spp.).

Comments: The first record of the Pipevine Swallowtail in Maine (Bar Harbor, Hancock County) was around 1900 (Aaron 1907). Another undated specimen from that locality (or nearby) was reported by Brower (1974). More recently, two adults were photographed on consecutive days in 2007 at Kennebunk Beach, York County (June Ficker), and on Lane's Island, Knox County (Kirk Gentalen), and a male in good condition was photographed in 2020 at Phippsburg, Sagadahoc County (Shaun Mains). This species strays far from its usual breeding areas, where it often occurs in gardens that grow cultivated pipevines. It is known to feed on cultivated Dutchman's Pipe in Massachusetts (Stichter 2015) and larvae were found on this plant on Mount Desert Island (Hancock County), Maine, in August 2022. This butterfly sequesters toxic chemicals from its larval food plants, thereby receiving protection from predators. As a result, it is believed to serve as a model for several other butterflies that mimic its pattern, including the Spicebush Swallowtail and the Red-spotted Purple (a more southerly subspecies of the White Admiral).

Threats: None documented in our region.

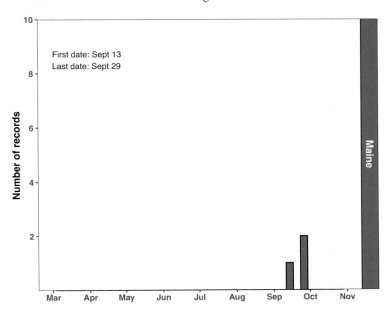

First date: Sept 13
Last date: Sept 29

SHORT-TAILED SWALLOWTAIL
Papilio brevicauda Saunders

Subspecies: Two subspecies occur in our region: *P. b. bretonensis* Mc-Dunnough occurs on the coasts of Cape Breton and the Gulf of St. Lawrence coastline of New Brunswick, and the subspecies *P. b. gaspeensis* McDunnough occurs inland in northern New Brunswick and northern Maine.

Distinguishing characters:
Size: Large.

Upperside: Black with rows of yellow spots near margins and across center of wings; hindwing with blue scaling between rows of yellow spots (often suffused with orange, especially in female), and black-centered orange dot near inner edge, the black center typically touching the wing margin.

Underside: Similar to upperside, yellow spots suffused with orange, especially on hindwing.

Additional: Hindwing with relatively short tail; forewing outer margin flat to convex (compare with Black Swallowtail).

Similar species: Black Swallowtail.

Status and distribution in Maine: Resident or rare colonist. Rare; known only from northern Aroostook County. State Special Concern. S1S2

Status and distribution in Maritimes: Resident. Rare in northern interior New Brunswick but widespread and common in localized colonies on New Brunswick's Gulf of St. Lawrence coast; rare and localized in Nova Scotia on the Cape Breton coast. NB-S3S4, NS-S1

Habitat: Dunes and salt marshes on the coast; riparian meadows, stream margins, and mesic openings in mixed and coniferous forest inland.

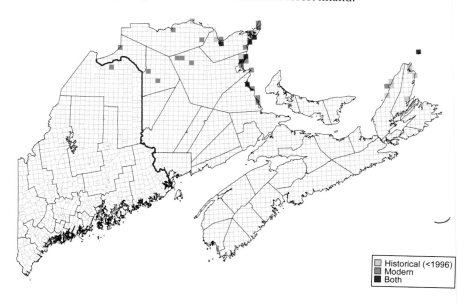

Historical (<1996)
Modern
Both

Male *P. b. bretonensis*—La Haute-Gaspésie, QC, July 2020 (Francis Pamerleau)

Male *P. b. gaspeensis*—Northumberland County, NB, 13 June 2020 (Peter Gadd)

Biology: Two generations per year. Adults emerge in early June and fly until late August, with peak flight in late June and early August. Larvae of subspecies *bretonensis* feed primarily on Scotch Lovage (*Ligusticum scoticum*) and have also been recorded on Sea Coast Angelica (*Angelica lucida*); larvae of subspecies *gaspeensis* are believed primarily to use American Cow-parsnip (*Heracleum maximum*). Overwinters as a pupa.

Adult behavior: This swallowtail has a strong flight, and males patrol for females on dunes and hilltops. It nectars frequently and has been observed at Blue Iris (*Iris versicolor*), Common Dandelion (*Taraxacum officinale*), Field Sow-thistle (*Sonchus arvensis*), Carolina Sea-lavender (*Limonium carolinianum*), Cow Vetch (*Vicia cracca*), and Scotch Bellflower (*Campanula rotundifolia*).

Comments: The Short-tailed Swallowtail has a small global range centered around the Gulf of St. Lawrence. The coastal subspecies, *bretonensis*, was described from specimens collected in Baddeck, Nova Scotia, and it has long been known to occur on Cape Breton and along the New Brunswick coast. The slightly smaller and more western subspecies, *gaspeensis*, was described from specimens collected on Quebec's Gaspé Peninsula, where it occurs in both boreal forest and coastal habitat on the St. Lawrence River estuary. *Papilio b. gaspeensis* was first documented in the Acadian region by Martin Turgeon on 2 June 2010 near Edmundston, New Brunswick. This discovery inspired Phillip and Emmett deMaynadier to search for this butterfly in northern Aroostook County, Maine, in late May 2012. They were rewarded with the

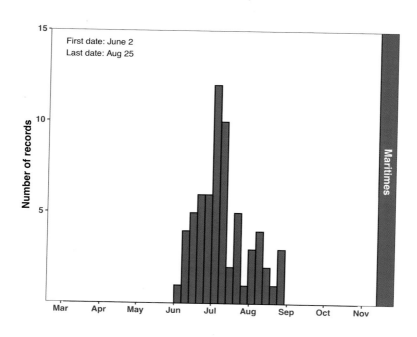

First date: June 2
Last date: Aug 25

first (and to date only) U.S. record of the species—a larva feeding on American Cow-parsnip, which was later reared to adult in captivity.

Threats: Subspecies *bretonensis* is restricted to dunes and salt marshes, making it vulnerable to habitat loss from rapid sea-level rise. Subspecies *gaspeensis* may also be threatened by the effects of climate change, as it is at the southern edge of its range in the Acadian region and at a lower altitude than much of its distribution in Quebec.

BLACK SWALLOWTAIL

Papilio polyxenes Fabricius

Subspecies: Only the sub-species *P. p. asterius* (Stoll) occurs in eastern North America.

Distinguishing characters:
Size: Large.
Upperside: Black with rows of yellow spots near margins and across center of wings (much reduced in females); hindwing with blue scaling between

Male—Kings, NS, 22 July 2014 (Phil Schappert)

rows of yellow spots (more extensive in females) and black-centered orange dot near inner edge.
Underside: Similar to upperside; yellow spots suffused with orange, especially on hindwing.
Additional: Hindwing with well-developed, narrow tail; forewing outer margin flat to concave (compare with Short-tailed Swallowtail).

Similar species: Short-tailed Swallowtail, Spicebush Swallowtail, and Pipevine Swallowtail.

Status and distribution in Maine: Resident. Common; widespread in south, seemingly absent from much of the northwest. S4S5

Status and distribution in Maritimes: Resident. Uncommon to common and widespread in New Brunswick and Nova Scotia, uncommon on Prince Edward Island. NB-S4, NS-S4, PE-S4

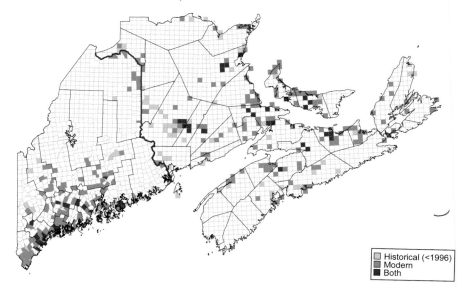

Historical (<1996)
Modern
Both

Habitat: A variety of open areas, including roadsides, gardens, old fields, coastal dunes, and salt marshes.

Biology: Two generations, with a partial third brood during some years. Adults emerge during May and fly until September (sometimes early October), with peak abundance from mid-June to late July. Larvae feed on plants in the carrot family (Apiaceae), including wild and cultivated carrot (*Daucus carota*), dill (*Anethum graveolens*), parsley (*Petroselinum crispum*), and Scotch Lovage (*Ligusticum scoticum*). Overwinters as a pupa.

Adult behavior: This swallowtail has a strong flight. Males patrol for females, particularly on hilltops, and occasionally visit damp soil or dung. Both sexes are frequently seen nectaring at a variety of flowers in our region, including thistles (*Cirsium* spp.), Common Dandelion (*Taraxacum officinale*), Common Lilac (*Syringa vulgaris*), and clovers (*Trifolium* spp.).

Comments: The Black Swallowtail's conspicuous black, green, and yellow-striped larvae are a relatively common sight in vegetable gardens, but they are rarely abundant enough to be considered a pest. This species' larval food plants are rare outside of inhabited areas, dunes, and salt marshes, resulting in a patchy distribution within the sparsely populated areas in the north of Maine and New Brunswick.

Threats: None documented in our region.

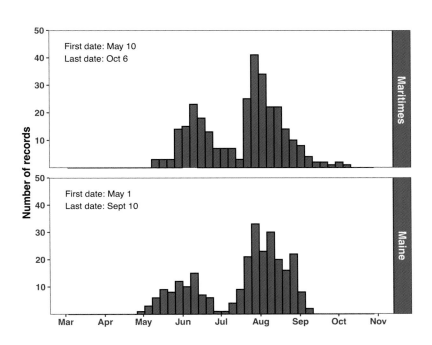

Maritimes
First date: May 10
Last date: Oct 6

Maine
First date: May 1
Last date: Sept 10

Number of records

Mar   Apr   May   Jun   Jul   Aug   Sep   Oct   Nov

EASTERN GIANT SWALLOWTAIL

Heraclides cresphontes (Cramer)

Subspecies: None.

Distinguishing characters:
    Size: Very large.
    Upperside: Dark brown (nearly black) with two rows of large yellow spots that merge near forewing apex; hindwing with black spot at inner edge, which is capped with orange and blue.

Sheffield, MA, 2 August 2011 (Frank S. Model)

    Underside: Forewing dark brown with large yellow spots; hindwing mostly yellow with dark central band enclosing blue crescents and usually two small orange patches in center.
    Additional: Hindwing with well-developed, spatulate tail; yellow teardrop in center.

Similar species: Canadian Tiger Swallowtail and Eastern Tiger Swallowtail (ventral only).

Status and distribution in Maine: Rare to frequent colonist; mostly southern. SNA

Status and distribution in Maritimes: Rare colonist. Four historical and several modern records in New Brunswick; one historical record in Nova Scotia. NB-SNA, NS-SNA

Habitat: Virtually any open area, including forest margins, roadsides, gardens, and orchards.

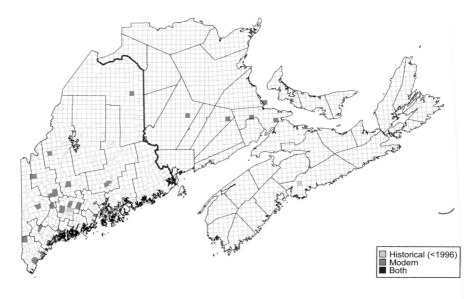

Historical (<1996)
Modern
Both

Biology: Multiple generations where resident; our records are from early August to mid-September. Larvae feed on members of the citrus family. In most of the butterfly's eastern range, the primary food plants are Common Prickly-ash (*Zanthoxylum americanum*), and Common Hoptree (*Ptelea trifoliata*), neither of which are native to our region. However, at two different locations in Maine, early stages have been found on ornamental Gasplant (*Dictamnus albus*), also of the citrus family. Southward, larvae are known as "orange dogs" to citrus growers. Overwinters as a pupa but not in our region.

Adult behavior: This species has a powerful, seemingly effortless flight. Males patrol along forest margins or watercourses in search of females and often visit damp soil. Both sexes nectar while rapidly beating their wings (flutter-feeding). In our region, they have been observed at a number of flowers, including Common Milkweed (*Asclepius syriaca*), Coastal Sweet-pepperbush (*Clethra alnifolia*), Orange-eye Butterfly-bush (*Buddleja davidii*), phlox (*Phlox* spp.), vervain (*Verbena* spp.), Eastern Purple Coneflower (*Echinacea purpurea*), and zinnias (*Zinnia* spp.).

Comments: Until recently, the Eastern Giant Swallowtail was mostly a very rare stray within the Acadian region. Larvae were reportedly found in Kings and Saint John Counties of New Brunswick as long ago as 1877 and 1878 (Heustis 1879a), which are the earliest known records from the region. The first Maine record was in 1886 (Scudder 1888–1889), and it was not recorded again in the state until 2011, when an influx of the species was also observed in Massachusetts (Stichter 2015). In 2012, many records of this butterfly were documented across our region, including the first reports in New Brunswick since 1879 and the first evidence of its breeding in Maine. Despite these recent incursions, there are no known records in Nova Scotia beyond a single worn, male specimen, probably captured in 1901 (Ferguson 1954). The expansion of this species into the Northeast appears to be correlated with climate change (Wilson et al. 2021).

Threats: None documented in our region.

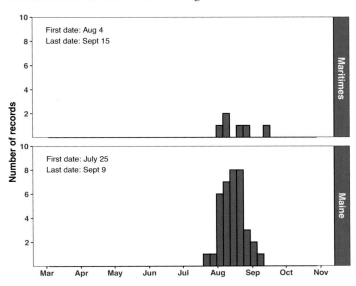

SPICEBUSH SWALLOWTAIL

Pterourus troilus (Linnaeus)

Subspecies: Only the nomino-
typical subspecies (*P. t. troilus*)
occurs in northeastern North
America.

Female—Eliot, ME, 14 July 2014 (Terry Chick)

Distinguishing characters:
  Size: Large.
  Upperside: Black with
    single row of white (fore-
    wing) to bluish-white
    (hindwing) marginal
    spots; a wash of blue-
    green (male) or blue
    (female) iridescence on
    hindwing (more extensive on males).
  Underside: Black with two rows of orange spots and broad band of irides-
    cent blue scaling.
  Additional: Hindwing with large spatulate tail.

Similar species: Black Swallowtail and Pipevine Swallowtail.

Status and distribution in Maine: Resident. Rare; restricted to York County.
State Special Concern. S2

Status and distribution in Maritimes: Not documented.

Habitat: Rich hardwood and mixed wood forests with slow streams or seep-
ages. Adults usually encountered in gardens and along woodland edges and
roads.

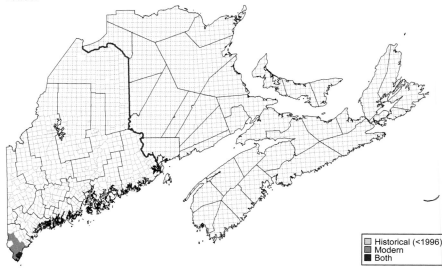

Historical (<1996)
Modern
Both

Biology: Two overlapping generations. Only a few adult records have been documented in July and August, but numerous observations of mid- to late-instar larvae in mid-July and August–September indicates the likely emergence of first-generation adults in June and second-generation adults in July and August, consistent with observations elsewhere in New England. Larvae feed on Sassafras (*Sassafras albidum*) and Northern Spicebush (*Lindera benzoin*), with most Maine populations found on the latter (Ward and deMaynadier 2012). Overwinters as a pupa.

Adult behavior: Males puddle, and both sexes visit flowers. Adults are strong flyers and frequent a variety of open habitats, but this is generally the only swallowtail in our region that regularly occurs within deep forest. Nectaring has been observed in Maine on Spreading Dogbane (*Apocynum androsaemifolium*), Steeplebush (*Spiraea tomentosa*), Purple Loosestrife (*Lythrum salicaria*), ornamental blazing stars (*Liatris* spp.), phlox (*Phlox* spp.), thistles (*Cirsium* spp.), and Joe-Pye weeds (*Eutrochium* spp.).

Comments: With only a single historical record, the residency status of the Spicebush Swallowtail in Maine remained uncertain until recent surveys revealed several populations in York County (deMaynadier 2009; Ward and deMaynadier 2012). This butterfly is most readily documented by searching food plants for the larvae, usually hidden in folded leaf nests. As in Massachusetts, surveys for larvae in Maine are most productive during the second generation in August and September (Jaffe 2009).

Threats: Rich seepage forests with Sassafras and Northern Spicebush are uncommon and mostly restricted to private lands in southern Maine, where these habitats are vulnerable to impacts from development and intensive forestry.

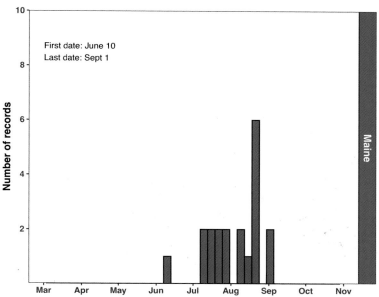

First date: June 10
Last date: Sept 1

Number of records

Mar Apr May Jun Jul Aug Sep Oct Nov

Maine

EASTERN TIGER SWALLOWTAIL

Pterourus glaucus (Linnaeus)

Subspecies: Only the nominotypical subspecies (*P. g. glaucus*) occurs in northeastern North America.

Distinguishing characters:

Size: Large to very large.

Upperside: Bright yellow with black barring; hindwing usually with relatively narrow black margin adjacent to the abdomen; female with expanded blue scaling within outer black hindwing border.

Male—Gardner, MA, 17 July 2014 (Garry Kessler)

Underside: Bright yellow with black barring; forewing margin usually with divided yellow spots; hindwing usually with relatively narrow black margin adjacent to the abdomen.

Additional: Hindwing with well-developed, spatulate tail.

Similar species: Canadian Tiger Swallowtail.

Status and distribution in Maine: Uncertain, stray or possible rare colonist; primarily in York County. SNA

Status and distribution in Maritimes: Not documented.

Habitat: A variety of open areas, including forest margins, rights-of-way, gardens, and old fields.

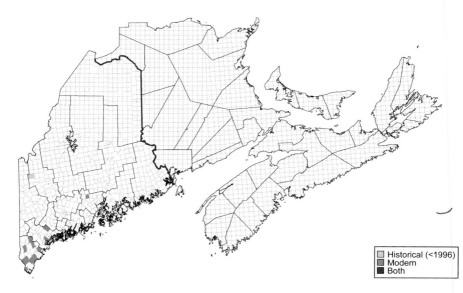

Historical (<1996)
Modern
Both

Biology: Two generations per year in the northern portion of its range. This species may not reproduce in our region, where the first brood is not known to occur. Adults appear irregularly in Maine, from late June through early September (rarely early October), with peak abundance in mid-August. Larvae feed on a variety of deciduous trees, including Tuliptree (*Liriodendron tulipifera*), cherries (*Prunus* spp.), and ashes (*Fraxinus* spp.). Overwinters as a pupa.

Adult behavior: As with many swallowtails, this species has a powerful flight. Like Canadian Tiger Swallowtails, males of the Eastern Tiger Swallowtail gather at patches of wet soil or on dung. Both sexes often nectar at flowers and have been observed in our region visiting Common Milkweed (*Asclepius syriaca*), bee-balms (*Monarda* spp.), phlox (*Phlox* spp.), thistles (*Cirsium* spp.), and Joe-Pye weeds (*Eutrochium* spp.).

Comments: The status of this species, which is widely distributed southward, is poorly understood in our region. In Maine, butterflies that are tentatively recognized as Eastern Tiger Swallowtails occur after the peak flight of Canadian Tiger Swallowtails, though there is considerable overlap in their flight periods. Southern Maine is at the northern edge of the contact zone between these species, and many butterflies within this area possess characteristics of both (Scriber et al. 2008). In Maine, very few individuals are entirely consistent with the Eastern Tiger Swallowtail, and the dark form female of this species (*glaucus*) has not been recorded. Most records of eastern tigers in our region may actually represent an unnamed species of hybrid origin, sometimes referred to as the "Midsummer Tiger Swallowtail." Acorn and Sheldon (2016) and Schmidt (2020) discuss the occurrence of such phenotypes in Ontario. The limited number of Maine records that appear to represent true eastern tigers suggests that they are merely strays, which rarely (if ever) reproduce in the state. Much more research is needed to understand the relationships and life histories within the tiger swallowtail complex in northeastern North America.

Threats: None documented in our region.

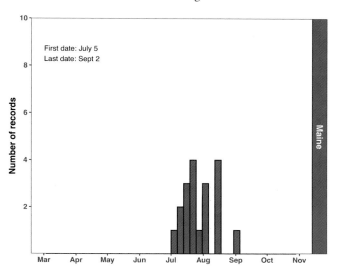

First date: July 5
Last date: Sept 2

CANADIAN TIGER SWALLOWTAIL

Pterourus canadensis
(Rothschild & Jordan)

Subspecies: None.

Distinguishing characters:

  Size: Large.

  Upperside: Bright yellow with black barring; hindwing with relatively broad black margin adjacent to the abdomen.

  Underside: Bright yellow with black barring; yellow forewing margin forming contiguous band (not interrupted by black veining); hindwing with relatively broad black margin adjacent to the abdomen.

  Additional: Hindwing with well-developed, spatulate tail.

Similar species: Eastern Tiger Swallowtail.

Status and distribution in Maine: Resident. Abundant; widespread. S5

Status and distribution in Maritimes: Resident. Abundant; widespread. NB-S5, NS-S5, PE-S5

Habitat: A variety of open areas, including forest margins, roadways, gardens, old fields, and bogs.

Male—China, ME, 10 June 2013 (W. Herbert Wilson Jr.)

Historical (<1996)
Modern
Both

Biology: One generation per year. Adults appear in mid to late May and fly until late July, with peak abundance in mid-June (they rarely occur into mid-August). Larvae feed on a variety of deciduous trees, including ashes (*Fraxinus* spp.), aspens (*Populus* spp.), birches (*Betula* spp.), and cherries (*Prunus* spp.). Overwinters as a pupa.

Adult behavior: This species has a strong, direct flight and is often coursing along woodland edges at a rapid pace. Males gather at patches of wet soil or on dung, often in large aggregations. During the Maine and Maritime surveys, this butterfly was recorded nectaring at the flowers of nearly eighty plant genera; most frequently lilacs (*Syringa* spp.), clovers (*Trifolium* spp.), milkweeds (*Asclepias* spp.), vetches (*Vicia* spp.), and hawkweeds (*Hieracium* spp.).

Comments: This is one of our most common and familiar butterflies, and its springtime appearance is a welcome harbinger of the warmer weather to come. The Canadian Tiger Swallowtail was previously considered a subspecies of the Eastern Tiger Swallowtail, but it was elevated to species level in 1991 based on genetic differences, number of generations, larval food preferences, and subtle morphological characteristics (Hagen et al. 1991). On rare occasions, fresh adults that look like Canadian Tiger Swallowtails occur as late as September, suggesting that their pupae broke diapause before overwintering. Observers in southern Maine must be careful not to confuse late-flying Canadian Tiger Swallowtails with Eastern Tiger Swallowtails. Complicating matters is the likely presence in southern Maine of at least one other summer-flying tiger swallowtail, which appears to be a discrete species (see Eastern Tiger Swallowtail).

Threats: None documented in our region.

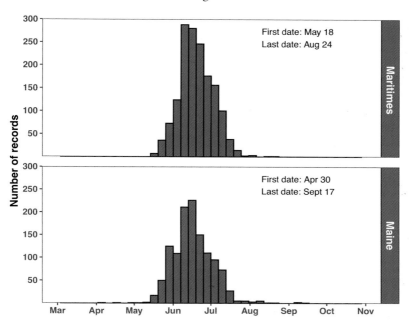

FAMILY PIERIDAE

As implied by their group name, butterflies of this family are mostly decorated in hues of white, yellow, and orange. Eighty species are known to occur in the United States and Canada, ranging in size from very small to large. To many, this family embodies the very notion of a butterfly. In fact, a butter-colored Eurasian sulphur may have been the inspiration for the name "butterfly." Most pierids prefer sunny areas and will quickly seek cover when the sun is obscured. Their cheerful coloration rarely fails to attract attention as they course through open woodlands, old fields, bogs, urban parks, and residential gardens. Whites and sulphurs owe their flashing colors to pigments known as pterins, which, unlike most other kinds of butterflies, are derived from waste uric acid stored during the larval stage. The name pterin comes from the Greek word *pteron*, meaning wing. The wing scales of some species also reflect ultraviolet light in distinct patterns that are recognized by potential mates during courtship. Many pierids produce seasonal phenotypes, with early and late season adults being darker ventrally. Some sulphurs have pale female forms that are reminiscent of whites. Most members of this family have a low, meandering flight, but larger species are powerful flyers that frequently rise above the treetops. Many whites and sulphurs, including small species, are known to migrate great distances and even have been observed far out at sea. Butterflies in this family are active flower visitors, and males often gather in large numbers at patches of damp earth. Larval food plants are diverse; in our region, whites feed on crucifers and most sulphurs eat legumes.

Nine species of Pieridae (three whites and six sulphurs) are known to occur in Maine and the Maritimes. They are divided into two subfamilies, the Pierinae (whites) and Coliadinae (sulphurs). Four are breeding residents that produce one or more generations per year, with two flying continuously from spring to autumn. Four species are strays, mostly southern migrants. One species (Orange Sulphur) regularly colonizes the region from the south, only to perish at the onset of winter.

Clouded Sulphurs—Fair Haven, VT (Bryan Pfeiffer). The most abundant sulphur in the Acadian region, this butterfly often gathers in large numbers at damp earth.

LITTLE YELLOW

Pyrisitia lisa (Boisduval &
Le Conte)

Mattapoisett, MA, 19 August 2012 (Garry Kessler)

Subspecies: Only the nomino-
typical subspecies (*P. l. lisa*)
occurs in the eastern United
States and Canada.

Distinguishing characters:
  Size: Small.
  Upperside: Male lemon
    yellow with broad black
    forewing border and nar-
    row black hindwing bor-
    der. Female pale yellow or white, with broken black hindwing border.
  Underside: Yellow (female sometimes white) with dark discal spot on
    forewing; hindwing with irregular pattern of dark spotting and usually
    a large red or orange spot on leading edge (sometimes reduced or absent
    in male).

Similar species: None.

Status and distribution in Maine: Stray. Rare; mostly historical records from
central and southern counties. SNA

Status and distribution in Maritimes: Stray. Rare; historical records from
scattered localities in New Brunswick and Nova Scotia. NB-SNA, NS-SNA

Habitat: Migrants can be found in virtually any sunny, open area, including
bogs, fields, city parks, and gardens.

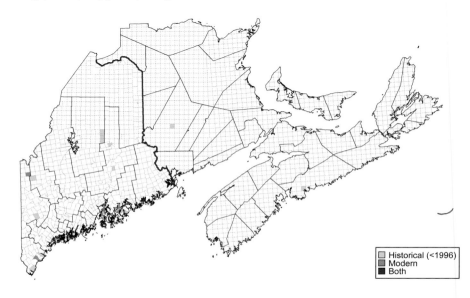

☐ Historical (<1996)
▨ Modern
■ Both

Biology: Multiple generations annually; adults are active all year where resident. Our records are from mid-June to early September. Larvae feed primarily on Partridge Sensitive-pea (*Chamaecrista fasciculata*) and Wild Sensitive-pea (*C. nictitans*). Overwinters as an adult in reproductive diapause but not in our region.

Adult behavior: This butterfly flies erratically and near the ground. Adults frequently visit flowers and damp earth. In our region, it has been observed nectaring at Sheep-laurel (*Kalmia angustifolia*).

Comments: The Little Yellow is an irregular migrant to our region, turning up where least suspected, from the tablelands of Mount Katahdin to the middle of a bog in Maine. Historically more frequent in the Northeast than today, this species was considered locally common as far north as Massachusetts during the late nineteenth and early twentieth centuries (Stichter 2015). Only two modern records (in 2002 and 2012) are known from the Acadian region. Although there is no evidence that this butterfly has ever established a temporary population here, it could potentially produce at least one brood if migrants arrived early enough and located a patch of Partridge Sensitive-pea, which is introduced in southern Maine.

Threats: None documented in our region.

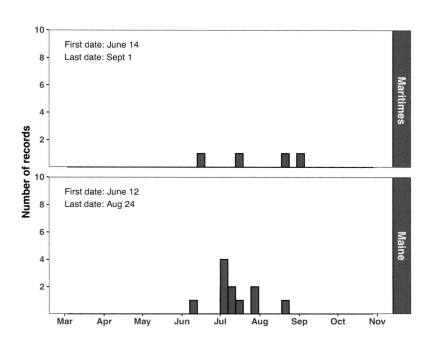

CLOUDED SULPHUR

Colias philodice Godart

Subspecies: None.

Distinguishing characters:

Size: Medium.

Upperside: Male lemon yellow with solid black wing margins; female yellow or white with pale spots in forewing margin; both sexes with black spot on forewing and orange spot on hindwing.

Female—Carrabassett Valley, ME, 1 June 2013 (Roger Rittmaster)

Underside: Lemon yellow (hindwing often greenish), usually with row of indistinct dark spots near margin; hindwing center with twin silvery eyespots (smaller upper spot sometimes missing); female may be white with little or no yellow.

Additional: Spring adults have narrower black wing margins and darker hindwings below.

Similar species: Pink-edged Sulphur and Orange Sulphur.

Status and distribution in Maine: Resident. Abundant; widespread. S5

Status and distribution in Maritimes: Resident. Abundant; widespread. NB-S5, NS-S5, PE-S5

Habitat: A variety of open areas, including roadsides, meadows, old fields, gardens, and lawns.

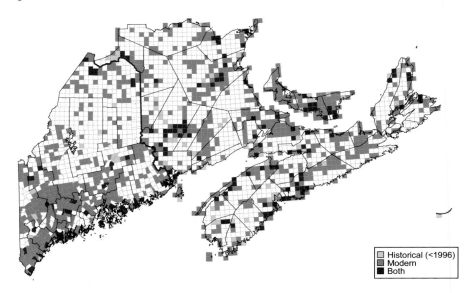

☐ Historical (<1996)
◼ Modern
■ Both

Biology: At least three overlapping generations per year. Adults emerge in early May, with peaks in abundance in late May/early June, late July, and late August/early September. Larvae feed on various legumes, particularly clovers (*Trifolium* spp.) and Alfalfa (*Medicago sativa*). Overwinters as a pupa.

Adult behavior: The Clouded Sulphur is a relatively strong flyer. Males erratically patrol for females near the ground throughout the day and are frequent puddlers, often forming large aggregations on dirt roads and other damp, open ground. During the Maine and Maritime atlas projects, adults were reported nectaring at the flowers of nearly fifty plant genera, with clovers (*Trifolium* spp.), American-asters (*Symphyotrichum* spp.), Common Dandelion (*Taraxacum officinale*), hawkweeds (*Hieracium* spp.), and goldenrods (*Solidago* spp.) being the most frequently visited.

Comments: All three of our *Colias* sulphurs have white-form ("alba") females, and they make up a significant proportion of the Clouded and Orange Sulphur populations by late summer. While white-form Pink-edged Sulphurs are identifiable by their lack of submarginal spots, those of Orange and Clouded Sulphurs can be impossible to separate with confidence. These two butterflies freely interbreed and produce hybrids, yet they are generally considered to represent discrete species.

Threats: None documented in our region.

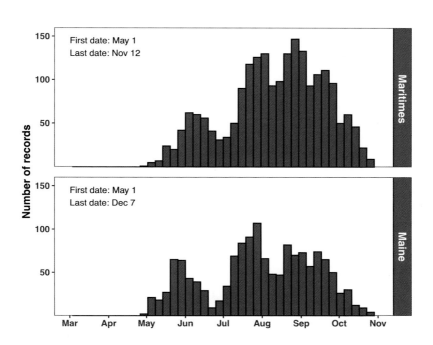

ORANGE SULPHUR

Colias eurytheme Boisduval

Subspecies: None.

Distinguishing characters:
Size: Medium.
Upperside: Male mostly orange with solid black wing margins; female orange or white with pale spots in forewing margin; both sexes with black spot on forewing and orange spot on hindwing.

Male—Northumberland County, NB, 3 August 2013 (Peter Ga…

Underside: Yellowish orange, usually with row of indistinct dark spots near margin; hindwing center with twin silvery eyespots (smaller upper spot sometimes missing); female may be white with little or no orange.
Additional: Spring adults often have reduced orange coloration and narrower black wing margins.

Similar species: Clouded Sulphur and Pink-edged Sulphur.

Status and distribution in Maine: Frequent colonist. Uncommon most years, occasionally common; widespread but most abundant in the south. S5B

Status and distribution in Maritimes: Frequent colonist. Rare most years, occasionally uncommon to common on the mainland; widespread. NB-S4B, NS-S5B, PE-S1S2B

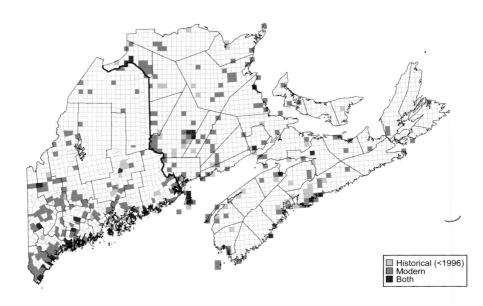

Historical (<1996)
Modern
Both

Habitat: A variety of open areas, including meadows, agricultural fields, gardens, and lawns.

Biology: At least two overlapping generations per year. Adults arrive from the south as early as mid-May, though occasionally as late as July. Numbers build over the summer and peak in late August and September. Larvae feed on legumes, particularly clovers (*Trifolium* spp.) and Alfalfa (*Medicago sativa*). There is no evidence of overwintering in our region.

Adult behavior: The Orange Sulphur is a strong flyer. Males patrol for females near the ground throughout the day. Adults nectar at a variety of flowers, particularly American-asters (*Symphyotrichum* spp.), Red Clover (*Trifolium pratense*), hawkweeds (*Hieracium* spp.), and Fall-dandelion (*Scorzoneroides autumnalis*) in our region.

Comments: Historically, the Orange Sulphur was a western species. It was rare east of the Mississippi River, with only a single sighting documented in the Acadian region during the late 1800s from Mount Desert Island, Maine (Scudder 1888–1889). In common with the rest of New England, it was not found in numbers in Maine until the 1930s (Farquhar 1934, Brower 1932, 1974). First recorded in the Maritimes at Digby, Nova Scotia, in 1906, this sulphur was present during most years in that province by the 1950s (Ferguson 1954). By late summer, white females (form *alba*) make up a significant proportion of the adult population in the Acadian region, where they can be nearly indistinguishable from white females of the Clouded Sulphur. These two butterflies freely interbreed and produce hybrids, yet they are generally considered to represent distinct species.

Threats: None documented in our region.

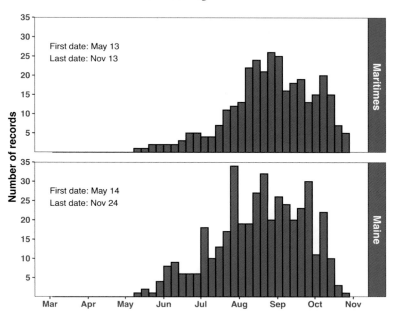

First date: May 13
Last date: Nov 13

Maritimes

First date: May 14
Last date: Nov 24

Maine

Number of records

PINK-EDGED SULPHUR

Colias interior Scudder

Subspecies: Only the nomino-typical subspecies (*C. i. interior*) occurs in north-eastern North America (see comments).

Male—Aurora, ME, 1 July 2009 (Bryan Pfeiffer)

Distinguishing characters:

Size: Medium.

Upperside: Male lemon yellow with solid black wing margins; female yellow (occasionally white) with black margin reduced on forewing and often absent on hindwing; both sexes with black spot on forewing and orange spot on hindwing.

Underside: Yellow or orange yellow; hindwing center with single silvery eyespot; lacks black spots found in other *Colias*.

Additional: Distinctive pink wing fringes (especially when fresh).

Similar species: Clouded Sulphur and Orange Sulphur.

Status and distribution in Maine: Resident. Uncommon to common; widespread. S4

Status and distribution in Maritimes: Resident. Common; widespread. NB-S5, NS-S5, PE-S5

Habitat: A variety of open areas, including forest roadsides, clearcuts, peatlands, and barrens.

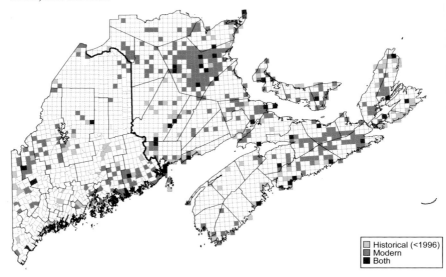

□ Historical (<1996)
▨ Modern
■ Both

Biology: One generation per year. Adults are active from mid-June until early to mid-August, with a few flying into September (some late flyers show little wear, suggesting a partial second brood). Numbers peak during mid-July. Larvae feed on blueberries, including Common Lowbush Blueberry (*Vaccinium angustifolium*), Velvet-leaved Blueberry (*V. myrtilloides*), and Highbush Blueberry (*V. corymbosum*). Overwinters as a partially grown larva.

Adult behavior: The Pink-edged Sulphur has a relatively weak flight, with males patrolling for females near the ground throughout the day. Males occasionally sip at damp soil, and adults nectar at a variety of flowers in our region, especially hawkweeds (*Hieracium* spp.), knapweeds (*Centaurea* spp.), clovers (*Trifolium* spp.), goldenrods (*Solidago* spp.), and Ox-eye Daisy (*Leucanthemum vulgare*).

Comments: Because of its close association with blueberry (its larval food plant), the Pink-edged Sulphur is much more habitat restricted than our other two *Colias* species. Southern Maine is close to the southern limit of this widespread, boreal species, and its occurrence there should be monitored as the impacts of climate change intensify. White-form females of the Pink-edged Sulphur are similar to those of the Clouded and Orange Sulphur but are identifiable by their lack of submarginal spots. In their recent study of the genus *Colias*, Hammond and McCorkle (2017) lumped *C. interior* with *Colias pelidne* Boisduval & Le Conte, and they hesitantly recognized the subspecies *laurentina* Scudder (described from Cape Breton Island, Nova Scotia) as occurring in the Acadian region. We tentatively agree with Pohl et al. (2018) and Pelham (2022) in treating our populations as *C. i. interior*.

Threats: None documented in our region.

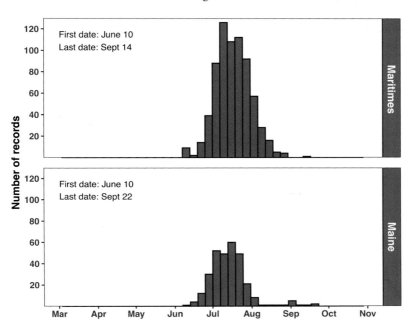

First date: June 10
Last date: Sept 14

First date: June 10
Last date: Sept 22

CLOUDLESS SULPHUR

Phoebis sennae (Linnaeus)

Subspecies: Only the subspecies *P. s. eubule* (Linnaeus) occurs in northeastern North America.

Distinguishing characters:
   Size: Large.
   Upperside: Male lemon yellow and virtually unmarked; female orange yellow to nearly white, with dark marginal spots and dark-ringed central spot on forewing.

Male—Monhegan, ME, 23 September 2012 (Douglas Hitchcox)

   Underside: Male yellow or greenish yellow with or without small reddish markings; central spot on forewing and one or two dark-ringed, silvery central spots on hindwing. Female orange yellow to nearly white, with scattered reddish markings; one or two dark-ringed, silvery central spots on both wings; markings usually more extensive than in male.

Similar species: Orange-barred Sulphur.

Status and distribution in Maine: Stray. Rare; recorded only from a few southern counties. SNA

Status and distribution in Maritimes: Not documented.

Habitat: Migrants can be found in virtually any sunny, open area, including city parks and gardens.

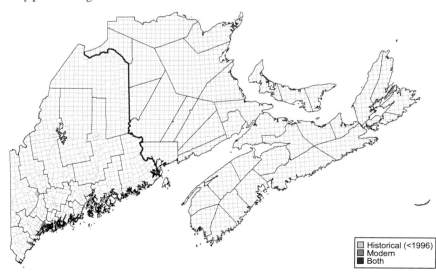

☐	Historical (<1996)
▨	Modern
■	Both

Biology: Multiple generations annually; adults are active all year where resident. Our records are from August and September. Larvae feed on various legumes (Fabaceae), especially Partridge Sensitive-pea (*Chamaecrista fasciculata*) and sennas (*Senna* spp., *Cassia* spp.). Overwinters as an adult in reproductive diapause but not in our region.

Adult behavior: A strong-flying sulphur that rarely pauses except to visit flowers or damp earth. Nectaring has been observed on Jewelweed (*Impatiens capensis*), Joe-Pye weeds (*Eutrochium* spp.), weedy mustards (Brassicaceae), and garden petunias (*Petunia* spp.).

Comments: A resident of the Southeast, the Cloudless Sulphur regularly emigrates northward. It rarely reaches as far north as Maine, where it was first recorded by John Parlin in 1922 at Canton, Oxford County (Parlin 1922). Farquhar (1934) reported a date of 19 August 1927 for Parlin's record, which Brower (1974) repeated, but this year is likely a typo for 1922. This species was more recently recorded at several south-coastal localities, including Monhegan Island, Lincoln County, 10 September 2002 (Bryan Pfeiffer) and 19 September 2012 (Doug Hitchcox); Freeport, Cumberland County, 30 August 2019 (Derek and Jeanette Lovitch); and Biddeford, York County, 15 September 2019 (M. Sprague). The Cloudless Sulphur is not known to reproduce anywhere in New England, but it is an increasingly regular immigrant in Massachusetts (Stichter 2015), and the first record in Vermont was documented on 16 June 2021. This species has the potential to breed in our region if migrants arrive early enough, though they would not survive the winter. Adults should be watched for later in the season, especially near the coast. Their large size is distinctive, even at a distance. Genomic evidence suggests that the subspecies *P. s. eubule* represents a separate species from more western and southern populations, which are currently recognized as the subspecies *P. s. marcellina* (Cramer) (Cong et al. 2016).

Threats: None documented in our region.

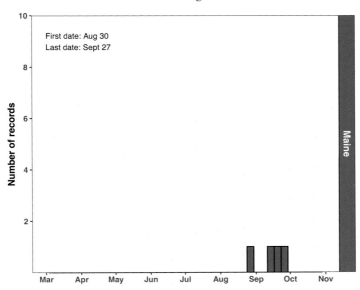

First date: Aug 30
Last date: Sept 27

ORANGE-BARRED SULPHUR

Phoebis philea (Linnaeus)

Subspecies: Only the nomino-typical subspecies (*P. p. philea*) occurs in the United States and Canada.

Distinguishing characters:

Size: Large.

Upperside: Male bright yellow with orange bar on forewing and orange outer hindwing mar-gin. Female varies from

Male—Brevard County, FL, 31 May 2019 (Edward Perry IV)

golden yellow to nearly white, with dark spots on forewing and along margin of hindwing; often with orange or red-orange hindwing margin.

Underside: Male golden yellow to orange yellow, usually with silvery spots in the center of both wings. Female reddish orange to nearly white, with scattered darker markings and silvery central spots; spotting often more extensive than in male.

Similar species: Cloudless Sulphur.

Status and distribution in Maine: Stray. Rare; one historical record from Kittery Point, York County. SNA

Status and distribution in Maritimes: Stray. Rare; one historical literature record from Saint John, New Brunswick. NB-SNA

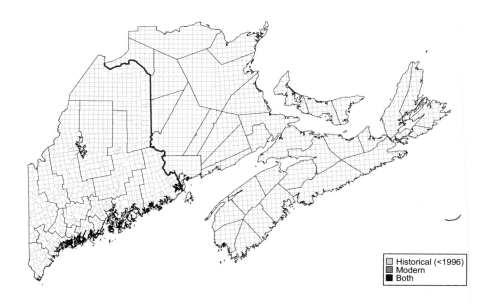

Historical (<1996)
Modern
Both

Habitat: Strays can be found in virtually any open, sunny area, including city parks and gardens. Where resident, it is less often found outside urban settings than the similar Cloudless Sulphur.

Biology: Three or more generations where resident all year. Our records are from early and mid-September. Larvae feed on various legumes (Fabaceae), especially woody sennas (*Senna* spp., *Cassia* spp.). There is no evidence of overwintering in our region.

Adult behavior: This species flies high and fast, rarely pausing except to visit flowers. The individual in New Brunswick reportedly visited blossoms of Dahlia (*Dahlia* sp.).

Comments: In eastern North America, the Orange-barred Sulphur is resident only in Florida, where it became established during the late 1920s. Adults sometimes stray northward, occasionally reaching as far as southern Canada. This species was first reported in Maine by Farquhar (1934), who listed an uncredited record from an unknown locality, dated August 1930. Brower (1974) repeated this report and credited it to Joseph R. Haskin. This statement is presumably in error, as Haskin (1931) reported a specimen from Connecticut, dated 26 August 1930, but he did not mention finding this butterfly in Maine. The only verifiable record from Maine is a battered, pale-form female, collected by Roland Thaxter at Kittery Point on 3 September 1930, which was previously misidentified and reported as a Large Orange Sulphur (*Phoebis agarithe*) (see Butterflies Dubiously or Erroneously Reported). In New Brunswick, a "perfect specimen" was reportedly collected at Saint John by William McIntosh on 17 September 1906, representing the first record of this species from Canada (Fletcher 1907; McIntosh 1907). Scott (1986) stated that it had been recorded in Nova Scotia, but we could not validate this claim.

Threats: None documented in our region.

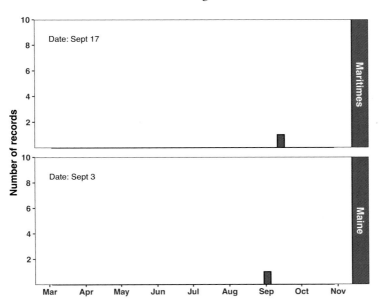

CHECKERED WHITE

Pontia protodice (Boisduval & Le Conte)

Subspecies: None.

Distinguishing characters:
  Size: Medium.
  Upperside: Male white with black spots on forewing; female more heavily marked with checkered pattern of black and dark gray (pattern reduced in early spring/autumn).

Female—Pittsburg, NH, 20 August 2007 (Erik Nielsen)

  Underside: Male white with dark gray spots on forewing and subtle gray or yellowish-gray markings on hindwing. Female forewing with heavier black or dark gray spots and yellowish veining at apex; hindwing with yellowish or yellowish-gray veining (darker in early spring/autumn).

Similar species: Mustard White and Cabbage White; white females of Clouded Sulphur and Orange Sulphur.

Status and distribution in Maine: Stray. Rare; a single literature record from Augusta, Kennebec County. SNA

Status and distribution in Maritimes: Not documented.

Habitat: Open, weedy habitats, including roadsides, old fields, vacant lots, and utility corridors; often in areas with very low-growing vegetation. The

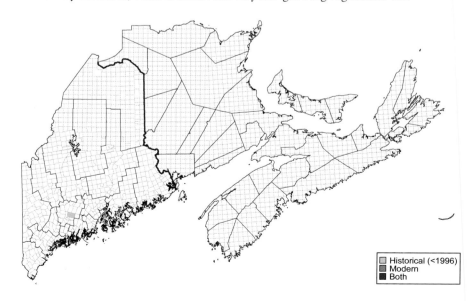

	Historical (<1996)
	Modern
	Both

single adult male recorded in Maine was encountered in an agricultural "truck garden" (Brower 1974).

Biology: Multiple generations possible where resident, spring to autumn. Larvae feed mostly on the buds and flowers of various crucifers (Brassicaceae), including Poor-man's Pepperweed (*Lepidium virginicum*), introduced Shepherd's-purse (*Capsella bursa-pastoris*), and cultivated species. The Maine individual was found in August or September in association with "wild mustard," a possible food plant. Overwinters as a pupa but probably not in our region.

Adult behavior: This species flies more rapidly and erratically than the Cabbage White or Mustard White. Males patrol for females throughout the day. Both sexes often visit flowers, such as Red Clover (*Trifolium pratense*) and Alfalfa (*Medicago sativa*).

Comments: The Checkered White is resident in the Deep South and a regular colonist into the Mid-Atlantic states. Believed to be declining in much of the east, it is known to erupt northward on occasion and migrants can establish multiyear colonies as far north as Ottawa, Ontario (Layberry et al. 1998). It is surprising that it has not been reported from Maine since 1951, as there are modern records from Massachusetts even within the last decade (Stichter 2015). This butterfly should be looked for during August and September in southern Maine, particularly along the coast, in clover fields and sparsely vegetated waste areas, such as vacant lots and roadsides. Strangely, two dates are associated with the Maine specimen that was reportedly collected by John H. Brower (A. E. Brower's son): 9 August 1951 (Hessel 1951) and 16 September 1951 (Brower 1974). Its association with "wild mustard" (Hessel 1951) may suggest that a small, temporary population existed at the collection site.

Threats: None documented in our region.

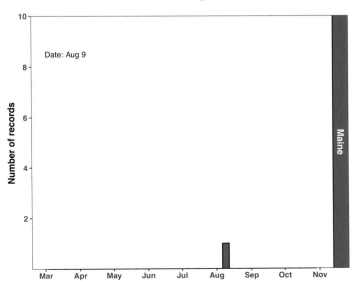

CABBAGE WHITE

Pieris rapae (Linnaeus)

Subspecies: Only the nomino-typical subspecies (*P. r. rapae*) occurs in North America.

Distinguishing characters:
  Size: Medium.
  Upperside: White; forewing with black tip, male with one black forewing spot, female with two; hindwing with black spot near leading edge.

Camden, ME, 18 August 2011 (Roger Rittmaster)

  Underside: Forewing white with cream-colored tip; hindwing cream colored; both sexes with two black forewing spots.
  Additional: Spring adults are smaller with reduced markings.

Similar species: Mustard White.

Status and distribution in Maine: Resident. Abundant; widespread. SNA

Status and distribution in Maritimes: Resident. Abundant; widespread. NB-SNA, NS-SNA, PE-SNA

Habitat: A variety of open weedy areas, including old fields, waste areas, roadsides, residential yards, and beaches.

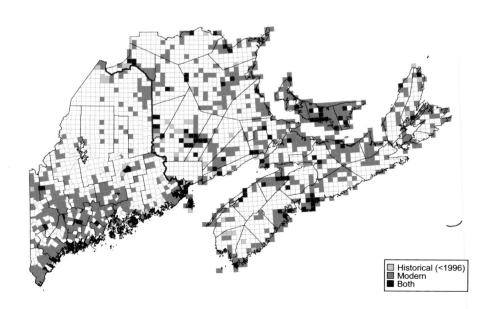

☐ Historical (<1996)
▨ Modern
■ Both

Biology: Three or more overlapping generations per year. Adults are active from April to late October, with numbers peaking during late July and August. Larvae feed on various wild and cultivated crucifers, such as mustards (*Brassica* spp.), Wild Radish (*Raphanus raphanistrum*), and American Sea-rocket (*Cakile edentula*). Overwinters as a pupa.

Adult behavior: The Cabbage White has a relatively strong flight. Males patrol for females and are most active in the morning. Adults nectar at a wide variety of flowers and have been observed visiting over ninety plants in the Acadian region. It is most often found at Red Clover (*Trifolium pratense*), hawkweeds (*Hieracium* spp.), knapweeds (*Centaurea* spp.), American-asters (*Symphyotrichum* spp.), and various crucifers (e.g., Wild Radish, American Sea-rocket).

Comments: A native to Eurasia and North Africa, the Cabbage White occurs widely as an introduced species and is now found virtually worldwide. It is a supreme generalist, capable of rapidly colonizing nearly any weedy, disturbed habitat and producing multiple generations per year. In addition, its larvae feed on—and are transported on—a wide variety of wild and cultivated food plants. The Cabbage White was first detected in North America in 1860 in Quebec (Scudder 1888–1889), and the first American record was a single specimen collected in 1865 by Sidney Smith at Norway (Oxford County), Maine. Because of its affinity for cole crops (cabbage, broccoli, kale, etc. [*Brassica oleracea*]), this butterfly can be an agricultural pest; gardeners often refer to it as the "cabbage worm."

Threats: None documented in our region (introduced species).

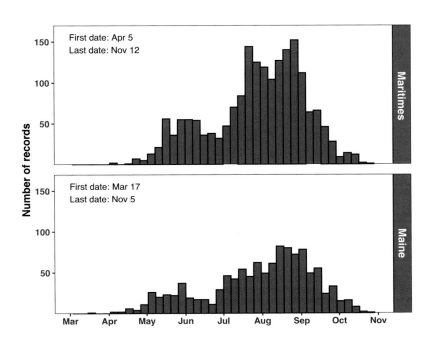

MUSTARD WHITE

Pieris oleracea (T. Harris)

Subspecies: Only the nominotypical subspecies (*P. o. oleracea*) occurs in the eastern United States and southeastern Canada.

Distinguishing characters:
   Size: Medium.
   Upperside: White with dark scaling at wing bases and forewing tip; summer form often unmarked.
   Underside: Forewing white, often with creamy tip; hindwing white to creamy white; distinctive gray-green scaling on veins of hindwing and forewing tip in spring form, lacking or much fainter in summer form.

Similar species: Cabbage White.

Status and distribution in Maine: Resident. Formerly uncommon to rare toward the coast (now potentially absent), common northward; widespread. S4

Status and distribution in Maritimes: Resident. Common in New Brunswick, uncommon to common elsewhere; widespread. NB-S5, NS-S4, PE-S4

Habitat: Edges and interiors of moist deciduous or mixed forest, often near streams.

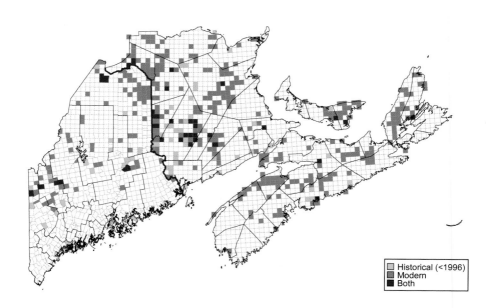

Historical (<1996)
Modern
Both

Spring form—York County, NB, 29 May 2014 (Jim Edsall)

Summer form—Carleton County, NB, 19 July 2011 (John Klymko)

Biology: Two generations per year. Adults are generally active from early May to early September, with peak flights in late May and mid-July. Larvae feed on crucifers, particularly toothworts (*Cardamine* spp.) and rockcresses (*Arabis* spp.). Overwinters as a pupa.

Adult behavior: This distinctive white has a relatively weak flight. Males patrol for females, and the species is frequently seen puddling, sometimes in large aggregations. Adults nectar at a wide variety of flowers, including knapweeds (*Centaurea* spp.), Common Strawberry (*Fragaria virginiana*), Common Dandelion (*Taraxacum officinale*), violets (*Viola* spp.), and hawkweeds (*Hieracium* spp.).

Comments: The Mustard White has declined in abundance across much of its southern range, and the absence of recent records from much of southern Maine suggests that it may be locally extirpated. Various causes of this decline have been proposed, including competition with the introduced Cabbage White and the fact that female Mustard Whites will lay eggs on unsuitable plants, namely introduced Garlic-mustard (*Alliaria petiolata*) and Garden Yellow-rocket (*Barbarea vulgaris*). These plants are toxic to the larvae, though there is some evidence that they may be adapting to Garlic-mustard (Keeler and Chew 2008). The decline in Nova Scotia was noted during the 1950s

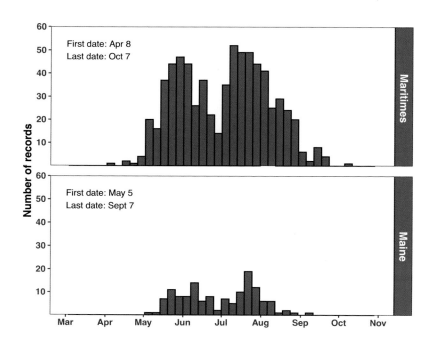

(Ferguson 1955), yet the abundance of modern records suggests it has at least partially recovered.

Threats: The continuing spread of Garlic-mustard is a potential threat. It is unclear if the Cabbage White still poses a risk through competition, as it has long been established in the region. Given the Mustard White's apparent decline in southern Maine, it should be monitored in our region.

FAMILY LYCAENIDAE

The Lycaenidae is a diverse assemblage of small, energetic butterflies. Over 200 species are found in the United States and Canada, and four subfamilies are represented in the Acadian region. With only a single species in the New World, the harvesters (Miletinae) have carnivorous larvae, which is an extraordinary trait among butterflies. As their name suggests, the coppers (Lycaeninae) include many orange or reddish-brown species. Among the hairstreaks (Theclinae) are numerous species that possess hairlike projections or tails on their hindwing (hence the name "hairstreak"). When at rest, these butterflies frequently rub their hindwings together, apparently to create the illusion that the tails are antennae to lure predators away from their defenseless bodies. This subfamily also includes the elfins, which are rather drab, secretive butterflies. The aptly named blues (Polyommatinae) typically display shining blue dorsal wings, particularly in males. Most lycaenids are avid flower visitors, and males of many species obtain nutrients from damp earth and animal dung. Members of this family are known for maintaining small, localized populations, often in specialized and vulnerable habitats. Larvae feed on an array of woody and herbaceous plants, though they are often restricted to certain genera or species. Harvester larvae feed exclusively on woolly aphids (Aphididae).

Twenty-nine species of Lycaenidae have been documented in Maine and the Maritimes. All but one (White M Hairstreak) are breeding residents. Most are single brooded, but several species produce up to three generations per year. In Maine, nine taxa are state listed as Endangered (Juniper Hairstreak, Hessel's Hairstreak, and Edwards' Hairstreak), Threatened (Clayton's Copper), or Special Concern (Hoary Elfin, Early Hairstreak, Coral Hairstreak, Crowberry Blue, and Northern Blue). One subspecies of Silvery Blue that is limited to Nova Scotia (*Glaucopsyche lygdamus mildredae*) may be extinct.

There is lingering disagreement about the generic placement of some species, especially those currently included in *Callophrys* (Pratt et al. 2011). Based on genomic evidence, Zhang et al. (2020) suggested that the subgenus *Tharsalea* be elevated to genus to include most of our coppers, and we tentatively follow this treatment in accordance with Pelham (2022). Despite continuing studies of the genus *Celastrina* (e.g., Schmidt and Layberry 2016), these butterflies are confusing, and species boundaries remain uncertain.

Eastern Tailed-Blues—Fair Haven, VT (Bryan Pfeiffer). This butterfly is a diminutive jewel among the nearly thirty gossamer wings inhabiting the Acadian region.

HARVESTER

Feniseca tarquinius (Fabricius)

Subspecies: None (see comments).

Distinguishing characters:
    Size: Small.
    Upperside: Dark brown with large orange patches.
    Underside: Orange brown with many white-margined dark spots; hindwing and apex of forewing suffused with white scaling, giving a frosted look.

Kent County, NB, 12 June 2008 (Denis Doucet)

Similar species: None.

Status and distribution in Maine: Resident. Uncommon; widespread. S4

Status and distribution in Maritimes: Resident. Uncommon to rare, apparently absent from Cape Breton. NB-S4S5, NS-S4, PE-S1S2

Habitat: Marshes, riparian meadows, swamp margins, and along trails through open woodlands.

Biology: At least two generations per year, with adults active from early May to late September, occasionally later. Larvae of the Harvester are carnivorous, feeding on several species of aphids, with one of the most common reported to be the Woolly Alder Aphid (*Prociphilus tessellatus*), which in our area is generally associated with alders (*Alnus* spp.). Overwinters as a pupa.

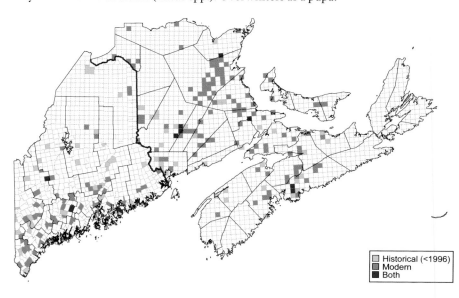

Historical (<1996)
Modern
Both

Adult behavior: This species has a very erratic flight. Adults are typically seen perched on sunlit foliage, sometimes high above the ground or on alder branches near Woolly Alder Aphid colonies. If disturbed, they often return to their original perches. The Harvester has a very short proboscis, which apparently limits its ability to nectar at flowers. As a result, these butterflies most often feed on honeydew secretions produced by aphids, but they also visit dung, sap, and damp earth.

Comments: Woolly Alder Aphid colonies are attended by ants that feed on the aphid's honeydew secretions and defend the colonies against predators. A study in Wisconsin showed that Harvester larvae go unmolested by these ants because they are chemically camouflaged. In other words, the larvae acquire the odor of Woolly Alder Aphids by rubbing against them, making the Harvester larvae undetectable to the chemoreceptors of the ants (Youngsteadt and Devries 2005). Adult Harvesters are inconspicuous and highly localized, so the species is probably far more common than records indicate. Surveys for larvae may be a more reliable way to find this butterfly. Larvae can be found by searching Woolly Alder Aphid colonies, particularly in autumn when the colonies become conspicuous after leaves fall (Wagner 2005). The subspecies *F. t. novascotiae* McDunnough was described from South Milford, Nova Scotia, but its defining characters are unreliable and occur in populations throughout the species' range. Like Layberry et al. (1998), we do not recognize this subspecies. An old, synonymous name for the Harvester, *Polyommatus porsenna* Scudder, was described from specimens collected in Maine (Aroostook and Oxford Counties).

Threats: None documented in our region.

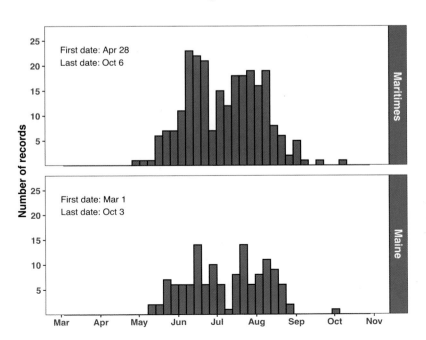

First date: Apr 28
Last date: Oct 6

Maritimes

First date: Mar 1
Last date: Oct 3

Maine

Number of records

AMERICAN COPPER

Lycaena phlaeas (Linnaeus)

Subspecies: Only the sub-species *L. p. hypophlaeas* (Bois-duval) occurs in eastern North America.

Distinguishing characters:
    Size: Small.
    Upperside: Forewing orange with large black spots and a brown-gray margin; hindwing brown gray with a broad orange margin.

Eliot, ME, 27 July 2018 (Bryan Pfeiffer)

    Underside: Forewing light orange with large black spots and a light gray margin; hindwing light gray with small black spots and a thin, orange submarginal band.

Similar species: Bronze Copper.

Status and distribution in Maine: Resident. Common in south, uncommon in north; widespread. S5

Status and distribution in Maritimes: Resident. Common to uncommon; widespread; NB-S5, NS-S5, PE-S4

Habitat: Generally open, disturbed areas such as meadows, pastures, road-sides, and utility corridors but also natural barrens, beaches, and dunes.

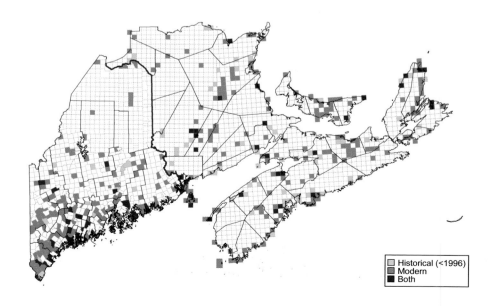

Historical (<1996)
Modern
Both

Biology: Likely two full generations and a partial third brood. One of the longest and latest flying lycaenids, adults are active from mid-May until late October. Larvae feed on docks (*Rumex* spp.), particularly Common Sheep Sorrel (*R. acetosella*). There are varying reports of the overwintering stage, but pupae seem likely given the early emergence of some adults.

Adult behavior: This species has a rapid, low flight. Males aggressively guard small territories from a perch on low vegetation. Adults frequent a variety of flowers in our region, particularly Common Dandelion (*Taraxacum officinale*), goldenrods (*Solidago* spp.), clovers (*Trifolium* spp.), and American-asters (*Symphyotrichum* spp.).

Comments: The eastern subspecies of the American Copper, *L. p. hypophlaeas*, is closely associated with disturbed habitats and a nonnative larval food plant. This has led to speculation that it is was introduced during colonial times from Europe, where it is also a common species (e.g., Opler and Krizek 1984). This theory has been questioned for various reasons, including that several butterflies endemic to North America, such as the Wild Indigo Duskywing, are now associated with similar habitats and introduced food plants (Glassberg 1999). It has also been suggested that this subspecies occurred naturally in parts of New England (e.g., White Mountains) and became more widespread with the introduction of *R. acetosella* (Pratt and Wright 2002). If so, it would be fascinating to know its original habitat preference, larval food plant, and distribution. Recent studies suggest that *hypophlaeas* is a discrete species.

Threats: None documented in our region.

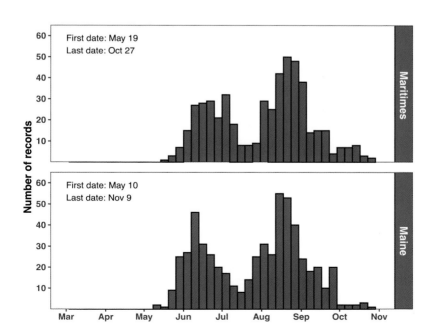

BRONZE COPPER

Tharsalea hyllus (Cramer)

Subspecies: None.

Distinguishing characters:
  Size: Small to medium.
  Upperside: Hindwing
    gray brown with broad
    orange margin; male
    forewing coppery brown
    with small dark spots;
    female forewing orange
    with large dark spots
    and broad gray-brown
    margin.

Steuben, ME, 30 June 2017 (Bryan Pfeiffer)

  Underside: Forewing light orange with prominent black spots and a chalky white margin; hindwing chalky white with bold black spots and broad orange margin.

Similar species: American Copper.

Status and distribution in Maine: Resident. Uncommon; widespread but localized in southern and central regions. S4

Status and distribution in Maritimes: Resident. Uncommon; widespread but localized in the southeastern half of New Brunswick, northern Nova Scotia, and Prince Edward Island. NB-S4, NS-S4, PE-S4

Habitat: Rich wetlands, including marshes, wet meadows, impoundments, the upper edge of salt marshes, and adjacent old fields.

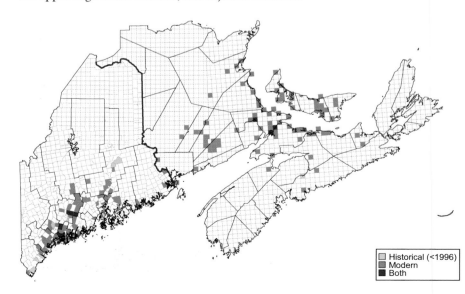

□ Historical (<1996)
▨ Modern
■ Both

Biology: Two generations. Adults are active from mid-June until early October, with peak flights in the first half of July and second half of August. Larvae feed on docks (*Rumex* spp.), particularly the wetland associated Greater Water Dock (*R. britannica*) and introduced Curly Dock (*R. crispus*); knotweeds (*Polygonum* spp.) are also reported. Overwinters as an egg.

Adult behavior: This copper has an erratic flight that is difficult to follow, but adults are typically found perched on wetland vegetation. Most active in the afternoon, they nectar at a variety of flowers, including Ox-eye Daisy (*Leucanthemum vulgare*) and Swamp Milkweed (*Asclepias incarnata*); the second generation is especially attracted to goldenrods (*Solidago* spp.).

Comments: The Bronze Copper has declined significantly in many parts of its range in the eastern United States (Cech and Tudor 2005). Data from our region paint a much different picture, especially in the Maritimes where there has a been a significant and recent range expansion. Historically, this species was recorded from just four atlas squares, but it is now known from seventy-four. It has taken advantage of new habitat, such as waterfowl impoundments, as well as permanent features like salt marshes. This colonization of the Maritimes may be the conclusion of an invasion of the Northeast that started over a century ago. Scudder (1888–1889) mentioned that the Bronze Copper had not been found east of the Connecticut River Valley. The earliest record from our region is a specimen collected in 1903 in York County, Maine, and it was not reported from farther east until 1937, when it was found in Sagadahoc County. The first records in New Brunswick, Nova Scotia, and Prince Edward Island are from 1972, 1979, and 2003, respectively.

Threats: The cumulative loss of wetland habitat and riparian buffers to development is a localized threat, especially for smaller wet meadows and marshes that are not adequately protected by existing wetland regulations.

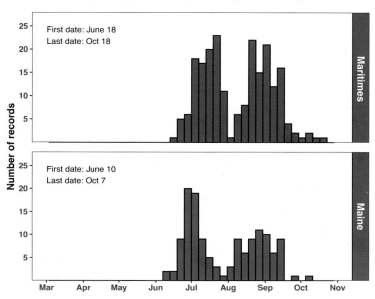

DORCAS COPPER
(CLAYTON'S COPPER)
Tharsalea dorcas (W. Kirby)

Subspecies: The subspecies *T. d. claytoni* (A. E. Brower) is endemic to the Acadian region. Nova Scotia populations are not well studied and are provisionally assigned to this subspecies (see comments).

Distinguishing characters:
    Size: Small to medium.
    Upperside: Rusty brown with scattered black spots; males with blue-purple iridescence

Mating pair—T8 R11 WELS, ME, 24 July 2014 (Donald Cameron)

    Underside: Forewing orange brown with prominent black spots; hindwing darker with indistinct black spots and a zigzag red-orange marginal band.

Similar species: Bog Copper and Salt Marsh Copper.

Status and distribution in Maine: Resident. Rare; limited to scattered localities in east central and northern regions. State Threatened. S2S3

Status and distribution in Maritimes: Resident. Rare; disjunct populations in western New Brunswick and Cape Breton, Nova Scotia. NB-S1, NS-S2

Habitat: Rich calcareous fens and wet meadows. Historically also from nearby upland fields with an abundance of the food plant, now lost to succession.

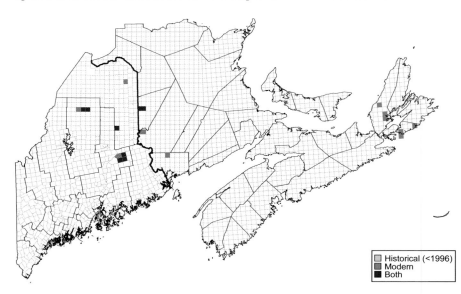

Historical (<1996)
Modern
Both

Biology: One generation, with adults flying from late July through mid-September (Webster and Swartz 2006), peaking during the first two weeks of August. The only known larval food plant is Shrubby-cinquefoil (*Dasiphora floribunda*). Overwintering occurs as eggs, which are laid on the foliage and stems of the food plant and drop to the ground in autumn.

Adult behavior: Adults rarely stray from natal wetlands, where males are often found patrolling low over shrubby vegetation. Flight is relatively fast and direct, with nectaring observed almost exclusively on Shrubby-cinquefoil.

Comments: The subspecies *T. d. claytoni* (Clayton's Copper) was originally described by Brower (1940) to define the disjunct populations of the Dorcas Copper in Maine. This subspecies was subsequently applied to a population across the border in western New Brunswick (Thomas 1991). Knurek (2009) found subtle morphological differences, but no genetic distinctions, between *T. d. claytoni* and the nominotypical subspecies (*T. d. dorcas*), which ranges from Newfoundland westward to at least central Canada. In 2010, a highly disjunct population of *T. dorcas* was discovered in Cape Breton, Nova Scotia (Klymko et al. 2012). We follow Pohl et al. (2018) in identifying all populations in our area as *T. d. claytoni*, but further study is needed to verify this treatment. Common Reed (*Phragmites australis*) is encroaching on one Maine population.

Threats: Populations in Maine are vulnerable to wetland and meadow succession, in which Northern White-cedar (*Thuja occidentalis*) is encroaching on areas previously dominated by shade-intolerant Shrubby-cinquefoil. Also, the water level of some sites is controlled by beavers or artificial impoundments, which affects growing conditions for the food plant: deep water floods the cinquefoil, while too little water encourages growth of less flood-tolerant trees and shrubs.

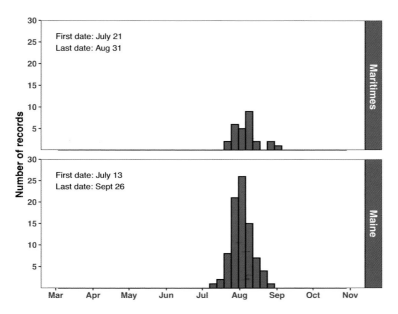

First date: July 21
Last date: Aug 31

Maritimes

First date: July 13
Last date: Sept 26

Maine

SALT MARSH COPPER (MARITIME COPPER)

Tharsalea dospassosi (McDunnough)

Subspecies: None.

Distinguishing characters:
    Size: Small.
    Upperside: Brownish gray with large black spots; males with bluish-purple iridescence.
    Underside: Forewing orange brown with prominent black spots; hindwing darker with small black spots and a zigzag red-orange marginal band.

Westmorland County, NB, 1 August 2011 (Maxim Larrivée)

Similar species: Bog Copper and Dorcas Copper.

Status and distribution in Maine: Not documented.

Status and distribution in Maritimes: Resident. Uncommon; widespread but localized on the Gulf of St. Lawrence coastline, absent elsewhere. NB-S3, NS-S2, PE-S3

Habitat: Salt marsh.

Biology: One generation per year. Adults are on the wing from mid-July to late August, with a peak flight in late July and early August. Larvae feed on Pacific Silverweed (*Argentina egedii*). Presumably overwinters as an egg like other summer-flying coppers.

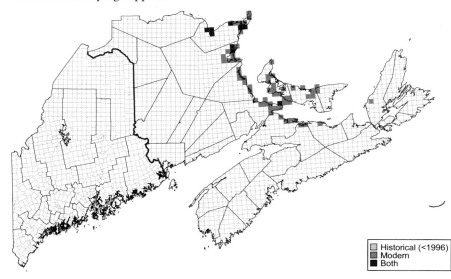

	Historical (<1996)
	Modern
	Both

Adult behavior: This copper has a low, weak flight and is generally seen perched on salt marsh vegetation. They usually nectar at Carolina Sea-lavender (*Limonium carolinianum*), Seaside Goldenrod (*Solidago semper-virens*), and Pacific Silverweed but will venture to the edges of salt marshes where other flowers are available.

Comments: This species is endemic to the Maritimes and Quebec's Gaspé Peninsula. Like the Maritime Ringlet, the Salt Marsh Copper may be a product of speciation that occurred within a glacial refugium during the last ice age. Discovered in 1939 with the Maritime Ringlet, in Bathurst, New Brunswick, it was originally described as a subspecies of the Dorcas Copper. Layberry et al. (1998) were the first to elevate it to the rank of species. Records suggest a recent and significant range expansion of this butterfly. Historical records are limited to northeastern New Brunswick, and it was not discovered in Prince Edward Island or Nova Scotia until 2002 and 2006, respectively, nor was it recorded in 1977 and 1978 in an extensive insect survey in Kouchibou-guac National Park, Kent County, New Brunswick (Miller and Lyons 1979), where it is now common. The species has been recorded more than a kilome-ter (0.6 mi) from suitable breeding habitat (Doucet 2009), suggesting a pro-pensity to wander that might facilitate the colonization of new salt marshes. Still, it is difficult to understand what would prompt a species with such a tiny global range suddenly to colonize so much available habitat, and, given that nearly all records for this habitat specialist are from within salt marshes, it is possible that the Salt Marsh Copper was simply overlooked in most of its range until recently. Baseline data gathered during the Maritimes Butterfly Atlas will reveal future changes in distribution with much more confidence.

Threats: This species is restricted to salt marshes in a small geographic area, which makes it vulnerable to habitat loss from rapid sea-level rise associated with climate change.

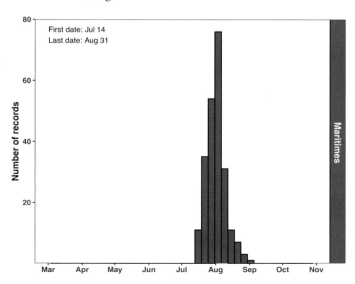

First date: Jul 14
Last date: Aug 31

BOG COPPER

Tharsalea epixanthe
(Boisduval & Le Conte)

Subspecies: Two subspecies occur in our region: the nominotypical subspecies (*T. e. epixanthe*) in southern and central Maine, and *T. e. phaedrus* (Hall) elsewhere, with a broad area of overlap (see comments).

Distinguishing characters:

Size: Small (our smallest copper).

Harrington, ME, 5 July 2011 (Jeff Pippen)

Upperside: Brownish gray with muted black spots and a variable amount of orange on the hindwing margin; males with bluish-purple iridescence.

Underside: Creamy yellow to chalky white, with black spots and zigzag orange marginal band on hindwing.

Similar species: Dorcas Copper and Salt Marsh Copper.

Status and distribution in Maine: Resident. Uncommon; widespread but localized. S4

Status and distribution in Maritimes: Resident. Common; widespread but localized. NB-S5, NS-S5, PE-S4

Habitat: Typically bogs and fens, occasionally other habitats if cranberry is present, including dune swales and roadside ditches.

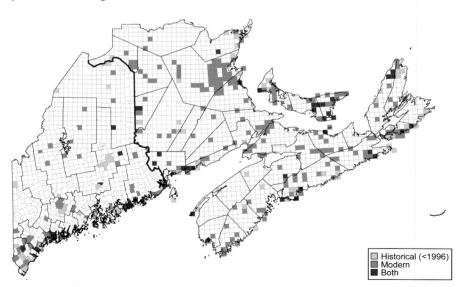

□ Historical (<1996)
▣ Modern
■ Both

Biology: One generation per year. Adults fly from late June until late August, with a peak in mid-July. Larvae feed on Small Cranberry (*Vaccinium oxycoccos*) and Large Cranberry (*V. macrocarpon*). Overwinters as an egg.

Adult behavior: Bog Coppers rarely stray far from their boggy environs and have a low, weak flight. Adults are generally seen perched on peatland vegetation. They typically nectar at cranberries but will visit a variety of other flowers, including upland species growing adjacent to bogs.

Comments: The Bog Copper's life history is closely tied to cranberries. Not only are these plants the larval food, they also serve as the primary adult nectar source and a substrate for egg-laying and pupation. Wright (1983) established that the emergence of adults in New Jersey corresponds with the flowering period of cranberry. The close association with cranberries limits the occurrence of this butterfly almost entirely to bogs, where it is often abundant. Most populations in Maine are intermediate, producing adults that resemble both the nominotypical subspecies (with more tawny ventral hindwings) and the subspecies *T. e. phaedrus* (with white ventral hindwings), the latter having been described from specimens collected in Lunenburg County, Nova Scotia.

Threats: While peat mining is a threat to habitat in New Brunswick, it is fortunate that the Bog Copper occurs in numerous small bogs where such operations are uneconomical.

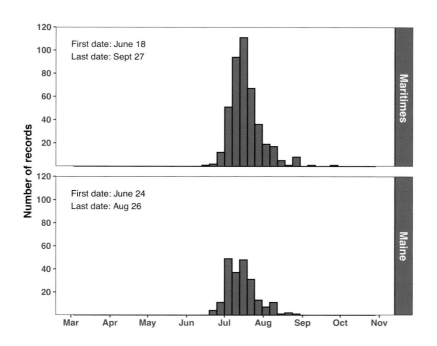

WHITE M HAIRSTREAK

Parrhasius m-album
(Boisduval & Le Conte)

Eliot, ME, 31 August 2018 (Josh Lincoln)

Subspecies: None.

Distinguishing characters:
    Size: Small.
    Upperside: Bright iridescent
        blue with wide black
        margins.
    Underside: Gray or brown-
        ish gray; hindwing with
        white postmedian line
        forming a clear M (or W)
        above the tails; a conspic-
    uous red-orange spot between M and tails; forewing with thin white
    postmedian line that is sometimes edged in dark gray.
    Additional: Two tails, one long and one short on each hindwing.

Similar species: Gray Hairstreak (underside only).

Status and distribution in Maine: Probably a rare colonist. Restricted to southern and coastal regions. SNA

Status and distribution in Maritimes: Not documented.

Habitat: Deciduous forests and adjacent meadows and shrublands.

Biology: The White M Hairstreak is a rare visitor that may produce one to two generations. There are only four records from our region, all from late July

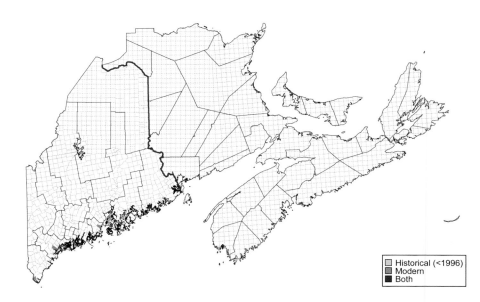

Historical (<1996)
Modern
Both

to late August. Larvae feed on oaks (*Quercus* spp.), likely Red Oak (*Q. rubra*) and White Oak (*Q. alba*), if the species is reproducing in southern Maine. It is unknown if this butterfly can overwinter in Maine; overwinters as pupae elsewhere.

Adult behavior: This hairstreak has a swift, erratic flight, with flashes of bright blue visible in the sunlight. Early generations are rarely seen, possibly because the butterfly spends much of its time in the forest canopy with only irregular visits to the ground. Most records in the Northeast are from late summer and fall when ground-level nectar is abundant. Adults have been observed feeding at White Meadowsweet (*Spiraea alba*) and Grass-leaved Goldenrod (*Euthamia graminifolia*) in Maine and many other species elsewhere.

Comments: The White M Hairstreak is an expected and welcome addition to our region's fauna, with the first adult observation in Falmouth by Doug Hitchcox on 24 July 2018, followed by two more records in 2018 from Rockport (Brian Willson) and Eliot (Bryan Pfeiffer and Josh Lincoln). A female was also recorded from Eliot on 5 August 2020 (John Calhoun), making this the first known repeat occurrence at a Maine locality. The good condition of these individuals suggests that local colonies were present. This primarily southeastern species was historically absent in New England (Scudder 1888–1889), but it is now established in Connecticut (O'Donnell et al. 2007). It was first recorded in Massachusetts in 1979, but since 2000 it has been recorded there annually, and regular springtime observations suggest that the species overwinters and/or repopulates the state each year (Stichter 2015). This northward advance may be the result of climate change.

Threats: None documented in our region.

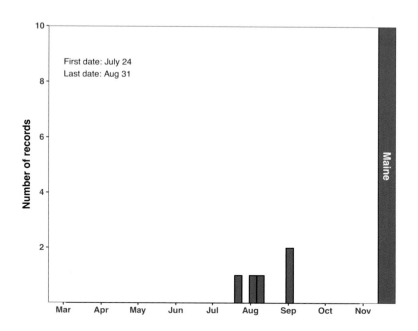

First date: July 24
Last date: Aug 31

GRAY HAIRSTREAK

Strymon melinus Hübner

Subspecies: Only the subspecies *S. m. humuli* (T. Harris) occurs in northeastern North America.

Distinguishing characters:
  Size: Small.
  Upperside: Dark bluish gray; hindwing margin with conspicuous orange and black spot
  Underside: Light gray with white, black, and orange medial band; hindwing margin with black spot

Eliot, ME, 28 July 2019 (Bryan Pfeiffer)

broadly capped with orange between tails and blue patch posterior to tails.
  Additional: Hindwing with two white-tipped tails, one long and one short; male abdomen is orange above (unique among our lycaenids).

Similar species: White M Hairstreak (underside only).

Status and distribution in Maine: Resident. Rare to uncommon; widespread in southern half of state. Additional migrant populations may also occur in some years, particularly in the southwest. S3

Status and distribution in Maritimes: Resident. Rare; widespread in southern and eastern half of New Brunswick, southern mainland Nova Scotia, and southern Cape Breton. NB-S2, NS-S3

Habitat: A variety of open, often disturbed areas, including clearings in dry forest, utility corridors, old fields, gardens, and coastal wetlands.

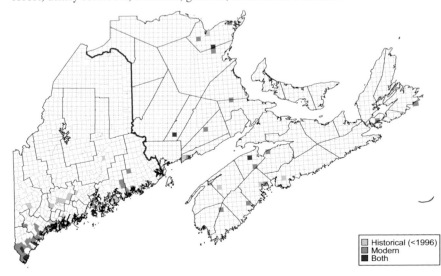

Historical (<1996)
Modern
Both

Biology: Two overlapping generations. Adults are active from early May until late September. In southwestern Maine, larvae probably feed on the flowers and fruit of a variety of plants, particularly legumes (Fabaceae) and mallows (Malvaceae). In the Maritimes, and possibly eastern Maine, larvae feed only on Sweet-fern (*Comptonia peregrina*). Overwinters as a pupa.

Adult behavior: This species has a rapid and erratic flight. The highly territorial males perch on vegetation awaiting passing females. Unlike other hairstreaks, it regularly basks with its wings open. The Gray Hairstreak is frequently found nectaring on flowers, with observations in our region on Cow Vetch (*Vicia cracca*), clovers (*Trifolium* spp.), goldenrods (*Solidago* spp.), Common Strawberry (*Fragaria virginiana*), and Common Lowbush Blueberry (*Vaccinium angustifolium*).

Comments: In most areas, the Gray Hairstreak occurs in various open habitats and has many larval food plants, including species in at least twenty families (Cech and Tudor 2005). However, in the Maritimes and elsewhere on the northern edge of its range, it is restricted to dry habitats, particularly openings in pine forest, where it appears to feed exclusively on Sweet-fern. This has led some to believe that this northern, habitat-restricted population is a distinct species (Acorn and Sheldon 2016). More research is also needed to determine if our southern population is resident or migratory or a mixture of the two. Like many hairstreaks, resting adults of this species rub their hindwings together in an alternating fashion. This behavior is thought to draw attention to the eyespots and tails, which resemble a head and antennae. Would-be predators attack this part of the wings, leaving the butterfly to escape without life-threatening injury.

Threats: None documented in our region, though its relative rarity here merits monitoring.

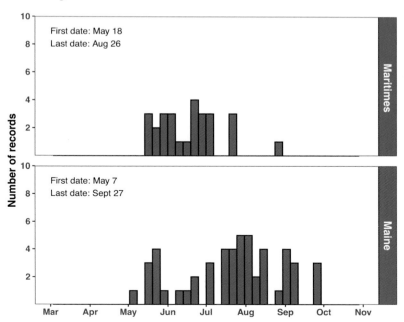

First date: May 18
Last date: Aug 26

Maritimes

First date: May 7
Last date: Sept 27

Maine

Number of records

JUNIPER HAIRSTREAK (OLIVE HAIRSTREAK)

Callophrys gryneus (Hübner)

Subspecies: Only the nomino-typical subspecies (*C. g. gryneus*) occurs in eastern North America.

Distinguishing characters:
    Size: Small.
    Upperside: Rarely seen, but variably tawny orange or dark brown, depending on sex and season.

Woburn, MA, 7 May 2017 (Michael Newton)

    Underside: Bright, apple green; white band of hindwing edged inward with brown; forewing lacking small white spot in discal area (usually present in Hessel's Hairstreak).
    Additional: Hindwing has thin tails.

Similar species: Hessel's Hairstreak.

Status and distribution in Maine: Resident. Rare; restricted to York and southern Oxford Counties. State Endangered. S1

Status and distribution in Maritimes: Not documented.

Habitat: Rocky outcrops and powerline corridors.

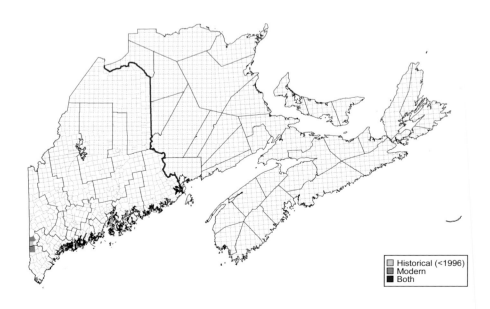

☐ Historical (<1996)
◼ Modern
◼ Both

Biology: Typically two or three generations, but at the northern edge of its range in Maine there may be only a single brood. All adult records from Maine are from mid to late May. Larvae feed on Eastern Red Cedar (*Juniperus virginiana*). Overwinters as a pupa at the base of food trees.

Adult behavior: Juniper Hairstreaks often perch on cedars, where they are well camouflaged, and can often be located by gently tapping the branches and trunks of trees. Nectaring has been observed on Lowbush Blueberry (*Vaccinium angustifolium*) and Highbush Blueberry (*V. corymbosum*) near cedar trees.

Comments: This species is known only from three localities in southwestern Maine, where it inhabits a powerline corridor and two south-facing hilltop outcrops. Reaching the northeastern edge of its range in our region, Eastern Red Cedar appears to be in decline due to natural succession, with most populations comprising a few aging trees and little regeneration. It is likely that the intense forest fires that swept through southern Maine in 1947 created open conditions conducive to regenerating populations of this pioneer tree species. No widespread disturbance of this magnitude has since occurred in the region. Historically, unimproved pastureland likely also supported the host plant, as it is unpalatable to livestock. Additional survey efforts are warranted for this elusive hairstreak, as there are likely undetected populations on isolated, remote hilltops in southwestern Maine, worthy of monitoring and protection.

Threats: Natural forest succession following agricultural abandonment and modern era forest fire suppression have both contributed to a decline in habitat for the Juniper Hairstreak. One colony was possibly lost to powerline vegetation clearing.

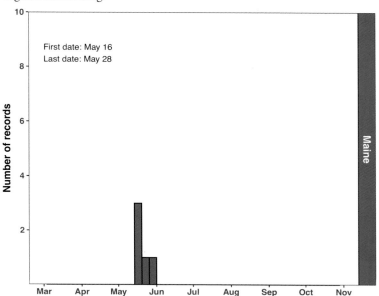

First date: May 16
Last date: May 28

HESSEL'S HAIRSTREAK

Callophrys hesseli (Rawson & Ziegler)

Subspecies: Only the nominotypical subspecies (*C. h. hesseli*) occurs in northeastern North America.

Distinguishing characters:
    Size: Small.
    Upperside: Dark brown or reddish brown (rarely seen).
    Underside: Reddish brown, overlaid with

Alfred, ME, 29 May 2018 (Bryan Pfeiffer)

bright green scales; hindwing white band brown edged on both sides; forewing often with small white spot in discal area (lacking in Juniper Hairstreak).
    Additional: Hindwing with white-tipped, hairlike tail.

Similar species: Juniper Hairstreak.

Status and distribution in Maine: Resident. Rare; restricted to York County. State Endangered. S1S2

Status and distribution in Maritimes: Not documented.

Habitat: Atlantic White Cedar swamps and bogs.

Biology: One generation; adults recorded from mid-May to early June. Larvae feed on Atlantic White Cedar (*Chamaecyparis thyoides*). Overwintering occurs as pupa.

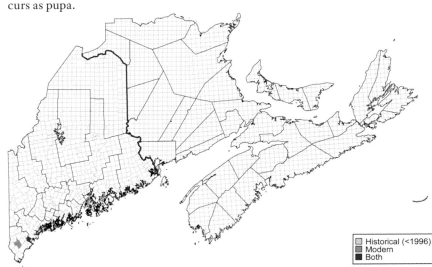

Historical (<1996)
Modern
Both

Adult behavior: Well camouflaged among the cedar foliage, perching adults are best located by tapping branches and trunks of host trees. Flushed individuals often return to the same tree or nearby perch. Adults are most often found in association with smaller cedars that grow along the margins of bog openings. Nectaring has been observed near cedar trees on Highbush Blueberry (*Vaccinium corymbosum*) and Black Huckleberry (*Gaylussacia baccata*). Leatherleaf (*Chamaedaphne calyculata*) is usually in bloom when adults are active and may also serve as a nectar source. Adults sometimes visit damp soil near suitable habitat.

Comments: Hessel's Hairstreak is known only from five localities in southwest Maine in swamp and peatland settings near the coast. Among our region's rarest and most beautiful insects, this hairstreak is notoriously difficult to observe, with adults spending much of their time high in the canopy of cedar trees. Additional surveys are needed in midcoast Maine, where potentially suitable habitat occurs in Knox and Waldo Counties at the extreme northern edge of the range of Atlantic White Cedar.

Threats: Atlantic White Cedar-dominated swamps and bogs preferred by Hessel's Hairstreak are rare and restricted to southern Maine. Many cedar stands are situated in proximity to residential development, where they are vulnerable to habitat fragmentation and water level changes from disturbance along their margins. Regeneration of the Atlantic White Cedar is thought to require disturbance, such as blowdowns and fire, in the absence of which cedars may be replaced by common shade-tolerant species, including Red Maple (*Acer rubrum*) and Eastern Hemlock (*Tsuga canadensis*) (Gawler and Cutko 2018). Historical impacts from Spongy Moth (*Lymantria dispar*) spraying and clearcutting may explain the apparent absence of populations in some isolated cedar swamps (NatureServe Explorer 2021).

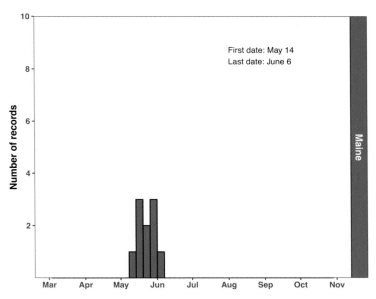

First date: May 14
Last date: June 6

Maine

BROWN ELFIN

Callophrys augustinus
(Westwood)

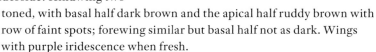

Subspecies: *Callophrys a. helenae* (dos Passos) occurs on Cape Breton; the nomino-typical subspecies (*C. a. augustinus*) occurs elsewhere in our region (see comments).

Distinguishing characters:
    Size: Small.
    Upperside: Dark brown (rarely seen).

Halifax County, NS, 10 May 2021 (Krista Melville)

    Underside: Hindwing two toned, with basal half dark brown and the apical half ruddy brown with row of faint spots; forewing similar but basal half not as dark. Wings with purple iridescence when fresh.

Similar species: Hoary Elfin and Henry's Elfin.

Status and distribution in Maine: Resident. Common; widespread but local-ized. S5

Status and distribution in Maritimes: Resident. Common but localized. NB-S5, NS-S5, PE-S4

Habitat: A wide variety of sites with poor soils and ericaceous ground cover, including peatlands, open coniferous woodlands, dry oak–pine forest, and sandy barrens.

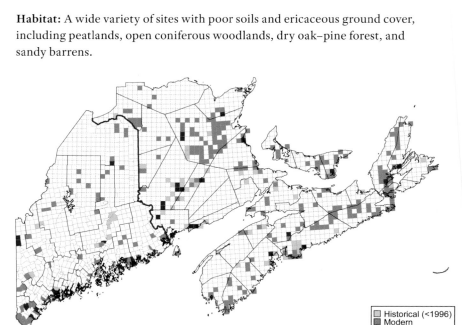

Historical (<1996)
Modern
Both

Biology: One generation per year; adults active from mid-April until early July. Larvae feed on shrubs of the heath family (Ericaceae), including blueberries (*Vaccinium* spp.), Red Bearberry (*Arctostaphylos uva-ursi*), Leatherleaf (*Chamaedaphne calyculata*), Labrador Tea (*Rhododendron groenlandicum*), and American-laurels (*Kalmia* spp.). Overwinters as a pupa.

Adult behavior: Seemingly less active than other elfins, adults of this species often perch low on vegetation, where they are difficult to detect unless disturbed. They nectar at various heaths, especially Leatherleaf and blueberries, and are frequently seen imbibing moisture along dirt roads and trails through open woodlands.

Comments: The Brown Elfin is variable in appearance in the Acadian region. The somewhat more contrasting phenotype associated with the subspecies *helenae* (originally described from Newfoundland) is common, especially in northeastern areas. It appears to be the only phenotype on Cape Breton, thus we apply the name *helenae* to that population in agreement with previous authors (e.g., Klots 1951; Layberry et al. 1998). The dark nominotypical subspecies occurs in mainland Nova Scotia and elsewhere in the Maritimes and is the dominant phenotype in most of Maine. Individuals in central and southern Maine often resemble the drabber subspecies *C. a. croesioides* (Scudder), which occurs from southern New England to the Florida Panhandle. These three subspecies are poorly defined, and individuals resembling one another often occur together. Jim Edsall collected a fresh specimen on 20 October 1975 in Peggy's Cove, Nova Scotia. It is unclear if this butterfly emerged exceptionally early (pupa failed to diapause over the winter) or exceptionally late (pupa failed to eclose in the spring).

Threats: None documented in our region.

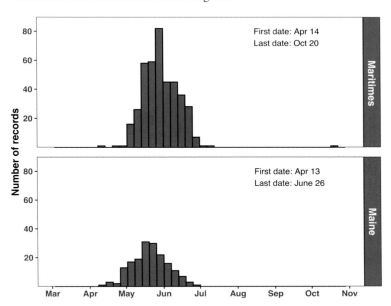

HOARY ELFIN

Callophrys polios (Cook & F. Watson)

Subspecies: Only the nominotypical subspecies (*C. p. polios*) occurs in eastern North America.

Distinguishing characters:
  Size: Small.
  Upperside: Rusty brown (rarely seen).
  Underside: Hindwing two toned, the basal half dark brown, the api-

Halifax County, NS, ME, 28 April 2021 (Krista Melville)

cal half mostly covered in gray scaling that gives it a frosted or hoary appearance; forewing brown with faint submarginal spots and a diffuse marginal band of purplish-grey scaling.

Similar species: Brown Elfin and Henry's Elfin.

Status and distribution in Maine: Resident. Rare; widespread but known in the southern half of the state almost exclusively from historical records. State Special Concern. S3?

Status and distribution in Maritimes: Resident. Uncommon to rare; widespread. NB-S4?, NS-S4, PE-S1

Habitat: Dry, open woodlands, sandplain barrens, and bog margins.

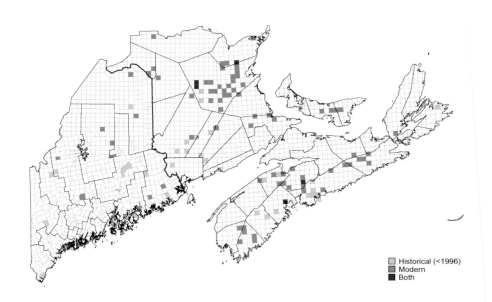

Historical (<1996)
Modern
Both

Biology: One generation per year, with adults flying from late April until mid-June. Larvae feed on Red Bearberry (*Arctostaphylos uva-ursi*) and Trailing Arbutus (*Epigaea repens*). Overwinters as a pupa.

Adult behavior: A low-flying species, adults often perch on the ground or low vegetation, where they are well camouflaged. They frequently visit damp soil and occasionally nectar in our region at Leatherleaf (*Chamaedaphne calyculata*), Common Strawberry (*Fragaria virginiensis*), Red Bearberry, and Trailing Arbutus.

Comments: Like its congeners, the Hoary Elfin usually occurs in low numbers and is easily overlooked. A single adult was documented in 2019 in southern Somerset County, Maine, but none were seen there again until 2022, in close association with Trailing Arbutus. At least in the Maritimes, where suitable habitat is common, it is likely more abundant than records indicate. For example, though the first Prince Edward Island record was in 2009, it is now known from five sites and undoubtedly occurs at several others. Elfins are often detected only when disturbed. Their small size, dark coloration, and erratic flight are reminiscent of moths to the untrained eye.

Threats: The Hoary Elfin has probably always been localized in occurrence in southern Maine, where residential development and ecological succession threaten its barren habitat. The only records from Oxford and York Counties (where the species is now absent or very rare) are from the 1860s (S. I. Smith) and 1934, respectively. This butterfly has apparently declined elsewhere in its northeastern range, with only historical records in New Hampshire, Connecticut, Pennsylvania, and West Virginia (NatureServe Explorer 2021).

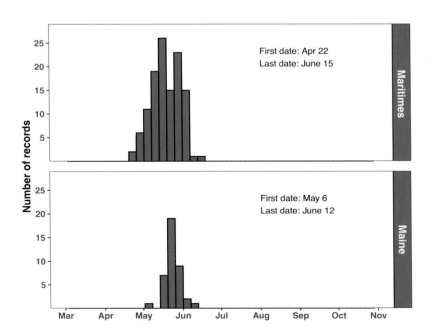

First date: Apr 22
Last date: June 15

First date: May 6
Last date: June 12

HENRY'S ELFIN

Callophrys henrici (Grote & Robinson)

Subspecies: Only the nominotypical subspecies (*C. h. henrici*) occurs in northeastern North America.

Distinguishing characters:
  Size: Small.
  Upperside: Dark brown, sometimes orange toned near margin (rarely seen).
  Underside: Strongly two-toned brown, basal half darker, with the two

Northumberland County, NB, 25 May 2020 (John Klymko)

halves separated by a median line edged with white at each end; hindwing with large patch of light scaling near margin, giving it a frosted appearance.
  Additional: Hindwing with small, blunt tail; margin scalloped.

Similar species: Brown Elfin and Hoary Elfin.

Status and distribution in Maine: Resident. Uncommon to rare; widespread. S4

Status and distribution in Maritimes: Resident. Uncommon to rare; widespread. NB-S4?, NS-S4?, PE-S1

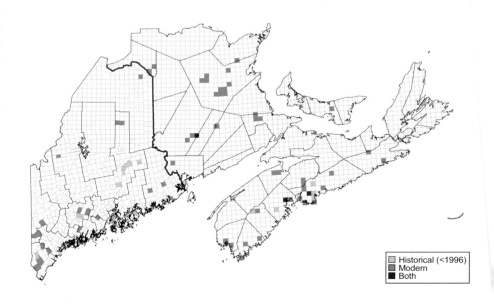

Habitat: Bog margins, forested wetlands, openings in wet spruce forest, and occasionally upland oak–pine forest.

Biology: One generation per year. Adults are active from late April until mid-June. Various larval food plants have been reported, though most populations use a single species (Cech and Tudor 2005). Reported larval hosts pertinent to our area include Mountain Holly (*Ilex mucronata*), blueberries (*Vaccinium* spp.), and the invasive exotic Glossy False Buckthorn (*Frangula alnus*). Overwinters as a pupa.

Adult behavior: This species has an erratic flight, and adults often perch on shrubby vegetation at the margins of woodlands. They occasionally nectar at flowers, including American-laurels (*Kalmia* spp.), blueberries, Leatherleaf (*Chamaedaphne calyculata*), Common Strawberry (*Fragaria virginiana*), and Red Bearberry (*Arctostaphylos uva-ursi*). Like other elfins, Henry's Elfins imbibe moisture from damp ground along dirt roads and trails.

Comments: Henry's Elfin is probably overlooked in our region, as it is relatively sedentary and usually encountered as singletons. In Ontario's Ottawa region, the butterfly has increased in abundance by colonizing abandoned farmland, where the larvae feed on Glossy False Buckthorn (Catling et al. 1998). In Massachusetts, where utilization of Glossy False Buckthorn and European Buckthorn (*Rhamnus cathartica*) was first documented (Winter 1986), these plants are probably now the primary food plants (Stichter 2015). Although buckthorns are aggressive invasive species that have become locally abundant in many parts of Maine and the Maritimes, this butterfly is not yet known to feed on them in the region.

Threats: None documented in our region.

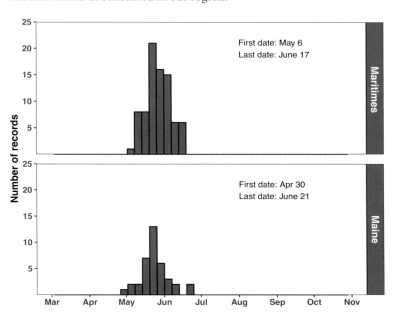

First date: May 6
Last date: June 17

First date: Apr 30
Last date: June 21

BOG ELFIN

Callophrys lanoraieensis
(Sheppard)

Subspecies: None

Distinguishing characters:
>   **Size:** Very small.
>   **Upperside:** Dark brown
>   (rarely seen).
>   **Underside:** Forewing
>   brown, inner half darker;
>   hindwing a complex but
>   smudged pattern of dark,
>   reddish-brown and light-
>   brown bands and patches, posterior margin with broad band of light
>   scaling, giving a frosted look.
>   **Additional:** Our region's smallest elfin.

Long A Township, ME, 22 May 2018 (Bryan Pfeiffer)

Similar species: Eastern Pine Elfin and Western Pine Elfin.

Status and distribution in Maine: Resident. Uncommon to rare; widespread but localized; no records from the northwest but likely occurs there. S4?

Status and distribution in Maritimes: Resident. Common and widespread but localized in New Brunswick; uncommon and widespread on mainland Nova Scotia; known from only two sites on Prince Edward Island. NB-S5, NS-S3, PE-S1

Habitat: Predominantly bogs and nearby habitats. Occasionally upland forests where Black Spruce (*Picea mariana*) is prevalent.

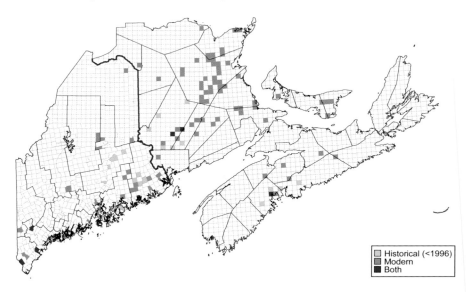

Historical (<1996)
Modern
Both

Biology: One generation per year. Adults are active from early May until mid-June. Larvae feed on the needles of Black Spruce. Overwinters as a pupa.

Adult behavior: Although they often perch on Black Spruce branches, Bog Elfins spend a great deal of time on lower vegetation, where they are difficult to see unless disturbed. These small butterflies have a bouncing, erratic flight, and their size makes them difficult to follow. They can often be found sipping moisture on the ground along logging roads and trails through Black Spruce-dominated uplands and sometimes nectar in open habitats that are adjacent to bogs. Adults have been observed at various peatland flowers, such as Leather-leaf (*Chamaedaphne calyculata*), Rhodora (*Rhododendron canadense*), and blueberries (*Vaccinium* spp.).

Comments: The Bog Elfin is typically reported as occurring almost exclusively in Black Spruce bogs. In our region, it sometimes occupies dry upland forest settings where Black Spruce is abundant. This species is a cryptic, early flying bog and woodland denizen, which undoubtedly is more widespread in the Acadian region than current records indicate.

Threats: This elfin appears reasonably adapted to intensive forest management as commonly practiced within its range. Peat mining is a localized threat to its bog habitats, especially in eastern New Brunswick.

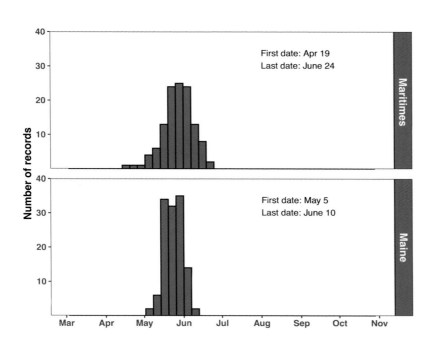

First date: Apr 19
Last date: June 24

Maritimes

First date: May 5
Last date: June 10

Maine

EASTERN PINE ELFIN

Callophrys niphon (Hübner)

Subspecies: Only the subspecies *C. niphon clarki* (T. Freeman) occurs in northeastern North America.

Distinguishing characters:
- **Size:** Small.
- **Upperside:** Males dark brown, females orange brown (rarely seen).
- **Underside:** Forewing reddish brown, usually with two dark dashes

Long A Township, ME, 22 May 2018 (Josh Lincoln)

on basal half and a series of submarginal chevrons; hindwing reddish brown and boldly patterned, black submarginal band, with a series of relatively shallow chevrons (typically less pronounced than in Western Pine Elfin); margins usually with narrow frosting topped by spots of lighter color that resemble glowing embers (in the Western Pine Elfin the margin has a row of dark spots surrounded by lighter color).

Similar species: Western Pine Elfin and Bog Elfin.

Status and distribution in Maine: Resident. Common; widespread but likely underreported in the north. S5

Status and distribution in Maritimes: Resident. Common and widespread in New Brunswick and southwestern Nova Scotia, uncommon to rare elsewhere; not recorded on Cape Breton. NB-S5, NS-S4, PE-S1

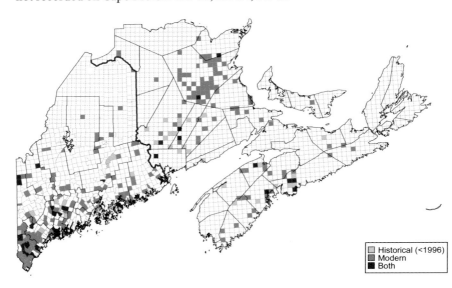

Historical (<1996)
Modern
Both

Habitat: Various woodland and barren habitats where pines (*Pinus* spp.) are common.

Biology: One generation per year. Adults are active from mid-April through June, with peak flight in late May. Larvae feed on needles of pines, including Eastern White Pine (*Pinus strobus*), Jack Pine (*P. banksiana*), and Pitch Pine (*P. rigida*). Overwinters as a pupa.

Adult behavior: Often perching on the tips of pine branches, the cryptic pattern of this species makes it very difficult to detect. They are most often encountered on moist ground along dirt roads and trails through pine woodlands. Common nectar sources include blueberries (*Vaccinium* spp.), strawberries (*Fragaria* spp.), Black Huckleberry (*Gaylussacia baccata*), Leatherleaf (*Chamaedaphne calyculata*), and Little Bluet (*Houstonia caerulea*).

Comments: Elfins are small, early flying, and relatively drab butterflies, making them easily overlooked. They undoubtedly occur in more areas than the data suggest. A significant increase in all elfin records during the modern era is likely the result of greater awareness and improved search efforts. The Eastern Pine Elfin, along with Western Pine Elfin, Bog Elfin, Juniper Hairstreak, and Hessel's Hairstreak, are the only butterfly species in eastern North America whose larvae feed exclusively on conifers.

Threats: None documented in our region.

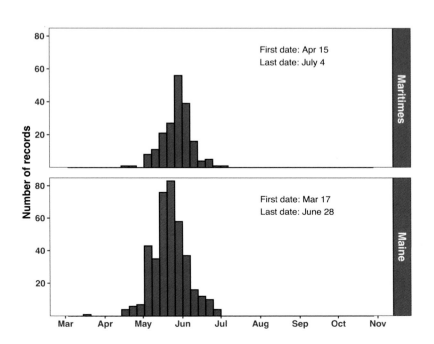

First date: Apr 15
Last date: July 4

Maritimes

First date: Mar 17
Last date: June 28

Maine

WESTERN PINE ELFIN

Callophrys eryphon
(Boisduval)

Subspecies: Only the nomino-
typical subspecies (*C. e. eryphon*) occurs in northeastern
North America.

Distinguishing characters:
    Size: Small.
    Upperside: Males dark
      brown, females orange
      brown (upperside rarely
      seen).
    Underside: Forewing

Thunder Bay District, ON, 22 May 2020
(Joshua Vandermeulen)

reddish brown, usually with one dark dash in basal half, and a series of
submarginal chevrons; hindwing reddish brown and boldly patterned,
black submarginal band is a series of deeply incised chevrons (more
pronounced than in Eastern Pine Elfin); margins unfrosted with a row
of dark spots surrounded by lighter color (in the Eastern Pine Elfin,
the frosted margins are topped by spots of lighter color that resemble
glowing embers).

Similar species: Eastern Pine Elfin and Bog Elfin.

Status and distribution in Maine: Resident. Rare; restricted to the northwestern third of the state. S3

Status and distribution in Maritimes: Resident. Uncommon to rare; widespread in northern New Brunswick. NB-S3

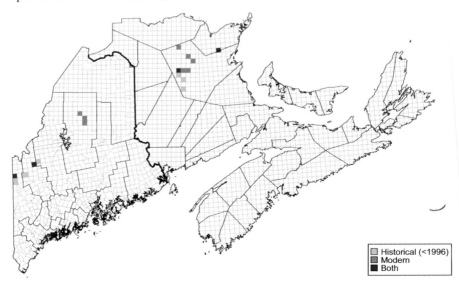

Historical (<1996)
Modern
Both

Habitat: Margins of woodlands with an abundance of pines, especially Eastern White Pine (*Pinus strobus*), often close to bogs.

Biology: One generation per year. Adults are on the wing from mid-May until mid-June, with peak flight in late May. Larvae feed on pines and have been observed ovipositing on White Pine in New Brunswick (Webster and deMaynadier 2005). Jack Pine (*P. banksiana*) may also be used, as it is a primary food plant in Quebec (Handfield 2011) and Ontario (Hall et al. 2014). Overwinters as a pupa.

Adult behavior: Behaving much like the Eastern Pine Elfin, adults of the Western Pine Elfin perch on branches of White Pine and frequently imbibe moisture on bare sandy ground at the edges of pine woodlands. They also nectar at Leatherleaf (*Chamaedaphne calyculata*) and probably also blueberries (*Vaccinium* spp.).

Comments: This species can be very difficult to distinguish from the more common and widespread Eastern Pine Elfin. Although the first published report of the Western Pine Elfin in Maine was by Kiel (1976) in Wilsons Mills Bog (Oxford County), we located several museum specimens that were collected as early as 1933 but were misidentified as the Eastern Pine Elfin. An early spring species that occurs in remote areas of the Acadian region, this elusive elfin awaits discovery at additional locations.

Threats: None documented in our region.

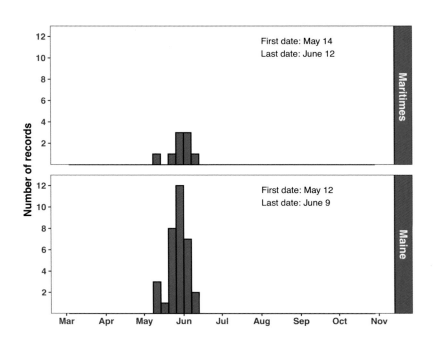

EARLY HAIRSTREAK

Erora laeta (W. H. Edwards)

Subspecies: None.

Distinguishing characters:
  Size: Small.
  Upperside: Dark gray in male and bright iridescent blue in female with dark wing borders.
  Underside: Both wings jade green with an irregular band of orange-red spots.

Similar species: None.

Stamford, VT, 30 May 2008 (Bryan Pfeiffer)

Status and distribution in Maine: Resident. Rare; only known from historical records scattered in the central part of the state. State Special Concern. SH

Status and distribution in Maritimes: Resident. Rare in all three provinces, with only historical records in Nova Scotia and Prince Edward Island; widespread. NB-S1, NS-S1, PE-SH

Habitat: Mature hardwood and mixed wood forests with abundant beech; most often encountered on dirt roads and trails.

Biology: Generally one generation in our region, though an old record from Maine on 22 July 1889 (S. I. Smith) suggests an occasional, partial second brood, as is common to our south. Most adults are active from late May to early June. Early instar larvae feed on the seed husk of American Beech (*Fagus grandifolia*), while older larvae burrow into the nut to eat developing seeds

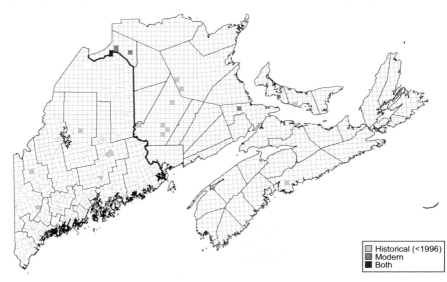

	Historical (<1996)
	Modern
	Both

(Layberry et al. 1998). Beaked Hazelnut (*Corylus cornuta*) may also be used as a food plant. Overwinters as a pupa.

Adult behavior: Thought to spend most of its brief adult life in the forest canopy, this species is rarely encountered throughout its range, and then only in specific circumstances—usually on dirt roads during hot sunny days, when adults may be found resting on damp soil. There are few observations of this species on flowers from anywhere in its range, though adults have been observed nectaring on Pin Cherry (*Prunus pensylvanica*) in New Brunswick (Webster and deMaynadier 2005).

Comments: It is unclear if the few records of Early Hairstreak in our region reflect true rarity or its inaccessibility due to a canopy-associated lifestyle. Nonetheless, it is one of our most sought after and mysterious butterflies.

Threats: American Beech, the larval food plant of the Early Hairstreak, faces multiple threats. Large, mature beeches have declined significantly on industrial forest lands of the Acadian region (McCaskill et al. 2016), where they are often targeted for removal because of low silvicultural value. Beech Bark Disease, caused by fungi that are spread by *Cryptococcus fagisuga*, a nonnative scale insect, affects a high proportion of trees, increasing tree mortality and reducing mast production (Cale et al. 2017). An emerging threat, which may be gravest of all, is beech mortality caused by the Beech Leaf-mining Weevil (*Orchestes fagi*). This European beetle was first documented in North America in Nova Scotia in 2012 (Sweeney et al. 2012) and by 2019 had spread to New Brunswick and Prince Edward Island (Klymko and Anderson 2022). American Beech trees suffered up to 88% mortality at study plots with weevil outbreaks monitored in Nova Scotia between 2014 and 2019 (Sweeney et al. 2020).

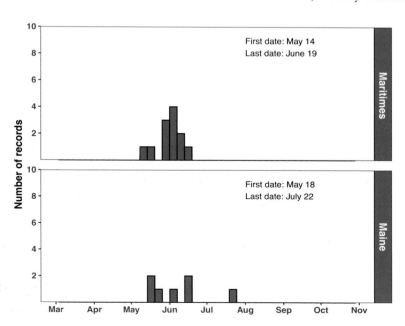

CORAL HAIRSTREAK

Satyrium titus (Fabricius)

Whitefield, ME, 19 July 2020 (Josh Lincoln)

Subspecies: Only the northeastern subspecies, *S. t. winteri* (Gatrelle), occurs in the Acadian region.

Distinguishing characters:
 Size: Small.
 Upperside: Dark brown; females sometimes with poorly defined row of marginal orange spots on hindwing (rarely seen).
 Underside: Brownish gray with central band of small, white-rimmed black spots; hindwing with well-defined marginal row of red-orange spots.
 Additional: Only tailless *Satyrium* hairstreak in our region.

Similar species: Acadian Hairstreak, Edwards' Hairstreak, and Banded Hairstreak.

Status and distribution in Maine: Resident. Uncommon and localized; restricted to southern and central regions. State Special Concern. S3S4

Status and distribution in Maritimes: McIntosh (1899b) reported two specimens from the Nerepis Valley, north of Saint John, New Brunswick, but they have not been located. We agree with Pohl et al. (2018) that the specimens were probably misidentified. There are no other records from the Maritimes.

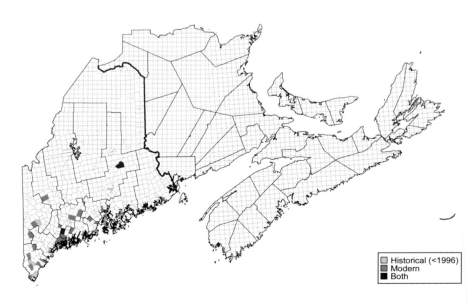

Historical (<1996)
Modern
Both

Habitat: Barrens, old fields, and other shrubby open areas, often with dry soils.

Biology: One generation with adults occurring from early July to early August. Larvae feed on various plums and cherries, including Choke Cherry (*Prunus virginiana*), Black Cherry (*P. serotina*), and possibly Beach Plum (*P. maritima*) coastally. In Michigan, the larvae were found during the day in small debris shelters constructed by ants at the base of host shrubs and saplings (Webster and deMaynadier 2005). Overwinters as eggs laid on the bark of small host trees and shrubs.

Adult behavior: Like other hairstreaks, this species has a swift and erratic flight. Males perch on the outer branches of trees and shrubs and fly out to investigate passing objects, including other males. Avid flower visitors, adults in our region have been observed at Common Milkweed (*Asclepias syriaca*), Spreading Dogbane (*Apocynum androsaemifolium*), Wood Lily (*Lilium philadelphicum*), and Canada Goldenrod (*Solidago canadensis*).

Comments: The Coral Hairstreak was thought to be exceptionally rare at the start of the Maine Butterfly Survey, when only three modern populations were known. Over twenty modern populations are now documented as a result of community science survey efforts. We recognize the subtly differentiated subspecies *S. t. winteri* pending additional research.

Threats: Residential development and ecological succession threaten the shrubby fields and barrens preferred by this butterfly. Aerial pesticide drift from adjacent oak–pine forests, targeted for Spongy Moth (*Lymantria dispar*) control, may be an additional threat.

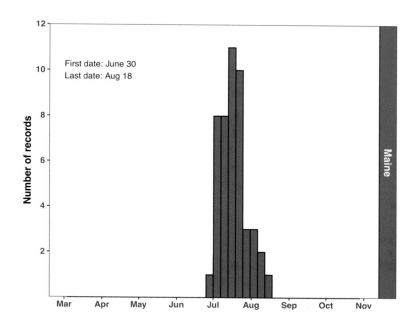

First date: June 30
Last date: Aug 18

STRIPED HAIRSTREAK

Satyrium liparops (Le Conte)

Subspecies: Only the subspecies *S. l. strigosa* (T. Harris) occurs in northeastern North America.

Distinguishing characters:
  Size: Small.
  Upperside: Dark grayish brown (rarely seen).
  Underside: Slate gray with multiple dark bands edged in white creating a striped appearance; hindwing with large submarginal blue patch capped with orange.

Kings County, NS, 17 July 2021 (Krista Melville)

  Additional: Hindwing with conspicuous tail and often a second, much-reduced tail above.

Similar species: Banded Hairstreak and Edwards' Hairstreak.

Status and distribution in Maine: Resident. Uncommon; widespread. S4

Status and distribution in Maritimes: Resident. Uncommon to rare; widespread. NB-S4, NS-S4?, PE-S3?

Habitat: Edges and openings of various deciduous and mixed woodlands and swamps.

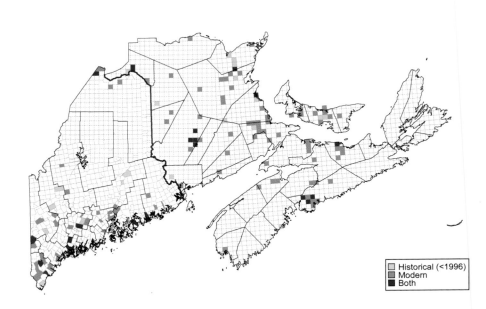

Historical (<1996)
Modern
Both

Biology: One generation per year. Adults are on the wing from early July until mid-August. Larvae feed on a variety of woody plants, particularly blueberries (*Vaccinium* spp.) and members of the rose family (Rosaceae), including cherries (*Prunus* spp.), and hawthorns (*Crataegus* spp.). Overwinters as an egg.

Adult behavior: Adults perch at various heights on the food plants, and males fly out to intercept passing females. This species uses a variety of nectar sources in our region, including Common Milkweed (*Asclepias syriaca*), Wild Carrot (*Daucus carota*), White Meadowsweet (*Spirea alba*), and Spreading Dogbane (*Apocynum androsaemifolium*).

Comments: Wagner and Gagliardi (2015) noted that in early June in Connecticut the larvae of this butterfly are among the most common species of Lepidoptera on apples (*Malus* spp.) and Highbush Blueberry (*Vaccinium corymbosum*), yet adults are infrequently seen around the plants. The authors suggest that adults appear scarce because they use flowers as a secondary food source; most calories are derived from the sugar-rich honeydew produced by sap-sucking insects, such as aphids, scales, and treehoppers. Instead of being conspicuous on roadside flowers, the butterflies are resting on foliage and branches, where they are more difficult to see. If so, the Striped Hairstreak may be much more common in our region than records indicate.

Threats: None documented in our region.

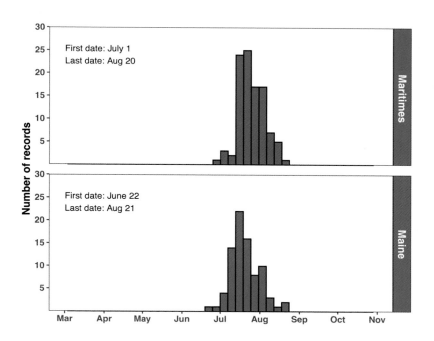

First date: July 1
Last date: Aug 20

Maritimes

First date: June 22
Last date: Aug 21

Maine

Number of records

BANDED HAIRSTREAK

Satyrium calanus (Hübner)

Subspecies: Only the sub-species *S. c. falacer* (Godart) occurs in northeastern North America.

Distinguishing characters:
    Size: Small.
    Upperside: Dark brown, male with scent patches on forewing (rarely seen).
    Underside: Slate gray with broken medial band of dark brown bars bor-

Waterford ME, 18 July 2020 (Bryan Pfeiffer)

dered by white on one or both sides; hindwing margin with conspicuous orange spot adjacent to a large blue patch that lacks an orange cap.
    Additional: Hindwing with conspicuous tail; often a second, highly re-duced tail is present.

Similar species: Striped Hairstreak.

Status and distribution in Maine: Resident. Common; widespread in south-ern third of state; records from New Brunswick suggest likely presence in Aroostook County. S4

Status and distribution in Maritimes: Resident. Rare; localized, primarily associated with the Saint John River Valley in New Brunswick; widespread in mainland Nova Scotia. NB-S2, NS-S3, PE-S1

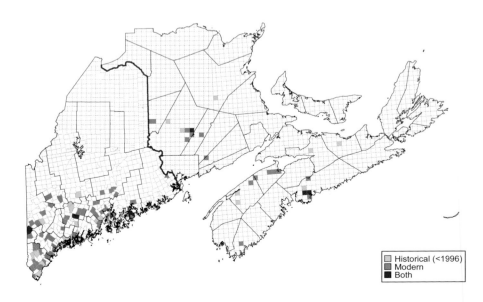

Historical (<1996)
Modern
Both

Habitat: Clearings and edges within woodlands where oak (*Quercus* spp.) or hickory (*Carya* spp.) is common, including urban forests; also rich floodplains with Butternut (*Juglans cinerea*).

Biology: One generation per year. Adults are active from late June until early September, with a peak flight in mid-July. Larvae feed on oaks, hickories, and Butternut. Overwinters as an egg.

Adult behavior: Males perch at various heights in host trees and fly out to investigate passing insects, often engaging in aerial "dogfights" with other hairstreaks. Adults are most often encountered at flowers. In our region, these include Common Milkweed (*Asclepias syriaca*), Wild Carrot (*Daucus carota*), Spreading Dogbane (*Apocynum androsaemifolium*), Bristly Sarsaparilla (*Aralia hispida*), and goldenrods (*Solidago* spp.).

Comments: On 20 July 2008, several observers reported a Banded Hairstreak in Charlottetown, Prince Edward Island, but no specimen or photo was taken. The hindwings of this and other hairstreaks display "false heads," with hairlike tails resembling antennae. Such complex pattern elements are thought to fool would-be predators into attacking a nonessential portion of the hindwing, allowing the butterfly to escape with its body intact. Indeed, hairstreaks can be found with triangular pieces of their hindwings missing, evidently the result of bird strikes. Studies by Sourakov (2013) suggest that jumping spiders that unsuccessfully capture hairstreaks with false heads may learn to ignore them as prey, thus conferring protection on multiple hairstreak species.

Threats: Because Butternut may be an important larval food plant in some areas, such as within the Saint John River Valley, the loss of this tree to Butternut canker (a disease caused by the introduced fungus *Ophiognomonia clavigignenti-juglandacearum*) may be a localized threat.

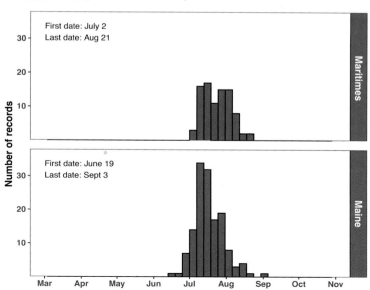

EDWARDS' HAIRSTREAK

Satyrium edwardsii (Grote & Robinson)

Subspecies: Only the nominotypical subspecies (*S. e. edwardsii*) occurs in northeastern North America.

Distinguishing characters:
  Size: Small.
  Upperside: Light brown, often with small, indistinct orange spot near hindwing tail, especially in female (rarely seen).
  Underside: Pale gray brown with

Wells, ME, 12 July 2017 (Trevor Persons)

postmedian band of separate, white-rimmed oval spots on both wings; hindwing with submarginal row of orange markings and prominent, orange, spear-shaped patch along the posterior hindwing margin; large blue spot near tail is narrowly orange capped (Banded Hairstreak is not orange capped).
  Additional: Hindwing with hair-like tail.

Similar species: Acadian Hairstreak and Banded Hairstreak.

Status and distribution in Maine: Resident. Rare; restricted to York and southern Oxford Counties. State Endangered. S2

Status and distribution in Maritimes: Not documented.

Habitat: Dry oak–pine woodlands and barrens.

Biology: One generation, with a short flight usually peaking in mid-late July. In most of the Northeast, Scrub Oak (*Quercus ilicifolia*) is the primary larval food

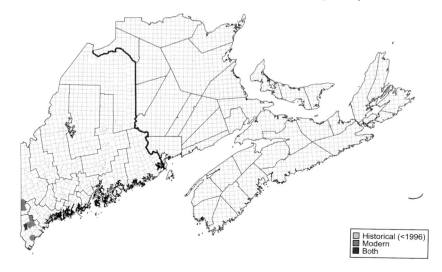

Historical (<1996)
Modern
Both

plant. Elsewhere in its range, Black Oak (*Q. velutina*), Scarlet Oak (*Q. coccinea*), and White Oak (*Q. alba*) are also used and are present in generally low abundance in southern Maine. Overwintering occurs as eggs laid in bark crevices of the food plant.

Adult behavior: Individuals often perch on the outer branches and leaves of open-grown Scrub Oak trees. With a fast, erratic flight, they frequently return to the same plant if flushed. Like other hairstreaks, males fly out from perches to inspect passing objects and sometimes engage in spiraling altercations with other males. Nectar sources are often limiting in barren habitats during July, when adults can be observed feeding at Common Milkweed (*Asclepias syriaca*) and Spreading Dogbane (*Apocynum androsaemifolium*).

Comments: A seminal study in Michigan (Webster and Nielsen 1984) revealed a strong relationship between the larvae of Edwards' Hairstreak and the mound-building ant, *Formica integra*. Late-instar larvae aggregate during the day in groups at the base of host trees within conical structures of detritus (byres) constructed by the ants. The larvae leave these structures in the evening to feed nocturnally and then return to them near dawn. The larvae are often surrounded by attending ants that feed on honeydew secreted from special abdominal glands. The ants presumably help protect the larvae from predators and parasitoids. Edwards' Hairstreak is similar in appearance to the Banded Hairstreak, which can lead to dubious records. For example, specimens reported by Brower (1974) from Augusta and Bar Harbor, Maine, were later determined to be Banded Hairstreaks.

Threats: Glacial outwash barrens and woodlands in southern Maine that provide habitat for this butterfly are threatened from development and ecological succession resulting from fire suppression. In addition, the larvae of both Edwards' Hairstreak and the introduced Spongy Moth (*Lymantria dispar*) (a pest that periodically defoliates large areas of oak) feed during approximately the same period in the spring, making the butterfly vulnerable to aerial pesticide application used to control the moth (Webster and deMaynadier 2005).

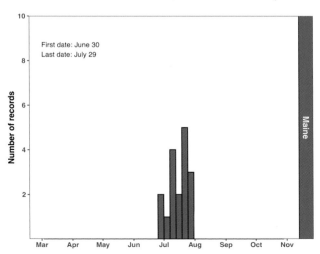

First date: June 30
Last date: July 29

ACADIAN HAIRSTREAK

Satyrium acadica
(W. H. Edwards)

Subspecies: Only the nomino-typical subspecies (*S. a. acadica*) occurs in eastern North America.

Distinguishing characters:
  Size: Small.
  Upperside: Brownish gray; hindwing with small orange patch above tail (rarely seen).
  Underside: Pale silver gray with medial row of white-rimmed, round, black spots and submarginal row of fused orange spots, widening toward the hindwing's posterior; hindwing with blue patch above tail.
  Additional: Hindwing with conspicuous tail; often a second, much-reduced tail is present.

Camden, ME, 11 July 2013 (Roger Rittmaster)

Similar species: Coral Hairstreak and Edwards' Hairstreak.

Status and distribution in Maine: Resident. Uncommon; widespread in southern half of state; in the north known only from eastern Aroostook County. S3S4

Status and distribution in Maritimes: Resident. Uncommon and widespread in New Brunswick; rare in eastern Prince Edward Island and northwestern Nova Scotia, apparently absent elsewhere. NB-S3, NS-S2, PE-S1

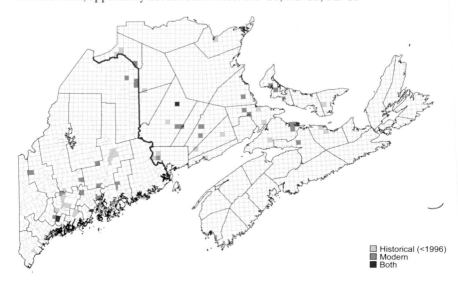

☐ Historical (<1996)
☐ Modern
■ Both

Habitat: A variety of open, generally wet sites with shrubby willows, including marshes, riparian meadows, and old fields.

Biology: One generation per year. Adults are active from early July until mid-August, with a peak flight in mid-July. Larvae feed on willows (*Salix* spp.). Overwinters as an egg.

Adult behavior: Adults are usually encountered when they are flushed from their low perches on vegetation or when they are visiting flowers, such as Spreading Dogbane (*Apocynum androsaemifolium*), White Meadowsweet (*Spiraea alba*), and milkweeds (*Asclepias* spp.).

Comments: When William H. Edwards named and described the Acadian Hairstreak in 1862, he offered no explanation as to why he chose the epithet *acadica*, which certainly sounds like a reference to the Acadian region. The only specimen mentioned in the description was taken near London, Ontario (Edwards 1862). Because Edwards also used the name *acadica* for a butterfly from Newfoundland, he must have equated it with meaning northern in distribution. Indeed, the Acadian Hairstreak has the most northeasterly distribution of any hairstreak in North America.

Threats: Declines have been reported in New Jersey (Gochfeld and Burger 1997) and Massachusetts (Stichter 2015), where authors suggest habitat loss (NJ) and climate change (MA) as potential causes. Only historical records are known from extreme southern Maine, where shrubby meadows and riparian areas have been subject to significant habitat loss and degradation.

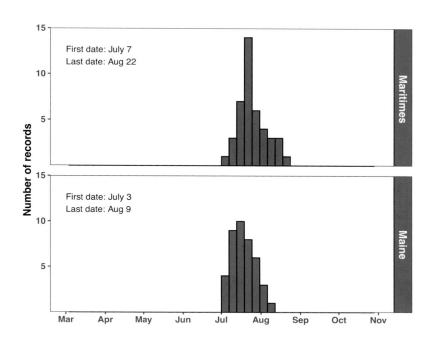

SILVERY BLUE

Glaucopsyche lygdamus
(E. Doubleday)

Subspecies: The subspecies *G. l. couperi* Grote occurs throughout our region and most of the Northeast; the current status of subspecies *G. l. mildredae* F. Chermock, described from Cape Breton, is unclear (see comments).

Albert County, NB, 27 May 2017 (Denis Doucet)

Distinguishing characters:
> **Size:** Small.
> **Upperside:** Bright metallic blue with narrow black margin (male) or grayish brown, usually with blue scaling on the basal half of wings (female); white fringe.
> **Underside:** Bluish gray with submarginal row of round, white-rimmed black spots on both wings.

Similar species: Greenish Blue.

Status and distribution in Maine: Resident. Abundant; widespread. S5

Status and distribution in Maritimes: Resident. Abundant; widespread. NB-S5, NS-S5, PE-S5

Habitat: A variety of open areas, including old fields, gardens, roadsides, and beaches.

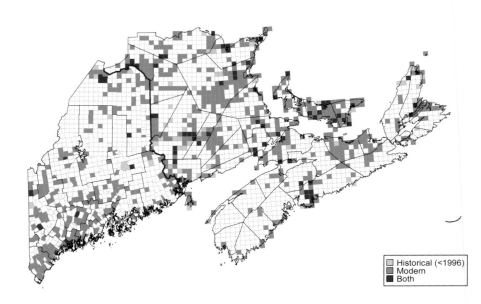

Historical (<1996)
Modern
Both

Biology: One generation per year. Adults appear from early to late May and fly until about mid-July. Larvae are ant attended and feed on legumes, primarily Cow Vetch (*Vicia cracca*), White Sweet-clover (*Melilotus albus*) and, along the coast, Beach Vetchling (*Lathyrus japonicus*). Overwinters as a pupa, often within ant nests.

Adult behavior: This species has a low, strong flight and is often seen nectaring or taking minerals from damp earth. Frequently used flowers include Cow Vetch, Common Strawberry (*Fragaria virginiana*), clovers (*Trifolium* spp.), blueberries (*Vaccinium* spp.), hawkweeds (*Hieracium* spp.), and buttercups (*Ranunculus* spp.).

Comments: Historically, the Silvery Blue was rare in the Acadian region. The isolated subspecies *mildredae*, associated with Beach Vetchling, occurred in Cape Breton. It was darker gray beneath, with larger black spots. The more northern subspecies *couperi* may have been present historically in northern New Brunswick (it was reported as common in 1886 at Jacquet River, Restigouche County [McIntosh 1899b]); however, it did not colonize other areas in the region until the 1900s: the first Maine, Nova Scotia, and Prince Edward Island records were during 1936, 1950, and 1964, respectively. The southward expansion of *couperi* was facilitated by the abundance of weedy legumes (particularly Cow Vetch) along road and rail corridors. It is unclear when *couperi* reached Cape Breton, where it is now abundant. Unfortunately, *mildredae* may now be extinct, having been genetically swamped by *couperi*. The persistence of *mildredae* should be investigated, particularly on coastal islands where *couperi* may not yet be established.

Threats: None documented in our region.

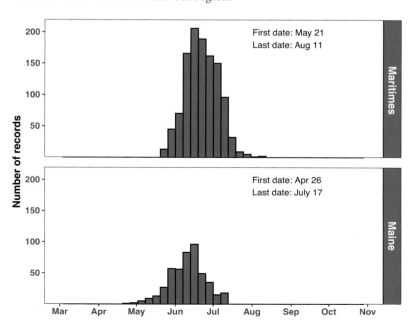

First date: May 21
Last date: Aug 11

First date: Apr 26
Last date: July 17

NORTHERN AZURE

Celastrina lucia (W. Kirby)

Subspecies: Only the nominotypical subspecies (*C. l. lucia*) occurs in eastern North America.

Distinguishing characters:

Size: Small.

Upperside: Pale metallic blue or purplish blue with black and white checkered fringe; female similar but with broad black margin on forewing and row of black marginal spots on hindwing.

Underside: Pale bluish gray to dusky gray with variable dark spotting and submarginal dark chevrons in three distinct morphs: "lucia" with darkened wing borders and a large dark central patch on the hindwing; "marginata" with darkened wing borders but no (or greatly reduced) patch; "violacea" without darkened borders or patch.

Similar species: Summer Azure.

Status and distribution in Maine: Resident. Abundant; widespread. S5

Status and distribution in Maritimes: Resident. Abundant; widespread. NB-S5, NS-S5, PE-S5

Habitat: Forest openings and edges, bogs, roadsides, old fields, and gardens.

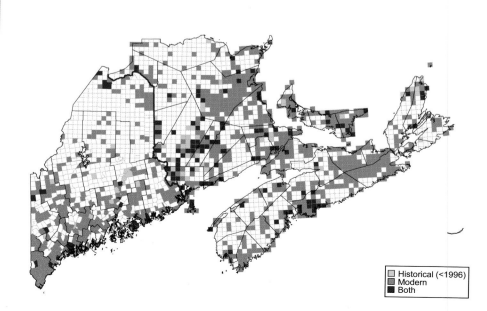

Historical (<1996)
Modern
Both

Form "lucia"—Westmorland County, NB, 15 May 2008 (Denis Doucet)

Form "marginata"—Kent County, NB, 20 May 2012 (Denis Doucet)

Form "violacea"—Magalloway Plantation, ME, 8 June 2018 (Bryan Pfeiffer)

Biology: Two overlapping generations. Adults are active from mid-April until September, with peak numbers from mid-May until early July. Larvae feed on the flowers and fruits of a wide variety of food plants, including Bush-honeysuckle (*Diervilla lonicera*), White Meadowsweet (*Spiraea alba*), Withe-rod (*Viburnum nudum*), Bristly Sarsaparilla (*Aralia hispida*), Labrador Tea (*Rhododendron groenlandicum*), and blueberries (*Vaccinium* spp.), as well as galls of cherry (*Prunus* spp.) leaves that are produced by mites. It has also been observed ovipositing on several other plant species in our region, some of which may not be suitable food plants. Overwinters as a pupa.

Adult behavior: This species has a rather weak, meandering flight. Males patrol for females low over vegetation and are frequent puddle visitors. Over its long flight period it visits a wide variety of flowers. During our surveys it was documented at flowers of over sixty plant genera, the most frequently reported including blueberries (*Vaccinium* spp.), vetches (*Vicia* spp.), buttercups (*Ranunculus* spp.), clovers (*Trifolium* spp.), and violets (*Viola* spp.).

Comments: The taxonomy of *Celastrina* has not been fully resolved, and disagreements remain about species boundaries and nomenclature. The Northern Azure was once considered a subspecies of the Spring Azure (*Celastrina ladon*), and another species, the Cherry Gall Azure (*Celastrina serotina*) was thought to occur in our region. Most recently, Schmidt and Layberry (2016) studied the Canadian *Celastrina* fauna and concluded that only Northern Azures are present in eastern Ontario, Quebec, and the Maritime Provinces. They also found that butterflies in Canada that were previously identified as the Cherry Gall Azure are applicable to the Northern Azure and Summer Azure. We tentatively follow this arrangement for the Acadian region. As the understanding of *Celastrina* taxonomy improves, other cryptic species (such as the Cherry Gall Azure) may be detected here. The lectotype (a specimen designated as the name-bearing type sometime after the original description was published) of form "marginata" is from Orono, Maine. See the account for the Summer Azure for more comments about the Northern Azure.

Threats: None documented in our region.

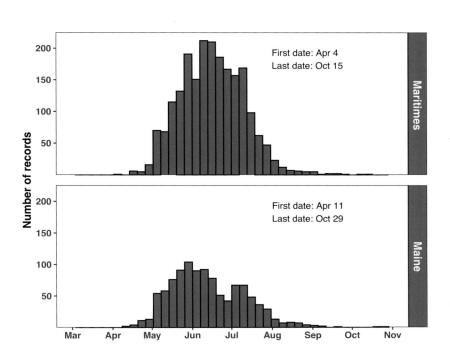

First date: Apr 4
Last date: Oct 15

Maritimes

First date: Apr 11
Last date: Oct 29

Maine

SUMMER AZURE

Celastrina neglecta
(W. H. Edwards)

Subspecies: None.

Distinguishing characters:
 Size: Small.
 Upperside: Male light blue
  or violet blue (hindwing
  often paler) with narrow,
  black forewing margin.
  Female forewing with
  variable amount of white
  scaling and broad, black
  border; hindwing mostly

Falmouth, ME, 6 August 2018 (Bryan Pfeiffer)

  white with row of black marginal spots.
 Underside: White with scattered, small black spots (sometimes reduced or
  absent); marginal row of indistinct black spots.
 Additional: Both sexes typically with solid white hindwing fringes.

Similar species: Northern Azure.

Status and distribution in Maine: Resident. Uncommon; widespread in southern third of state. S4

Status and distribution in Maritimes: Not documented (previous records are misidentifications).

Habitat: Mostly woodland clearings and edges but also old fields, marshes, urban parks, and gardens.

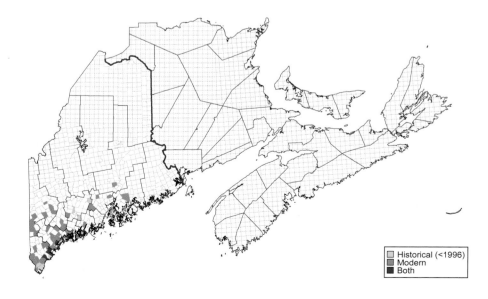

☐ Historical (<1996)
▨ Modern
■ Both

Biology: One generation per year in our region, with a peak flight in mid-July. Larvae feed on floral buds of many plants, including dogwoods (*Benthamidia* [= *Cornus*] spp.), meadowsweets (*Spiraea* spp.), and Ninebark (*Physocarpus opulifolius*). Overwinters as a pupa.

Adult behavior: This butterfly has a rather weak, meandering flight. Males are frequently encountered at moist ground and both sexes readily visit flowers, such as clovers (*Trifolium* spp.), Cow Vetch (*Vicia cracca*), Common Milkweed (*Asclepias syriaca*), fleabanes (*Erigeron* spp.), and meadowsweets (*Spiraea* spp.).

Comments: Like the Northern Azure, this species was once treated as a subspecies of the Spring Azure (*Celastrina ladon*), which has not been confirmed in our region. Summer broods of the Northern Azure are similar to the Summer Azure, but they have checkered hindwing fringes and more developed ventral spotting. Additional wing characters used to separate these species are summarized by Schmidt and Layberry (2016). Like males of the Northern Azure, those of the Summer Azure do not typically possess the distinctive, overlapping forewing scales of Spring Azure males. However, some males of the Northern Azure from farther south, and just entering York County, Maine (sometimes referred to as the "New England Azure"), express this trait, perhaps as a result of past genetic introgression from the Spring Azure at the northern limits of its range (D. M. Wright, pers. comm. to J. Calhoun). A similar situation was recently discovered in Summer Azures at the southern edge of the Spring Azure's range in Florida (Wright et al. 2019). In addition, there is evidence in nearby areas of the Northeast (e.g., Massachusetts and Rhode Island) that the Summer Azure is now producing two generations as they do farther south. If so, this same bivoltine flight may eventually occur in our region. Species limits within the genus *Celastrina* have been debated for nearly two centuries; DNA research, especially genomics, may provide a clearer understanding of this group's taxonomy.

Threats: None documented in our region.

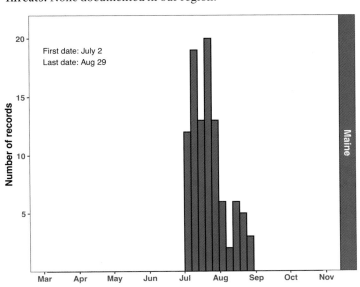

First date: July 2
Last date: Aug 29

EASTERN TAILED-BLUE

Cupido comyntas (Godart)

Subspecies: Only the nomino-typical subspecies (*C. c. comyntas*) occurs in eastern North America.

Distinguishing characters:
  Size: Small.
  Upperside: Metallic purplish blue with black margins (male) or dark brownish gray with sparse blue scaling that is more prominent in spring brood (female); hindwing with black marginal spots, the spot adjacent to tail with small orange cap.

Mating pair—Eliot, ME, 6 August 2018 (Bryan Pfeiffer)

  Underside: Grayish white with well-defined black spots edged in white; hindwing with two orange-capped black spots near tail.
  Additional: Hindwing with short, thin, white-tipped tail.

Similar species: Western Tailed-Blue.

Status and distribution in Maine: Resident. Common in the southern half of the state, rare elsewhere. S5

Status and distribution in Maritimes: Resident. Uncommon in southwestern New Brunswick, rare elsewhere. NB-S4S5, NS-S3S4, PE-SU

Habitat: A variety of open disturbed areas, including old fields, vacant lots, gardens, and roadsides.

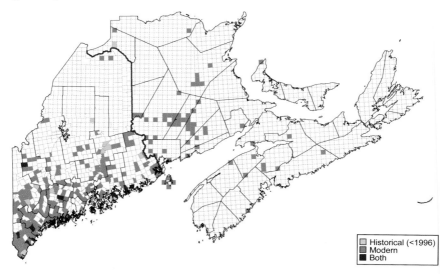

□ Historical (<1996)
▨ Modern
■ Both

Biology: Two or three overlapping generations per year. Adults appear in early May or June and fly into October, with a peak flight during August. Late summer abundance may be augmented by immigrants from farther south. Larvae feed on the flowers and seeds (sometimes new leaves) of plants in the pea family, including vetches (*Vicia* spp.), clovers (*Trifolium* spp.), and tick-trefoils (*Desmodium* spp.). Overwinters as a nearly fully grown larva.

Adult behavior: This blue has a weak, low flight, and adults frequently bask with wings partially open. When resting with wings closed, they often rub their hindwings together like a hairstreak. It is a frequent flower visitor, and common nectar sources in our region include clovers (*Trifolium* spp.), American-asters (*Symphyotrichum* spp.), goldenrods (*Solidago* spp.), and Cow Vetch (*Vicia cracca*).

Comments: Until recently, the Eastern Tailed-Blue was a very rare species in the Maritime Provinces. The first known record was a specimen from Edmundston, New Brunswick, in 1984. Prior to the Maritimes atlas project, only a handful of additional sites had been discovered, all in the western half of New Brunswick. Its status in the Maritimes began to change in 2012, when it was found in a number of new areas in New Brunswick, including two eastward. In 2013, it was found at many new localities, including four in Nova Scotia. It was recorded for the first time in Prince Edward Island in 2014. Interestingly, the first records from northern Maine were not documented until 2020. These discoveries potentially indicate a recent and rapid range expansion into northern and eastern portions of our region.

Threats: None documented in our region.

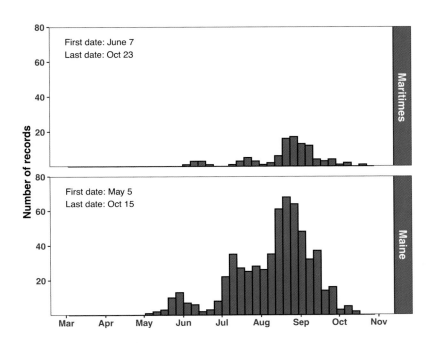

WESTERN TAILED-BLUE

Cupido amyntula (Boisduval)

Male—Northumberland, NB, 14 June 2007 (Denis Doucet)

Subspecies: Only the subspecies *C. a. maritima* (Leblanc) occurs in northeastern North America (see comments).

Distinguishing characters:
  Size: Small.
  Upperside: Metallic purplish blue with black margins (male) or dark brownish gray with sparse blue scaling basally (female); hindwing with black marginal spots, the spot adjacent to tail rarely with small orange cap.
  Underside: Grayish white (typically paler than Eastern Tailed-Blue) with black spots edged in white (nearly absent in some individuals); hindwing with one or two orange-capped black spots near tail (typically larger in Eastern Tailed-Blue).
  Additional: Hindwing with short, thin, white-tipped tail.

Similar species: Eastern Tailed-Blue.

Status and distribution in Maine: Resident. Rare in northern Aroostook County, absent elsewhere. S3

Status and distribution in Maritimes: Resident. Common in northern New Brunswick, uncommon in central and southeastern New Brunswick, absent elsewhere. NB-S5

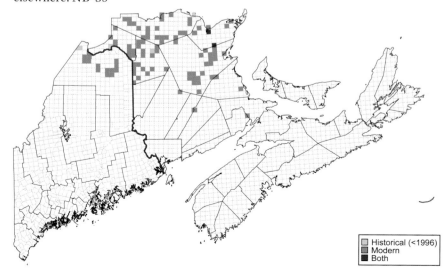

☐ Historical (<1996)
▨ Modern
■ Both

Habitat: A variety of open areas including roadsides, forest clearings, dunes, and salt marshes.

Biology: One generation per year. Adults appear in early June and fly until mid-July, with peak numbers during mid-June. Larvae feed on the flowers and seed pods of plants in the pea family, including Cow Vetch (*Vicia cracca*) and vetchlings *(Lathyrus* spp.). Overwinters as a nearly fully grown larva, often in seedpods.

Adult behavior: This species has a weak, low flight and adults often bask with partially open wings. It regularly visits flowers, and in our region has been observed nectaring at Cow Vetch, Common Cinquefoil (*Potentilla simplex*), Beach Vetchling (*Lathyrus japonicus*), and White Clover (*Trifolium repens*).

Comments: Historically, the diversity and abundance of suitable food plants were limited for Western Tailed-Blues in the Acadian region. Native members of the pea family are not nearly as abundant or widespread as introduced weedy species, such as Cow Vetch. Nonnative food plant options are probably contributing to the range expansion of this disjunct subspecies of *amyntula* (Cech and Tudor 2005). Recent records from northern Maine (including five new township records in 2022, as well as from central and southeastern New Brunswick, suggest that expansion is ongoing. It was first recorded in Maine in 1995. The baseline data sets of the Maine and Maritime surveys will help detect future changes in this species' distribution. The lectotype (a specimen designated as the name-bearing type sometime after the original type description was published) of the subspecies *maritima* is from Jacquet River, New Brunswick.

Threats: None documented in our region.

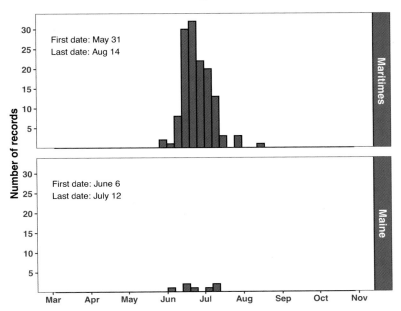

GREENISH BLUE

Icaricia saepiolus (Boisduval)

Subspecies: Only the subspecies *I. s. amica* (W. H. Edwards) occurs in eastern North America.

Distinguishing characters:

Size: Small.

Upperside: Pale metallic blue with black margins (male) or dusky brown with sparse blue scaling near base of wings (female); forewing typically with small black dash in the center; hindwing with faint orange-capped black spots along margin.

Madawaska County, NB, 1 July 2018 (John Klymko)

Underside: Chalky gray with greenish scaling near body and numerous white-rimmed black spots; hindwing usually with one or more orange spots near the margin.

Similar species: Silvery Blue, Summer Azure, Northern Azure.

Status and distribution in Maine: Resident. Rare; historically widespread, now absent or limited to the extreme north. S1

Status and distribution in Maritimes: Resident. Rare; historically widespread in New Brunswick, now potentially limited to the north; in Nova Scotia it is known from two historical records. NB-S1S2, NS-SH

Habitat: A variety of open and disturbed areas, especially roadsides.

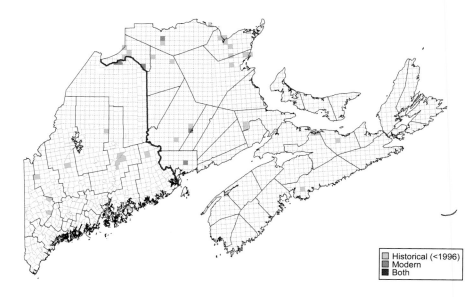

Historical (<1996)
Modern
Both

Biology: One generation per year. Adults appear in early to mid-June and fly for about 1 month, with peak abundance in late June. Larvae feed on the flowers and seed pods of clovers, particularly introduced Alsike Clover (*Trifolium hybridum*) and White Clover (*T. repens*). Overwinters as a nearly fully grown larva.

Adult behavior: This species has a weak flight and is typically seen flying low over vegetation or basking with its wings partially open. Males often aggregate on wet soil along forest roads, sometimes in large numbers. It regularly nectars at clovers and Cow Vetch (*Vicia cracca*).

Comments: The Greenish Blue's history in eastern North America is one of boom and bust. There are no records of this northern butterfly in the Acadian region, or even Quebec, before 1900. By midcentury it was common in northern New Brunswick and present in northern Maine (Ferguson 1954). It was fairly common in Maine by 1950 (Klots 1951), but only two decades later its abundance was waning (Brower 1974). Thomas (1996) considered it to be rare in New Brunswick, and it has disappeared from other areas in the southern part of its range, such as central Ontario (Hall et al. 2014). The first record for the Acadian region was a female collected in Digby County, Nova Scotia, in 1908 (Ferguson 1954). The only other record from Nova Scotia is a specimen collected at Halifax in 1984, but it is unclear if this species was ever well established in the province. Since 2005, the Greenish Blue has been found only at two localities in our region: Fort Kent, Maine (2012), and Saint-Leonard, New Brunswick (2015). It is assumed the butterfly spread southeast via road corridors, where introduced clovers used as food plants are common. Why this blue declined as quickly as it appeared remains a mystery. The status of this subspecies needs more study.

Threats: The causes of the decline of Greenish Blue from the Acadian region and elsewhere in the East are unknown. The *couperi* subspecies of Silvery Blue also colonized the region in the mid-twentieth century by using weedy legumes along travel corridors, yet it remains abundant.

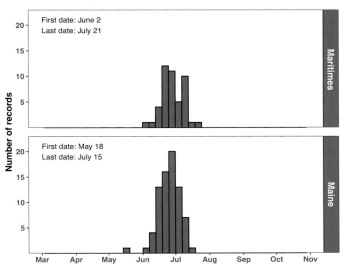

CROWBERRY BLUE AND NORTHERN BLUE

Plebejus idas (Linnaeus)

Subspecies: Two of three subspecies in northeastern North America are found in our region: *P. i. empetri* T. Freeman (**Crowberry Blue**) and *P. i. scudderii* (W. H. Edwards) (**Northern Blue**).

Distinguishing characters:

    Size: Small.

    Upperside: Males bright iridescent blue with a very narrow black border; females brownish gray with variable blue scaling basally; hindwing with row of marginal black spots (more developed in female), sometimes capped with orange.

    Underside: Whitish gray with prominent black spots; hindwing has marginal row of black spots with metallic blue scaling, usually capped by orange and black.

    Additional: The underside black spots of the Crowberry Blue tend to be larger than those of the Northern Blue, but differences are slight.

Similar species: None.

Status and distribution in Maine: Resident. Rare; Crowberry Blue is restricted to coastal Washington County; Northern Blue is restricted to northern Aroostook County. State Special Concern (both subspecies). S3

Status and distribution in Maritimes: Only the Crowberry Blue is documented. Resident. Widespread coastally; common in eastern Nova Scotia, uncommon to rare elsewhere. NB-S3, NS-S5, PE-S2

Habitat: The Crowberry Blue is found in coastal peatlands, headlands, and shrub-dominated dunes; the Northern Blue is restricted to openings within dry, coniferous woodland.

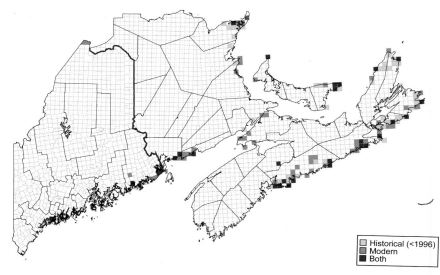

Historical (<1996)
Modern
Both

P. i. empetri—Harrington, ME, 12 July 2017 (Bryan Pfeiffer)

P. i. scudderi—Thunder Bay District, ON, 23 June 2019 (Colin Jones)

Biology: One generation. Adults are on the wing from late June to mid-August, with peak flight in early to mid-July. The Crowberry Blue appears limited to habitats with Black Crowberry (*Empetrum nigrum*), the known larval food plant, though Pink Crowberry (*E. eamesii*) may also be used in Nova Scotia and Prince Edward Island. The only record of the Northern Blue in our region is a worn female from 13 July 2005. The larval food plants of this subspecies in Maine are unknown, but elsewhere it is reported to feed on various heaths, including Dwarf Blueberry (*Vaccinium caespitosum*), Labrador Tea (*Rhododendron groenlandicum*), and Sheep Laurel (*Kalmia angustifolia*). Overwinters as an egg.

Adult behavior: Adult flight is relatively weak, with males often patrolling low above the heath vegetation. The Crowberry Blue can be active early in the day and under cool, cloudy, even foggy conditions, which are common in the coastal fog belt where it is found. The Crowberry Blue has been observed nectaring on the flowers of peatland plants, such as Bog Laurel (*K. polifolia*), Sheep Laurel, and Labrador Tea, as well as weedy species like clovers (*Trifolium* spp.).

Comments: Originally described from specimens collected at Baddeck, Nova Scotia, the Crowberry Blue is endemic to the coastal ecoregions of Maine, the Maritimes, and likely Quebec's Gaspé Peninsula. Further study is needed in Maine to determine if coastal headlands with Black Crowberry are also inhabited by the Crowberry Blue as they are in the Maritimes. The Northern Blue was only recently documented by Reeves and Christopher Livesay in Maine's

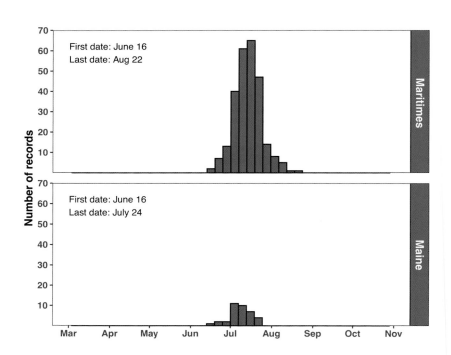

northern-most municipality (Big Twenty Township). This subspecies may be found to occur elsewhere in northern Maine and New Brunswick. More study is needed to understand the relationship between the Crowberry Blue and the Northern Blue.

Threats: In New Brunswick, peat mining is a threat to Crowberry Blue habitat. In Maine, peat mining was a greater threat historically, as many sites are now in conservation ownership. Climate change is a potential global concern for the Crowberry Blue and a regional threat to the Northern Blue, which occurs at the southern extremity of its eastern range in the Acadian region.

FAMILY NYMPHALIDAE

The Nymphalidae is the largest family of butterflies, with nearly 230 species in the United States and Canada. The group comprises a diverse assortment of species ranging in size from very small to very large. The front pair of legs in most species are reduced and resemble small brushes, leading to the name "brushfoots." Six subfamilies are represented in the Acadian region: snouts (Libytheinae), milkweed butterflies (Danainae), longwings and fritillaries (Heliconiinae), admirals and sisters (Limenitidinae), true brushfoots (Nymphalinae), and satyrs (Satyrinae). Many are cryptically colored below, resembling leaves or bark when at rest. Others display bright eyespots or other flashy, aposematic designs to avoid predation. Most nymphalids nectar at flowers, though some prefer sap, dung, and carrion. They also frequently obtain nutrients from damp earth. These butterflies occupy a broad range of habitats, from shady forests to open fields. Larval food plants include trees, herbs, vines, grasses, and sedges.

Forty-one species of Nymphalidae have been recorded in Maine and the Maritimes. Thirty-two are breeding residents, most of which produce a single generation per year. Seven species are rare or frequent colonists, and one is a stray. In Maine, six species are state listed as Endangered (Frigga Fritillary and Katahdin Arctic), Threatened (Arctic Fritillary), or Special Concern (Monarch, Satyr Comma, and Silvery Checkerspot). Two species are listed in the Maritimes as Endangered (Monarch [Nova Scotia] and Maritime Ringlet [New Brunswick]) or Special Concern (Monarch [New Brunswick]). The Monarch is federally listed as a species of Special Concern in Canada, and the Maritime Ringlet is listed as Endangered in Canada. One species, the Regal Fritillary, is now extirpated in Maine and virtually all of eastern North America.

Genomics research by Zhang et al. (2020) suggests the placement of the greater fritillaries in the genus *Argynnis* (formerly *Speyeria*), while the genera *Aglais* and *Polygonia* would be included in the genus *Nymphalis*. We follow Pelham (2022), who tentatively recognizes the former but not the latter for reasons of nomenclatural stability (stabilized name usage). More research is needed to understand relationships within the genus *Phyciodes*, which may include additional, unrecognized species in our region.

Baltimore Checkerspots—East Montpelier, VT (Bryan Pfieffer).
This checkerspot is one of the most eye-catching of the forty-one
brushfoot species gracing Maine and the Maritimes.

AMERICAN SNOUT

Libytheana carinenta (Cramer)

Subspecies: Only the subspecies *L. c. bachmanii* (Kirtland) occurs in eastern North America.

Distinguishing characters:

Size: Medium.

Upperside: Orange and dark brown; forewing with white apical spots.

Underside: Hindwing violet-gray or tan, with varying degrees of mottling.

Veazie, ME, 30 May 2011 (Jonathan Mays)

Additional: Labial palps elongated, forming distinctive "snout"; forewing apex is squared off.

Similar species: None.

Status and distribution in Maine: Stray. Rare; recorded from six scattered locations, mostly coastal. SNA.

Status and distribution in Maritimes: Stray. Rare; recorded from five scattered locations in New Brunswick and Nova Scotia. NB-SNA, NS-SNA

Habitat: Strays in our region may be found in virtually any habitat, especially open areas such as old fields, roadsides, utility corridors, and gardens. Where resident, the species primarily occurs in and around moist bottomland forests.

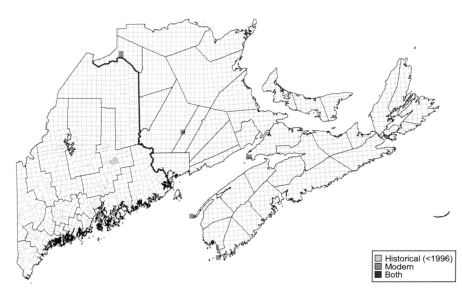

Historical (<1996)
Modern
Both

Biology: This species is a rare stray in our region. There are generally two generations within its normal range to the south, where larvae feed on hackberry trees *(Celtis* spp.). There is no evidence of overwintering in our region.

Adult behavior: These butterflies have an erratic, bouncing flight and almost always rest with their wings closed, giving them the appearance of a dead leaf, with their elongated palps (snout) resembling the petiole. Adults frequently nectar, and in our region it has been recorded visiting the flowers of Ox-eye Daisy (*Leucanthemum vulgare*) and White Meadowsweet (*Spirea alba*). Other plants found in our region that are frequently visited by this species farther south include Spreading Dogbane (*Apocynum androsaemifolium*), goldenrods (*Solidago* spp.), and American-asters (*Symphyotrichum* spp.). They are also attracted to wet soil.

Comments: It appears that this species is becoming more common in the Northeast. Scudder (1888–1889) knew of only one or two records in New England, but it is now considered a "regular visitor" in Massachusetts (Stichter 2015). Of the eleven records from the Acadian region, one is from 1974, and the rest are from 2004 or later. Despite how frequent they may become in the future, breeding opportunities for this butterfly are limited in Maine and the Maritimes. Although hackberries are sometimes planted in urban areas, they are not native to our area.

Threats: None documented in our region.

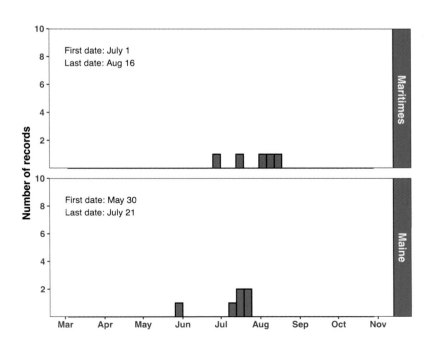

First date: July 1
Last date: Aug 16

First date: May 30
Last date: July 21

MONARCH

Danaus plexippus (Linnaeus)

Subspecies: Only the nominotypical subspecies (*D. p. plexippus*) occurs in North America.

Distinguishing characters:
> Size: Large.
> Upperside: Bright orange with black veins, black margin with two rows of pale spots.
> Underside: As above, hind wing a lighter orange.

Male—Monhegan, ME, 10 August 2019 (Bryan Pfeiffer)

Similar species: Viceroy.

Status and distribution in Maine: Frequent colonist. Numbers vary annually from common to uncommon; widespread, most common in southern and central regions. State Special Concern. S3S4B

Status and distribution in Maritimes: Frequent colonist. Numbers vary annually; some years common in southern New Brunswick and southern Nova Scotia; generally rare on Prince Edward Island and Cape Breton. Special Concern in Canada; Special Concern in New Brunswick; Endangered in Nova Scotia. NB-S3B, NS-S2?B, PE-S1B

Habitat: Various open spaces, including old fields, gardens, riparian meadows, roadsides, and utility corridors.

Biology: Most adults of the late summer and autumn generations migrate to central Mexico, where they overwinter in a small number of massive roosts in high elevation Oyamel Fir (*Abies religiosa*) forests. In late winter, adults in Mexico mate before a partial northward migration. The progeny of overwintering individuals arrive in the Acadian region each year. The species usually arrives in late June and has two to four overlapping generations depending on the arrival of

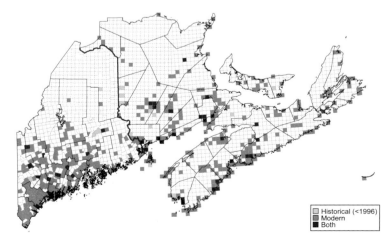

Historical (<1996)
Modern
Both

spring migrants; most vacate by October. Larvae feed on milkweeds (*Asclepias* spp.). There is no evidence of overwintering in our region.

Adult behavior: The Monarch is a strong flyer with a distinctive gliding flight pattern. They are often seen flying purposefully through an area, or cruising meadows where milkweed and nectar sources are available. Monarchs nectar at a variety of native and cultivated flowers in our region, including goldenrods (*Solidago* spp.), American-asters (*Symphyotrichum* spp.), knapweeds (*Centaurea* spp.), coneflowers (*Echinacea* spp.), thistles (*Cirsium* spp.), and milkweeds (*Asclepias* spp.).

Comments: The Monarch is likely the most famous butterfly in the world, largely because of its spectacular multigenerational migration of the eastern North American population between breeding sites in the United States and Canada and a narrow band of overwintering forest in Mexico. This event is especially noteworthy in the Acadian region, where one-way migratory distances (for example from Cape Breton, Nova Scotia) can exceed 5800 km (3600 mi), rivaling those documented anywhere else on the continent. However, this migratory population is imperiled. The size of overwintering colonies declined by over 80% between the mid-1990s and mid-2010s (Semmens et al. 2016). It has been more stable since but remains much reduced (Monarch Watch 2022). This decline recently prompted the U.S. Fish and Wildlife Service to list the Monarch as a candidate for protection under the Endangered Species Act (USFWS 2020). The butterfly has been listed as Special Concern under Canada's Species at Risk Act since 2003, and in 2016 the Committee on the Status of Endangered Wildlife in Canada (COSEWIC) recommended that it be listed as Endangered. Most recently, in 2022, the International Union for Conservation of Nature (IUCN) classified migratory populations of the Monarch as Endangered in its Red List of Threatened Species.

Threats: One of the most significant threats to the Monarch's eastern North American population is the incremental loss and fragmentation of its montane forest overwintering sites in central Mexico due to illegal logging and climate change (COSEWIC 2016). Degradation of milkweed habitat throughout the Monarch's breeding range, from development and agricultural intensification (particularly in the midwestern United States), has likely also contributed to population declines (Brower et al. 2012; but see Crossley et al. 2022).

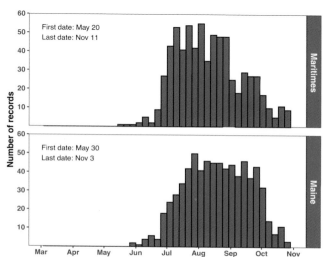

VARIEGATED FRITILLARY

Euptoieta claudia (Cramer)

Subspecies: None.

Distinguishing characters:

Size: Medium.

Upperside: Tawny orange with black venation, bands, and submarginal spots.

Underside: Forewing orange basally; otherwise both wings mottled light brown, hindwing with pale central and marginal

Queens County, NB, 5 August 2015 (John Klymko)

band, and submarginal row of indistinct eyespots.

Additional: Outside margin of forewing concave.

Similar species: Other lesser fritillaries (*Boloria* spp.).

Status and distribution in Maine: Rare colonist, not recorded in most years. Most frequent in southern and coastal regions. SNA

Status and distribution in Maritimes: Rare colonist in New Brunswick, where breeding has been confirmed once. Rare stray in Nova Scotia. NB-SNA, NS-SNA

Habitat: Various open spaces, including old fields, roadsides, utility corridors, and gardens. Adults are particularly fond of areas with low vegetation interspersed with patches of bare earth.

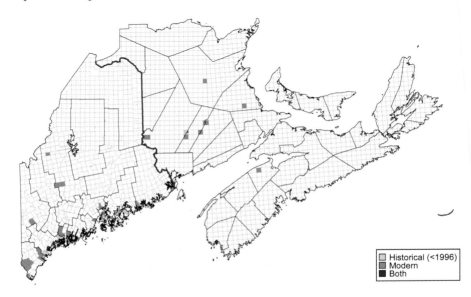

Historical (<1996)
Modern
Both

Biology: This species is a migrant that arrives as early as late May, though most records are from mid-July to late October. Multiple generations are possible when adults arrive early, but the frequency of local breeding is unknown. Important larval food plants farther south are passionflowers (*Passiflora* spp.), plantains (*Plantago* spp.), flaxes (*Linum* spp.), and violets (*Viola* spp.). Only violets are known to be fed upon in our region. There is no evidence of overwintering in the Acadian region, but a longstanding overwintering colony has been recorded near Quebec City (Layberry et al. 1998; Handfield 2011).

Adult behavior: Like most migrants, the Variegated Fritillary is a strong flyer. It feeds from a variety of native and cultivated flowers, including American-asters (*Symphyotrichum* spp.), coneflowers (*Echinacea* spp.), milkweeds (*Asclepias* spp.), hawkweeds (*Hieracium* spp.), goldenrods (*Solidago* spp.), and Carolina Sea-lavender (*Limonium carolinianum*). Males occasionally puddle.

Comments: In our region, this species has been recorded in numbers only in 2011, when there were twelve reports, all in Maine. There are typically no more than one or two records per year. The paucity of observations during most years is probably partly due to the skittish behavior of adults, which can be very difficult to approach. In fact, the genus name, *Euptoieta,* may be derived from a Greek word meaning "easily scared" (Opler and Krizek 1984), though an alternative interpretation suggests that it means "truly terrifying," in reference to the species name *claudia*, which was named after infamous emperor Claudius I (Guppy and Shepard 2001).

Threats: None documented in our region.

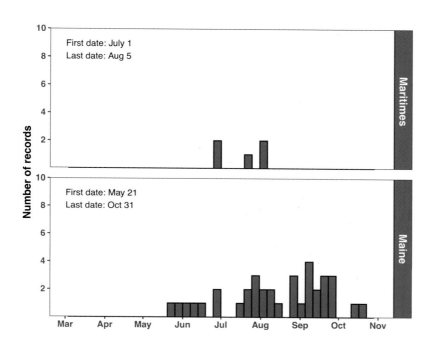

BOG FRITILLARY

Boloria eunomia (Esper)

Subspecies: The only subspecies in our region is *B. e. dawsoni* (W. Barnes & McDunnough).

Distinguishing characters:
  Size: Small to medium.
  Upperside: Dark orange with complex black markings and a submarginal row of black spots.
  Underside: Hindwing bright orange; central row of conjoined, irregularly shaped white spots; postmedian row of black-bordered, pearly-white spots; marginal row of triangular white spots.

Magalloway Plantation, ME, 17 June 2019 (Bryan Pfeiffer)

Similar species: Other lesser fritillaries (*Boloria* spp.).

Status and distribution in Maine: Resident. Uncommon and extremely localized; widespread in north. S3

Status and distribution in Maritimes: Resident. Uncommon and extremely localized; widespread in northern New Brunswick. NB-S3

Habitat: Bogs; typically those with wet open areas of reduced heath cover dominated by sphagnum moss and sedges.

Biology: One generation per year, with adults on the wing from early June to early July. Activity is usually limited to 2 weeks or less at any given locality. A

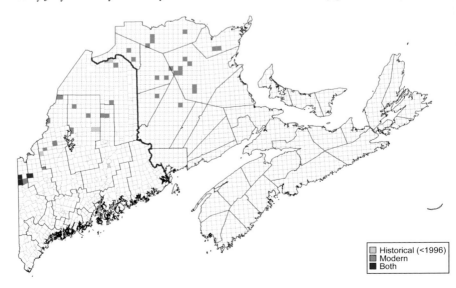

Historical (<1996)
Modern
Both

variety of larval food plants have been reported, including willows (*Salix* spp.), violets (*Viola* spp.), Small Cranberry (*Vaccinium oxycoccos*) and Creeping Spicy-wintergreen (*Gaultheria hispidula*) (e.g., Layberry et al. 1998). Small Cranberry is likely the primary food plant in the Acadian region, as it is common in occupied bogs, and females in Maine have been observed ovipositing on it. Bog-rosemary (*Andromeda polifolia*) may also be used, as butterflies from Maine have been reared on this plant in captivity (R. Boscoe, pers. comm. to J. Calhoun). Overwinters as a partially grown larva.

Adult behavior: This species has a strong, meandering flight. It is often common near the perimeter of bogs, where males patrol for females over low vegetation. Adults remain active until nearly sunset on warm evenings. The most frequented nectar source is Labrador-tea (*Rhododendron groenlandicum*).

Comments: Although it has been known to occur in Maine since 1937 (Klots 1939), the Bog Fritillary was not recorded in New Brunswick until 1999. This is a testament to how difficult this species is to detect, owing to its localized nature and short flight period. Nonetheless, it can be abundant where found. Emergence is typically explosive, followed by a sharp decline in numbers after the first week. Targeted surveys, both professional and amateur, have produced most of the region's records.

Threats: The Acadian region is at the very southern edge of the Bog Fritillary's range in eastern North America, making local populations especially sensitive to global climate change. Peat harvesting is a potential threat to the large bogs of northern New Brunswick. Fortunately, many bogs known to harbor colonies of this butterfly are too small to harvest.

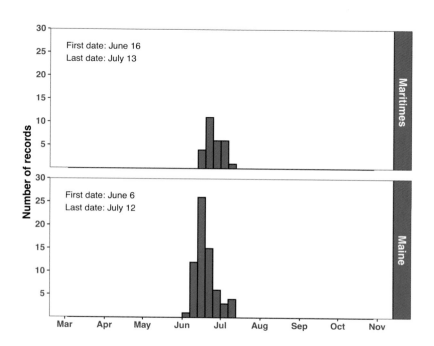

ARCTIC FRITILLARY
(PURPLE LESSER FRITILLARY)
Boloria chariclea (Schneider)

Subspecies: The only subspecies in our region is *B. c. grandis* (W. Barnes & McDunnough).

Distinguishing characters:
  Size: Small to medium.
  Upperside: Dark orange with complex black markings, some dark scaling basally.
  Underside: Hindwing boldly patterned with abundant purplish brown scaling; black and white spots with marginal band of small, elongate white marks, each capped with inward-pointing black triangle.

Big Twenty Township, ME, 18 July 2019 (Bryan Pfeiffer)

Similar species: Other lesser fritillaries (*Boloria* spp.).

Status and distribution in Maine: Resident. Rare; only in the north. State Threatened. S1

Status and distribution in Maritimes: Resident. Uncommon in northern New Brunswick and Cape Breton Island; known from historical records in southeastern New Brunswick and northwestern Nova Scotia. Brower (1974) mentions an historical record from Charlotte County, New Brunswick. NB-S3, NS-S1S2

Habitat: Open boreal coniferous woodlands with ericaceous understory, in both dry upland and wet boggy sites; sometimes found in powerline rights-of-way and other openings within coniferous forest.

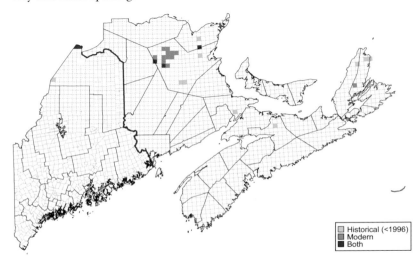

☐ Historical (<1996)
☐ Modern
■ Both

Biology: One generation per year. Adults fly from early to late July in Maine and late July to late August in the Maritimes. The food plant used in our region has not been determined, but elsewhere larvae reportedly feed on willows (*Salix* spp.), knotweeds (*Polygonum* spp.), violets (*Viola* spp.), and blueberries (*Vaccinium* spp.). Overwinters as a newly hatched larva, which may require 2 years to develop.

Adult behavior: Males patrol the understory shrub layer with a low, rapid flight. Adults frequently visit flowers, including Sheep-laurel (*Kalmia angustifolia*), dogbanes (*Apocynum* spp.), goldenrods (*Solidago* spp.), meadowsweets (*Spiraea* spp.), Spotted Hawkweed (*Hieracium maculatum*), and Pearly Everlasting (*Anaphalis margaritacea*).

Comments: This butterfly occurs in localized colonies but can be quite common in suitable habitat. Populations currently recognized as *B. chariclea* were previously considered to represent two different species: *B. chariclea* (confined to tundra) and *B. titania* (Esper) (confined to boreal forests). The name *titania* is now generally restricted to Palearctic populations, while the name *chariclea* is applied to all those in North America (Shepard 1998), though more than one species may be involved in the New World. The flight period in Maine is markedly different from that in the Maritimes, possibly indicating the occurrence of two discrete species in the region. More research is needed.

Threats: Acadian populations are located mostly in boreal woodlands at the southern limits of this species' range, conceivably making them more susceptible to the impacts of climate change (Tang and Beckage 2010). Our southernmost records are historical, suggesting that some range contraction has already occurred. In addition, the unique open, mature woodlands preferred by this species may be lost if intensively logged and replanted to create dense, young forests.

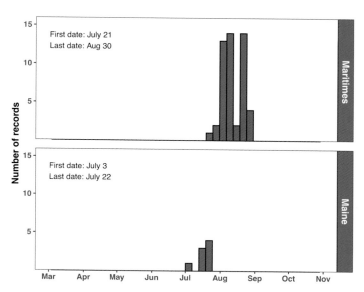

MEADOW FRITILLARY

Boloria bellona (Fabricius)

Subspecies: The nominotypical subspecies (*B. b. bellona*) occurs in our region and most of eastern North America (see comments).

Norwich, VT, 4 July 2020 (Bryan Pfeiffer)

Distinguishing characters:
  Size: Small to medium.
  Upperside: Orange with complex black markings and a submarginal row of black spots; margin mostly orange.
  Underside: Hindwing mottled orange brown on basal half, apical half purple brown with a frosted cast.
  Additional: Forewing squared at apex; long palps.

Similar species: Other lesser fritillaries (*Boloria* spp.), especially Frigga Fritillary.

Status and distribution in Maine: Resident. Common to uncommon in the south, rare in the north; widespread. S3S4

Status and distribution in Maritimes: Resident. Uncommon in southwestern and central New Brunswick, rare in northern New Brunswick. NB-S3

Habitat: A variety of open, often moist habitats, including pasture, road margins, old fields, powerline rights-of-way, and sometimes peatlands.

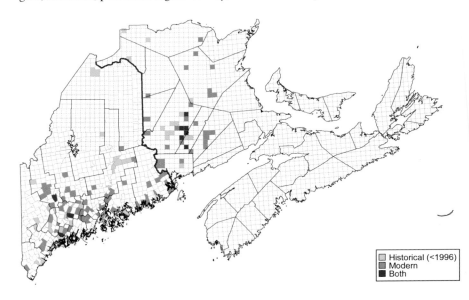

Historical (<1996)
Modern
Both

Biology: Three generations per year in southern Maine, likely two generations with a partial third brood in northern Maine and New Brunswick. Adults fly from early May until mid-September, with peaks in abundance in late May and early June, July, and late August. Larvae feed on various violets (*Viola* spp.). Overwinters as a partially grown larva.

Adult behavior: This fritillary has a fast, erratic flight. Males patrol throughout the day for females. Adults visit flowers, including coneflowers (*Rudbeckia* spp.), hawkweeds (*Hieracium* spp.), bluets (*Houstonia* spp.), blueberries (*Vaccinium* spp.) and many other open field species. Males occasionally visit damp soil and decaying matter.

Comments: Of our lesser fritillaries, the Meadow Fritillary has been most successful in exploiting disturbed habitats (such as pastures and old fields) and has expanded at the southern edge of its range in Missouri, Kentucky, and Virginia (Scott 1986). No apparent expansion has occurred in our region. It has not colonized vast areas of the Maritimes, where seemingly suitable habitat exists, and there are similar gaps in distribution across southern and northeastern Maine. Some butterflies in Maine and New Brunswick resemble the darker *B. b. toddi* (W. Holland), but that subspecies is generally believed to occur north and west of our region.

Threats: This fritillary has experienced unexplained declines in Connecticut and Massachusetts (O'Donnell et al. 2007; Stichter 2015). In our region, it is likely that some colonies have been lost as abandoned farmland has reverted to forest.

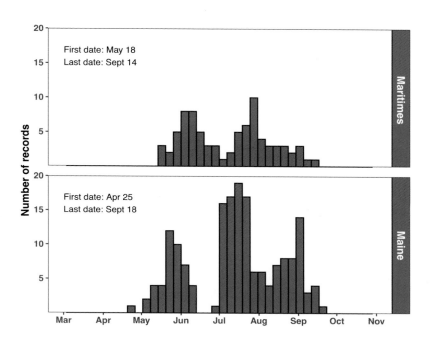

FRIGGA FRITILLARY

Boloria frigga (Thunberg)

Subspecies: Only the subspecies *B. f. saga* (Staudinger)
occurs in northeastern North
America.

Distinguishing characters:
  Size: Small to medium.
  Upperside: Orange with
    black markings and
    considerable dark scaling
    toward base of the wings.
  Underside: Hindwing with
    frosted pale violet band

Northeast Carry Township, ME, 15 June 2018 (Bryan Pfeiffer)

along the margin and prominent basal white patch along the leading
edge.

Similar species: Meadow Fritillary.

Status and distribution in Maine: Resident. Rare; only known from one
locality. State Endangered. S1

Status and distribution in Maritimes: Not documented.

Habitat: Sphagnum bog margins with groundwater influence.

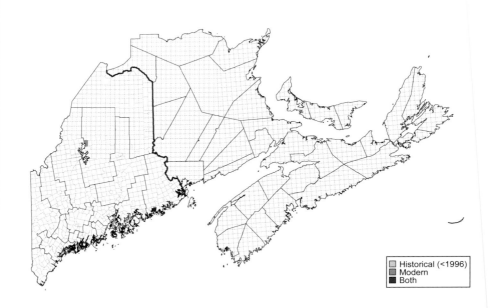

	Historical (<1996)
	Modern
	Both

Biology: One generation with a peak flight generally in mid-June, usually lasting less than 2 weeks. Only Bog Willow (*Salix pedicellaris*) is confirmed as a larval food plant in Maine, though Bog Birch (*Betula pumila*) should be considered, as it is widespread in our region's northern peatlands and is an alternate host elsewhere. Overwinters as a late-instar larva.

Adult behavior: This species flies rapidly near the ground in lightly forested margins of the bog where Bog Willow is abundant. Nectaring has been observed on Three-leaved False Solomon's-seal (*Maianthemum trifolium*) and several bog shrubs, including Labrador Tea (*Rhododendron groenlandicum*), Rhodora (*R. canadense*), Black Chokeberry (*Aronia melanocarpa*), Bog Laurel (*Kalmia polifolia*), and Bog-rosemary (*Andromeda polifolia*).

Comments: Until recently, the southeastern most documented locality for the Frigga Fritillary was in central Quebec. The collection of a single very worn female on 24 June 2002 by Phillip deMaynadier and Beth Swartz provided a Maine state record and a significant range extension for northeastern North America (deMaynadier and Webster 2009). After two decades with no additional breeding colonies discovered, it is possible that only a few glacial relict populations of this species persist within the Acadian region.

Threats: Restricted to just one or a few isolated populations, this species is vulnerable to stochastic disturbances, such as wildfire, disease, parasitism, and drought. At the extreme southern edge of its eastern range in our region, the Frigga Fritillary is also potentially threatened in Maine by global climate change.

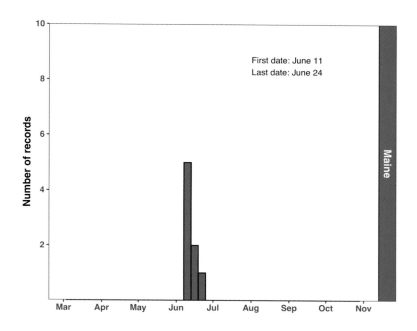

First date: June 11
Last date: June 24

SILVER-BORDERED FRITILLARY

Boloria selene (Denis & Schiffermüller)

Subspecies: Two poorly differentiated subspecies occur in our region: *B. s. myrina* (Cramer) in southern Maine and *B. s. atrocostalis* (Huard) elsewhere. A broad area of overlap occurs across central Maine.

Distinguishing characters:

Buxton, ME, 27 August 2019 (Bryan Pfeiffer)

Size: Small to medium.

Upperside: Orange, marked with black, with a submarginal row of black spots.

Underside: Hindwing brick orange with yellow patches, series of bold white spots with silvery reflections, a single white-rimmed black spot near the base, and a submarginal row of black spots.

Additional: Subspecies *atrocostalis* has a broader dark border and more extensive dark basal scaling above.

Similar species: Upperside resembles other lesser fritillaries (*Boloria* spp.); underside similar to the greater fritillaries (*Speyeria* spp.).

Status and distribution in Maine: Resident. Common; widespread. S5

Status and distribution in Maritimes: Resident. Common; widespread. NB-S5, NS-S5, PE-S5

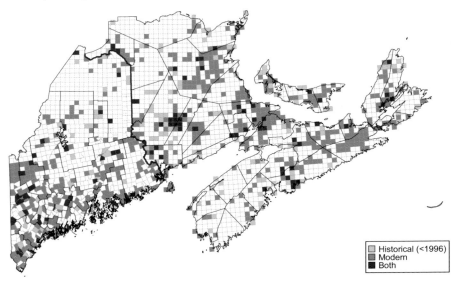

Historical (<1996)
Modern
Both

Habitat: A variety of open, moist habitats, including marshes, bogs, wet meadows, and roadside ditches.

Biology: Two overlapping generations per year, with the first beginning in mid to late May (peaking in mid-June) and the second beginning in late July (peaking in August). Larvae feed on wetland violets (*Viola* spp.). Overwinters as a partially grown larva.

Adult behavior: A strong flyer, this species is rarely seen far from its breeding habitat. Adults frequently nectar, visiting over fifty plant species in our region, including hawkweeds (*Hieracium* spp.) and buttercups (*Ranunculus* spp.) early in the season and goldenrods (*Solidago* spp.), red clover (*Trifolium pratense*), and American-asters (*Symphyotrichum* spp.) later in the season.

Comments: Major declines have been noted in many places at the edge of this species' range, including New Jersey and Ohio (Iftner et al. 1992; Gochfeld and Burger 1997). In Maryland a single-brooded race may be extirpated (Cech and Tudor 2005). Such declines are mostly due to loss of habitat in areas where the species is more localized in distribution. No apparent declines have been documented in the Acadian region, and the species frequently colonizes wet meadow habitat created by forest harvesting. New World *myrina* is probably a different species from Old World *selene*.

Threats: Loss of wetland and riparian habitat has been identified as a threat elsewhere in the species' range, but it is not yet considered a serious problem for this species in our region.

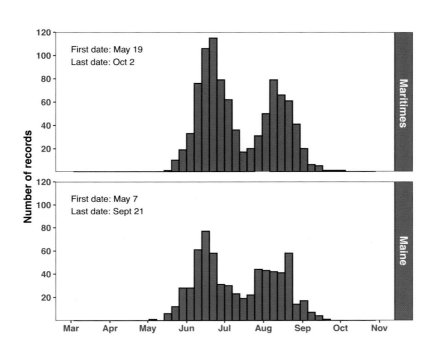

REGAL FRITILLARY

Argynnis idalia (Drury)

Subspecies: Only the nomino-
typical subspecies (*A. i. idalia*)
occurs in northeastern North
America.

Distinguishing characters:
    Size: Large.
    Upperside: Forewing
        orange with black spots
        (male) or black and white
        spots (female); most of
        hindwing heavily suf-
        fused with bluish black,

Fort Indian Gap, PA, 13 July 2018 (Steve Collins)

with orange and white spots (male) or all white spots (female).
    Underside: Forewing orange with black and white spots; hindwing dark
        brown with large silver spots.

Similar species: Other greater fritillaries (*Argynnis* spp.), but the Regal Fritillary is distinctive.

Status and distribution in Maine: Extirpated; only known from historical
records scattered in the southern and central parts of the state. SX

Status and distribution in Maritimes: Stray; only one historical record from
southern New Brunswick. NB-SNA

Habitat: Large, open grassy areas with an abundance of nectar, including wet
meadows, old fields, and weedy pastures. Occurs more commonly in tall grass
prairie westward.

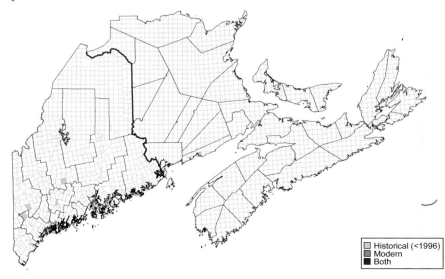

☐ Historical (<1996)
▨ Modern
■ Both

Biology: Formerly one summer generation. Maine records are from early July to early September. Males typically emerge about 1 week before females, and some females can be found quite late in the season. Eggs are laid near violets (*Viola* spp.) in late summer, and newly hatched larvae overwinter without feeding, until the following spring when larvae forage on violets.

Adult behavior: Regal Fritillaries are rapid flyers, with males often patrolling circuitous routes just above the vegetation in search of females. They nectar often, especially on Common Milkweed (*Asclepias syriaca*), Swamp Milkweed (*A. incarnata*), and thistles (*Cirsium* spp.) elsewhere in their range.

Comments: Once described in Maine as locally "somewhat plentiful" and "not very common" (Lyman 1876; Scudder 1888–1889), this striking butterfly has not been seen in the state since 1941. It was recorded in New Brunswick only in 1880, when "quite a number" were found in the vicinity of Saint John (Fletcher 1897). Last seen in New England in 1991 (Wagner et al. 1995), the Regal Fritillary disappeared from virtually all of its former eastern range by the early 1990s. Today, only a single isolated population, in south-central Pennsylvania, is believed to survive in the East. It represents the last population of the nominotypical subspecies.

Threats: The reasons for the disappearance of this species in the East are poorly understood, but they probably include habitat loss and fragmentation, natural succession, changes in agricultural practices, pesticide drift, and the spread of parasitoids and/or pathogens (Wagner et al. 1995). Attempted re-introductions in Massachusetts have been unsuccessful due to complexities of the species' biology (Stichter 2015).

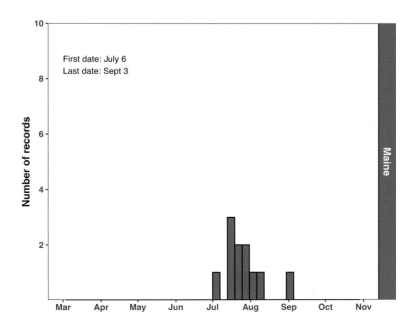

First date: July 6
Last date: Sept 3

GREAT SPANGLED FRITILLARY

Argynnis cybele (Fabricius)

Falmouth, ME, 6 August 2018 (Josh Lincoln)

Subspecies: Two subspecies occur in our region, *A. c. cybele* and *A. c. novascotiae* McDunnough (see comments).

Distinguishing characters:
   Size: Large.
   Upperside: Bright orange or yellowish orange with complex black markings; darker basally, especially in female; forewing characteristically without basal spot (rarely present; compare Aphrodite and Atlantis fritillaries).
   Underside: Hindwing orange brown with bold silvery spots and a wide, pale submarginal band.
   Additional: Eyes amber brown

Similar species: Atlantis Fritillary and Aphrodite Fritillary.

Status and distribution in Maine: Resident. Common; widespread. S5

Status and distribution in Maritimes: Resident. Common; widespread. NB-S5, NS-S5, PE-S5

Habitat: A variety of open areas, including old fields, gardens, roadsides, and riparian meadows.

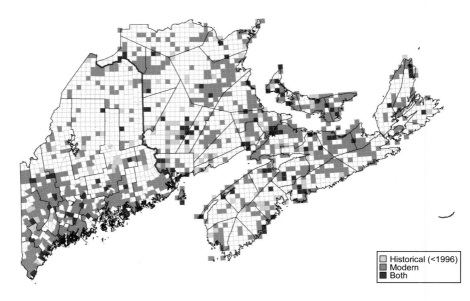

☐ Historical (<1996)
☐ Modern
■ Both

286

Biology: One generation per year. Adults are on the wing from mid-June through September, with a peak flight in late July. Males emerge before females (in Maine, the earliest female record is 25 June). Females oviposit in late summer on or near violets (*Viola* spp.), the larval food plants. Overwinters as a newly hatched larva.

Adult behavior: This species is a strong flyer, and males patrol throughout the day for females. Adults visit a wide variety of native and ornamental flowers, particularly those with large or multiple flower heads, including milkweeds (*Asclepias* spp.), Orange-eye Butterfly-bush (*Buddleja davidii*), coneflowers (*Echinacea* spp.), Spotted Joe-Pye Weed (*Eutrochium maculatum*), goldenrods (*Solidago* spp.), and thistles (*Cirsium* spp.). Adults also feed at moist soil and dung.

Comments: Males of the Great Spangled Fritillary possess numerous specialized pheromone-producing scales (androconia) along the forewing veins, making them appear swollen. During courtship, males perch next to females and extend their forewings while flapping them to waft pheromones over the female's antennae (Scott 1986). Characters used to define the subspecies *A. c. novascotiae* (described from Queen's County, Nova Scotia) include smaller size and narrower pale band on the ventral hindwing. Butterflies attributed to this subspecies, which are generally darker and paler orange than *A. c. cybele*, are mostly limited to the Maritimes, though they also occur in northern and eastern coastal Maine, with intermediates elsewhere.

Threats: None documented in our region.

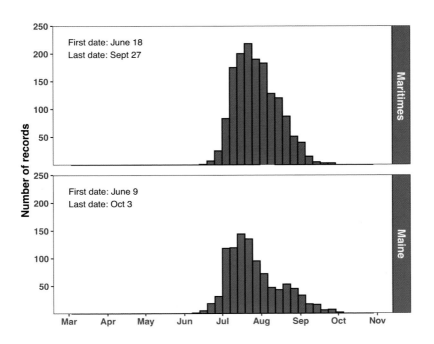

First date: June 18
Last date: Sept 27

Maritimes

First date: June 9
Last date: Oct 3

Maine

Number of records

Mar   Apr   May   Jun   Jul   Aug   Sep   Oct   Nov

APHRODITE FRITILLARY

Argynnis aphrodite (Fabricius)

Subspecies: Two subspecies occur in our region: *A. a. aphrodite* in southern Maine, and *A. a. winni* Gunder elsewhere, with a broad area of intergradation.

Distinguishing characters:
  Size: Medium to large.
  Upperside: Bright orange with complex black markings; forewing with basal spot; forewing mar-

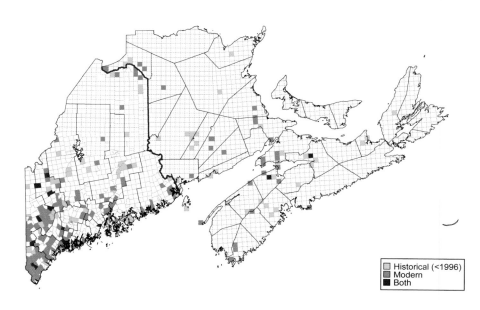

Camden, ME, 30 July 2017 (Roger Rittmaster)

gin mostly orange in subspecies *aphrodite* and mostly dark in subspecies *winni* (though many appear intermediate); male forewing veins without heavy black scaling.
  Underside: Hindwing reddish brown (redder than Atlantis Fritillary) with bold silvery spots and a narrow, pale submarginal band.
  Additional: Eyes amber brown.

Similar species: Atlantis Fritillary and Great Spangled Fritillary.

Status and distribution in Maine: Resident. Common in the southwest, uncommon elsewhere; widespread. S4

Status and distribution in Maritimes: Resident. Widespread but uncommon in New Brunswick and Nova Scotia. NB-S3, NS-S3S4

☐ Historical (<1996)
◻ Modern
◼ Both

Habitat: A variety of open areas, including old fields, gardens, roadsides, and riparian meadows.

Biology: One generation per year. Adults are on the wing from late June through September, with a peak flight in mid to late July. Typical of greater fritillaries, males emerge before females. Larvae feed on violets (*Viola* spp.) Overwinters as a newly hatched larva.

Adult behavior: This species is a strong flyer, and males patrol throughout the day for females. Adults frequent a wide variety of flowers, particularly those with larger flower heads, such as milkweeds (*Asclepias* spp.), goldenrods (*Solidago* spp.), and coneflowers (*Echinacea* spp.).

Comments: The Aphrodite Fritillary is very similar to the Great Spangled and Atlantis fritillaries in both appearance and habits. The three can be seen together, but the Aphrodite is usually the least abundant. Many historical records are based on misidentified specimens, usually those of the Great Spangled Fritillary. The Aphrodite Fritillary has been reported from Prince Edward Island (e.g., Layberry et al. 1998), but we could find no verifiable records from the province. Noticeably smaller than *A. a. aphrodite*, the subspecies *A. a. winni* is often confused with the Atlantis Fritillary in northern Maine and the Maritimes.

Threats: Some colonies in old fields and pastures have been lost to woody succession.

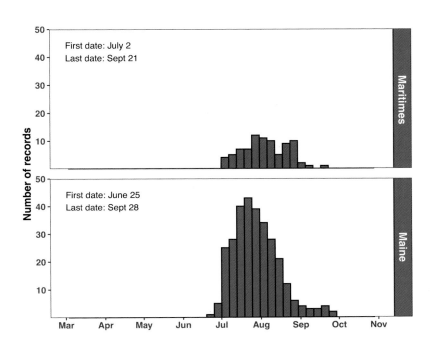

ATLANTIS FRITILLARY

Argynnis atlantis
W. H. Edwards

Subspecies: Only the nomino-typical subspecies (*A. a. atlantis*) occurs in our region.

Distinguishing characters:
  Size: Medium to large.
  Upperside: Bright orange with complex black markings; forewing with basal spot, margins mostly dark; male fore-wing veins with heavy black scaling.

Stockton Springs, ME, 11 July 2014 (Roger Rittmaster)

  Underside: Hindwing reddish brown with bold silvery spots and a narrow, pale submarginal band.
  Additional: Eyes gray blue in living individuals.

Similar species: Aphrodite Fritillary and Great Spangled Fritillary.

Status and distribution in Maine: Resident. Uncommon in extreme south, common elsewhere; widespread. S5

Status and distribution in Maritimes: Resident. Common and widespread. NB-S5, NS-S5, PE-S5

Habitat: A variety of open areas, including old fields, gardens, roadsides, riparian meadows, and bogs.

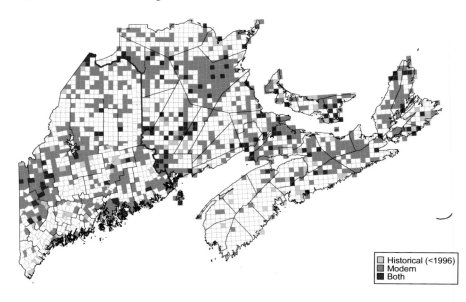

Historical (<1996)
Modern
Both

Biology: One generation per year. Adults are on the wing from June through September, with a peak flight in early July. As with other greater fritillaries, males emerge before females. Larvae feed on violets (*Viola* spp.). Overwinters as a newly hatched larva.

Adult behavior: Like other greater fritillaries this species is a strong flyer, and males patrol throughout the day for females. They visit a wide variety of flowers, particularly those with larger flower heads, such as Common Milkweed (*Asclepias syriaca*), goldenrods (*Solidago* spp.), thistles (*Cirsium* spp.), Orange-eye Butterfly Bush (*Buddleja davidii*), and Spotted Joe-Pye Weed (*Eutrochium maculatum*). Adults also visit damp earth, dung, and decaying matter. One was even photographed at a hummingbird feeder.

Comments: Uncommon in Maine's southernmost counties, this species is the most boreal of the greater fritillaries in our region. Among this group, it is also the most likely to be found in peatlands. Its smaller size, more pointed forewing apex, darker forewing margins, and gray-blue eyes help separate this species in the field from the Great Spangled and Aphrodite fritillaries. Unfortunately, its striking eyes turn brown after death. A darkly suffused aberration of this species, aberration named "chemo," was described from a female specimen collected in 1887 at "Lake Chemo" (Chemo Pond), Penobscot County, Maine.

Threats: None documented in our region.

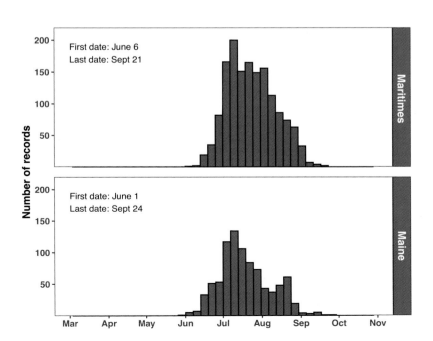

Number of records

First date: June 6
Last date: Sept 21

Maritimes

First date: June 1
Last date: Sept 24

Maine

Mar   Apr   May   Jun   Jul   Aug   Sep   Oct   Nov

VICEROY

Limenitis archippus (Cramer)

Subspecies: Only the nominotypical subspecies (*L. a. archippus*) occurs in northeastern North America.

Distinguishing characters:

Size: Medium to large.

Upperside: Tawny orange to brownish orange with black veins; thick black margin with submarginal row of white spots; hindwing with thin black transverse band (sometimes narrow or partial).

Camden, ME, 31 July 2013 (Roger Rittmaster)

Underside: Like upperside except forewing tip and hindwing lighter orange; white submarginal spots larger.

Similar species: Monarch.

Status and distribution in Maine: Resident. Common; widespread. S5

Status and distribution in Maritimes: Resident. Rare in southwestern Nova Scotia and on Cape Breton Island, common elsewhere; widespread. NB-S5, NS-S5, PE-S4

Habitat: Marshes, riparian meadows, roadsides, forest edges, and openings in wet forest.

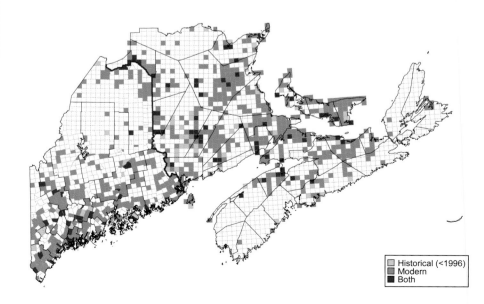

☐ Historical (<1996)
☐ Modern
■ Both

Biology: Two generations per year. Adults are on the wing from late May until late September, with peak flights in June and August. Larvae feed on willows (*Salix* spp.) and poplars (*Populus* spp.). Overwinters as a partially grown larva within a hibernaculum made of a rolled leaf.

Adult behavior: This species has a flap-and-glide flight pattern like that of the Monarch, though the wings are held flat in the glide (versus an open "V" in the Monarch). Adults frequently perch on low vegetation and nectar at a variety of flowers, such as goldenrods (*Solidago* spp.), Wild Carrot (*Daucus carota*), and White Meadowsweet (*Spiraea alba*). They also are attracted to damp earth, dung, carrion, and rotting fruit.

Comments: It was long believed that the Viceroy was a Batesian mimic, a palatable species that gains protection from predators by resembling an unpalatable look-alike, in this case the Monarch. However, at least some populations of the Viceroy are as unpalatable as Monarchs (or nearly so), a relationship that better fits a Müllerian mimicry model, wherein two or more unpalatable species resemble one another and mutually benefit from their similarity (Ritland and Brower 1991; Ritland 1995). It is not known if the Viceroy is unpalatable in our region, but this would make sense in places like Prince Edward Island and Cape Breton, where Monarchs are rare during most years. There are at least two historical records in Maine of rare hybrids between the Viceroy and the White Admiral (form *arthechippus*) (Platt et al. 1978). More recently, one was captured in Knox County, Maine, on 13 August 2020.

Threats: None documented in our region.

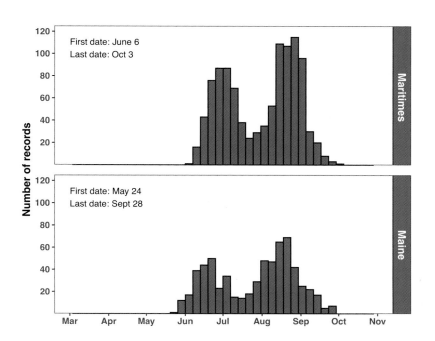

WHITE ADMIRAL

Limenitis arthemis (Drury)

Subspecies: Only the nomino-typical subspecies (*L. a. arthemis*) occurs in our region (see comments).

Distinguishing characters:
    Size: Large.
    Upperside: Black with bold white band; margin with blue dashes.
    Underside: Dark brown with basal red spots and a bold white median band; with series of red submarginal spots and blue marginal dashes.

Lincoln Plantation, ME, 21 June 2018 (Bryan Pfeiffer)

    Additional: Some individuals lack a white band or have a partial band. Known as form *proserpina*, they closely resemble the subspecies commonly called the Red-spotted Purple, *L. a. astyanax* (Fabricius).

Similar species: None.

Status and distribution in Maine: Resident. Common; widespread. S5

Status and distribution in Maritimes: Resident. Common; widespread. NB-S5, NS-S5, PE-S5

Habitat: Forest edges and openings, especially along trails, roadsides, and utility corridors.

Biology: Two generations per year. The second brood is only partial, except perhaps in southern Maine. Adults fly from late May or early June through

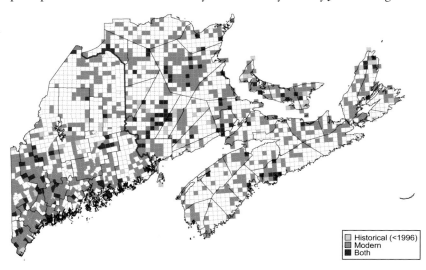

Historical (<1996)
Modern
Both

September, with peak flights in early July and late August. Larvae feed on a wide variety of trees and shrubs, including willows (*Salix* spp.), poplars (*Populus* spp.), and birches (*Betula* spp.). Overwinters as a partially grown larva within a hibernaculum made of a rolled leaf.

Adult behavior: Adults of this species are strong flyers, but they often perch on bare ground or on low vegetation. They nectar at a wide diversity of flowers, such as milkweeds (*Asclepias* spp.), Wild Carrot (*Daucus carota*), Spreading Dogbane (*Apocynum androsaemifolium*), Ox-eye Daisy (*Leucanthemum vulgare*), and White Meadowsweet (*Spiraea alba*). White Admirals also feed at damp earth and a wide variety of other nonfloral nutrient sources, including the following impressive list from within our region: mammal and bird dung, sap, rotting fruit, Yellow-bellied Sapsucker (*Sphyrapicus varius*) holes, hummingbird feeders, fire pit charcoal, a dead frog, a porcupine (*Erethizon dorsatum*) hide, a Common Snapping Turtle (*Chelydra serpentina*) skeleton, and a roadkill Smooth Green Snake (*Opheodrys vernalis*) (shared with a Viceroy).

Comments: Southern Maine is at the northern edge of a blend zone, where the White Admiral (with white bands) transitions into the more southern Red-spotted Purple (without white bands). White Admiral individuals lacking bands (or with reduced bands) occur rarely as far north as central Maine (they can be locally frequent in extreme southern Maine). Although they are similar to the Red-spotted Purple, they have a darker black ground color and lack the shimmering blue on the dorsal hindwing of that subspecies. Some butterflies have been documented in our region that appear to be hybrids between the White Admiral and the Viceroy, something that occurs relatively often within the genus *Limenitis* (Guppy and Shepard 2001) (see Viceroy account).

Threats: None documented in our region.

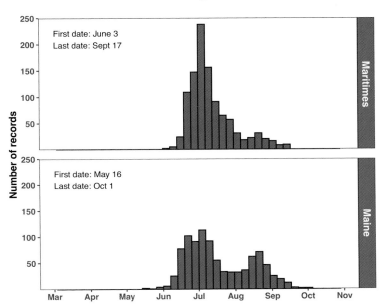

MILBERT'S TORTOISESHELL

Aglais milberti (Godart)

Subspecies: Only the nomino-typical subspecies (*A. m. milberti*) occurs in our region.

Distinguishing characters:
   Size: Medium.
   Upperside: Dark brown with broad submarginal orange band; single white spot at forewing tip.
   Underside: Two toned: dark brown on basal half, tan on apical half.

Rangeley, ME, 26 June 2015 (Roger Rittmaster)

Similar species: None.

Status and distribution in Maine: Resident. Common; widespread. S4

Status and distribution in Maritimes: Resident. Uncommon and widespread in New Brunswick; rare in Nova Scotia and Prince Edward Island. NB-S4, NS-S2S3, PE-S1

Habitat: Forest openings, edges, and old fields. Often associated with wet areas.

Biology: There are contradictory reports of the number of generations that this species produces in the Northeast. One generation is reported in Ontario (Hall et al. 2014), while two are suspected in northern New Hampshire (Kiel 2003) and Vermont (McFarland and Pfeiffer 2022). Up to three generations

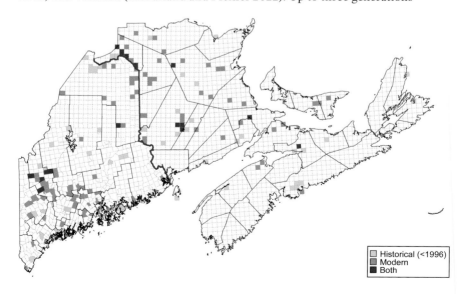

Historical (<1996)
Modern
Both

are reported in Massachusetts (Scudder 1888–1889; Stichter 2015). The flight histograms for Maine and the Maritimes do not provide a clear indication of how many generations are actually produced in this region, but two is likely the maximum, and only one may be typical northward. Adults can be found from early spring to autumn, but they are most common during July and August. Larvae feed on nettles (*Urtica* spp.). The adult is the most often reported overwintering stage, but Scudder (1888–1889) claimed that in New England a proportion also overwinter as pupae.

Adult behavior: This species has a rapid, energetic flight. Feeding mostly on sap, dung, and carrion, adults also visit flowers much more frequently than related species like the Compton Tortoiseshell and the Mourning Cloak. A variety of plants are visited, including Common Milkweed (*Asclepias syriaca*), goldenrods (*Solidago* spp.), and ornamentals like Orange-eye Butterfly-bush (*Buddleja davidii*).

Comments: Usually encountered as one or two individuals, the Milbert's Tortoiseshell occasionally occurs in unusually large numbers. For example, it was reported in 1952 to be abundant at sites in Colchester County, Nova Scotia (Ferguson 1954), though it is generally rare in the province. The lack of recent records from large areas of Nova Scotia for which there are historical records (such as Digby County and Cape Breton) may be the result of inadequate sampling rather than local extirpation. The weakly differentiated subspecies *A. m. viola* (dos Passos), described from Newfoundland, is characterized by having little or no yellow at the base of the orange forewing band. Similar individuals are often found in our region, but they are merely individual variants.

Threats: Potential declines are reported elsewhere in the Northeast, possibly as a result of climate change (Gochfeld and Burger 1997; Stichter 2015).

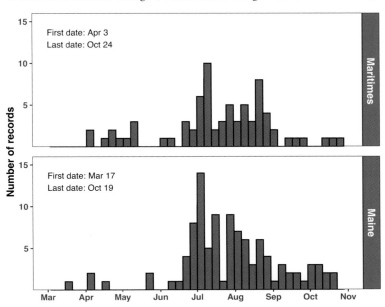

EUROPEAN PEACOCK
Aglais io (Linnaeus)

Subspecies: Only the nomino-
typical subspecies (*A. i. io*)
has been confirmed in North
America.

Distinguishing characters:
    Size: Medium.
    Upperside: Rusty red;
        forewing with bold black,
        blue, and yellow eyespot
        on both wings.

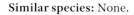

Halifax, NS, 13 August 2014 (Phil Schappert)

    Underside: Dark brown to
    black with fine striations; hindwing with purple sheen.
    Additional: Irregular wing edges.

Similar species: None.

Status and distribution in Maine: Not documented.

Status and distribution in Maritimes: Likely resident. Rare; known only
from Halifax County, Nova Scotia. NS-SNA

Habitat: Mostly disturbed habitats, particularly urban and rural gardens.

Biology: Presumably one generation per year. The handful of records to date
in our region are from August, September, and October. Where established,
adults are active from midsummer until fall and again in the spring. Adults
overwinter in sheltered spots, such as tree hollows and outbuildings, sometimes in groups. Larvae feed primarily on nettles (*Urtica* spp.), though this is
not yet confirmed in our region.

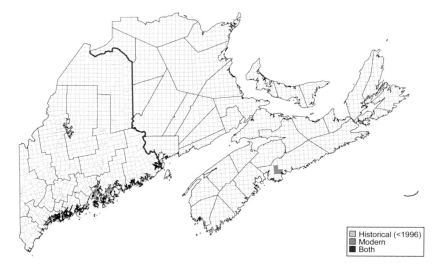

	Historical (<1996)
	Modern
	Both

Adult behavior: Adults are wary and have a rapid flight. They often bask and nectar with wings open. If disturbed while resting with wings closed, they may flash their wing eyespots in defense. Adults have been observed in our region nectaring at New York Aster (*Symphiotrichum novi-belgii*) and several ornamentals, including daylilies (*Hemerocallis* spp.) and Eastern Purple Cone-flower (*Echinacea purpurea*). They are known to visit tree sap and rotten fruit elsewhere.

Comments: The European Peacock was first discovered in our region by Duff and Donna Evers, who photographed one in their flower garden near Halifax, Nova Scotia, in August 2014. This is an Old World species, so it was assumed the butterfly had been imported from Europe, likely as a cargo stowaway. Five more have since been found in the Halifax region, one in 2020 and four in 2021. Such a cluster of records has been documented only at one other location in North America: in and around Montreal, Quebec, where there has been an established population since at least the late 1990s (Handfield 2011). Although there is no definitive evidence that this species is breeding around Halifax, the number of records documented over several years suggests it is locally established. The European Peacock has been reported from a few other localities in eastern Canada and the northeastern United States, as well as from the West Coast of North America in Oregon and British Columbia. The Quebec population is genetically consistent with the widespread European subspecies, *A. i. io* (Nazari et al. 2018). The Nova Scotia butterflies also appear to represent the nominotypical subspecies. This species is sometimes placed in the genera *Inachis* or *Nymphalis*.

Threats: None documented in our region (introduced species).

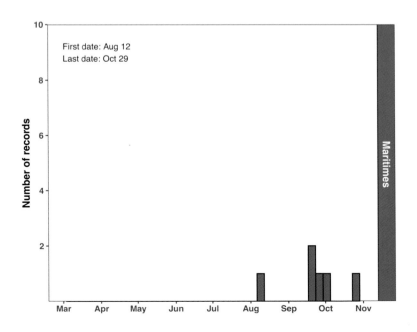

First date: Aug 12
Last date: Oct 29

COMPTON TORTOISESHELL

Nymphalis l-album (Esper)

Subspecies: Only the subspecies *N. l-album j-album* (Boisduval & Le Conte) occurs in eastern North America.

Distinguishing characters:
- **Size:** Medium to large.
- **Upperside:** Brown basally and orange apically, with black and white spotting and golden-brown margin.

Searsmont, ME, 9 April 2017 (Fyn Kynd)

- **Underside:** Dark brown on basal half and light brown on apical half, with intricate striations and a very weak white "comma."

Similar species: *Polygonia* species; in fact, the Compton Tortoiseshell is called the False Comma in Europe.

Status and distribution in Maine: Resident. Uncommon; widespread. S3S4

Status and distribution in Maritimes: Resident. Generally rare; widespread. NB-S3, NS-S2S3, PE-S1

Habitats: Wooded areas and adjacent openings, such as unpaved roads through hardwood and mixed forests.

Biology: One generation per year, with adults emerging during July and August. After a period of activity, some adults enter a state of dormancy

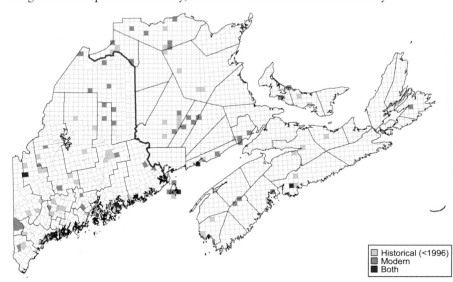

Historical (<1996)
Modern
Both

(aestivation) in a protected location until autumn, when they again become active and fly into October or November. Late in the season, they take refuge in sheltered spots in which to spend the winter, such as under loose bark, in wood piles and hollow logs, and in primitive buildings. They again become active when the weather warms the following spring, but (like the Mourning Cloak) they can fly during midwinter if the temperature exceeds 10 °C (50 °F). Adults of this species can survive up to 8 or 9 months, much longer than most butterflies. Larval food plants are elms (*Ulmus* spp.), birches (*Betula* spp.), willows (*Salix* spp.), and poplars (*Populus* spp.).

Adult behavior: This species is a strong flyer and very difficult to approach. Adults are usually found resting on dirt roads, patrolling trails, or perched in openings with wings outstretched. They are much more frequently seen feeding at dung, carrion, fermented fruit, and sap than at flowers.

Comments: This butterfly is also widely distributed in Europe and Asia. Typically uncommon to rare in our region, its numbers vary from year to year. During occasional "boom" years it can become quite common, and individuals sometimes stray far from their normal range. Although no seasons of exceptional abundance were documented during the atlas periods, the Compton Tortoiseshell was unusually frequent in 2020. This species was long known as *N. vaualbum* (or *N. vau-album*), and some authors still use that name. There is some evidence that North American populations of this butterfly may represent a different species from those in the Old World. If so, our populations would be known by the name *N. j-album*.

Threats: None documented in our region.

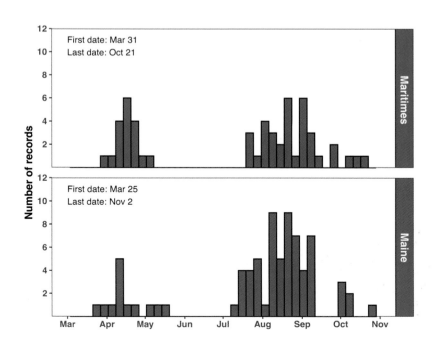

MOURNING CLOAK
Nymphalis antiopa (Linnaeus)

Subspecies: Only the nominotypical subspecies (*N. a. antiopa*) occurs in North America.

Distinguishing characters:
  Size: Large.
  Upperside: Maroon brown with submarginal blue spots and cream-colored margins.
  Underside: Brown with fine striations, margins buff colored.

Chesuncook Township, ME, 30 September 2019 (Phillip deMaynadier)

Similar species: None.

Status and distribution in Maine: Resident. Common; widespread. S5

Status and distribution in Maritimes: Resident. Common and widespread in New Brunswick and Nova Scotia; uncommon and widespread in Prince Edward Island. NB-S5, NS-S5, PE-S3S4

Habitats: Typically associated with forests but can be found in a wide variety of other habitats.

Biology: In our region there is one principal generation and possibly a partial second brood. Often the first butterfly encountered in early spring, overwintering adults are on the wing as early as March, and they are occasionally found flying on warm winter days. Although the species flies from spring to autumn, there is a dip in adult abundance in midsummer before new adults

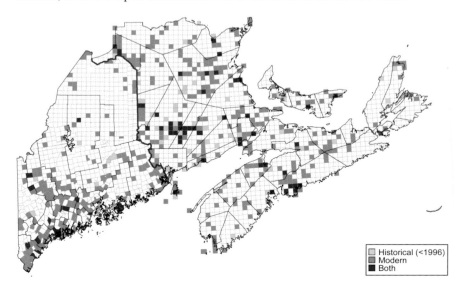

Historical (<1996)
Modern
Both

emerge during mid-July and August. After a period of activity, summer adults enter a state of dormancy (aestivation) until autumn, when they again become active before hibernating during the winter under logs, loose bark, or in other protected locations. Larvae feed on a wide variety of woody species, including willows (*Salix* spp.), elms (*Ulmus* spp.), and poplars (*Populus* spp.).

Adult behavior: This species is a strong flyer, frequently alternating between flapping and gliding. It can be found nectaring at a variety of flowers, including Common Lilac (*Syringa vulgaris*), Common Dandelion (*Taraxacum officinale*), Ox-eye Daisy (*Leucanthemum vulgare*), and goldenrods (*Solidago* spp.), but it is more often seen feeding on sap and dung.

Comments: The Mourning Cloak is also a resident in northern Eurasia and is a rare migrant into Britain, where it is called the Camberwell Beauty. It is known in Europe for its long-distance dispersals and mass influxes into areas where it does not normally occur. The dynamics of its movements in our region have not been studied, but there is evidence that it is capable of long-distance flights here as well. The species was recorded on Machias Seal Island, disputed territory in the Gulf of Maine more than 16 km (10 mi) from the mainland. This tiny island apparently has no resident butterflies—all other species recorded there are well-known long-distance dispersers that do not overwinter in our region, including the American Lady and the Monarch. There is evidence that New World populations of this butterfly differ from the nominotypical subspecies, which was described from Sweden. The subspecies names *N. a. hyperborea* and *N. a. linterni* were long ago proposed for North America, but a thorough study of their status and distribution has yet to be published. We tentatively identify our subspecies as *N. a. antiopa* in accordance with Pohl et al. (2018) and Pelham (2022) pending additional research.

Threats: None documented in our region.

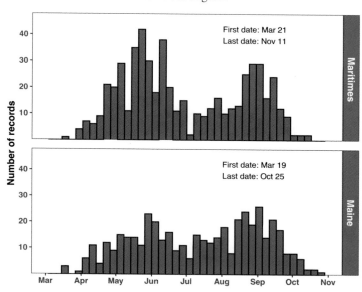

QUESTION MARK

Polygonia interrogationis
(Fabricius)

Subspecies: None.

Distinguishing characters:
  Size: Large (largest
    anglewing).
  Upperside: Margins with
    purple scaling; forewing
    orange with bold black
    spots, with rectangular
    spot anterior to the series of
    round dots; hindwing heav-
    ily suffused dark brown.

Buxton, ME, 27 August 2019 (Bryan Pfeiffer)

  Underside: Medium to light brown, boldly patterned; hindwing with silver
    question mark.
  Additional: Wing margins slightly scalloped; hindwing with relatively
    long tail. **Summer form** described above; **overwintering form** similar,
    except underside has weaker patterning and hindwing upperside is
    orange with black spots.

Similar species: The commas, especially Eastern Comma.

Status and distribution in Maine: Frequent colonist. Generally rare to uncommon, but in some years abundant; widespread but most frequent along
the coast. S4B

Status and distribution in Maritimes: Frequent colonist. Generally rare to

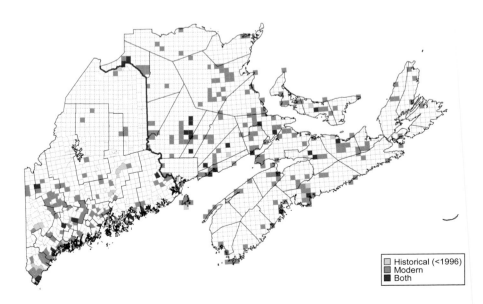

Historical (<1996)
Modern
Both

uncommon, but in some years abundant; widespread but most frequent along the coast. NB-S4B, NS-S3B, PE-S3B

Habitat: Deciduous and mixed forests and nearby open areas, including streamsides, trails, and old fields.

Biology: This species is a migrant with two generations. The summer form generation immigrates from farther south, usually during May and June. The overwintering generation, produced locally, appears in mid to late summer. In some years there is a pronounced southwestward coastal flight during autumn (Hildreth 2013). Larval food plants include elms (*Ulmus* spp.), nettles (Urticaceae), and hops (*Humulus* spp.). There is no evidence of overwintering in our region.

Adult behavior: Like other migrant species, the Question Mark is a strong flyer. It is often found feeding on carrion, sap, and dung and only occasionally nectaring at a variety of flowers in our region, including Choke Cherry (*Prunus virginiana*), Common Dandelion (*Taraxacum officinale*), Common Milkweed (*Asclepias syriaca*), Orange-eye Butterfly Bush (*Buddleja davidii*), and Huckleberry (*Gaylussacia baccata*) in eastern Maine (Hildreth 2016).

Comments: The Question Mark is an irregular migrant to the Acadian region, where records reflect significant variation in its abundance from year to year. For example, there were 402 records of the species submitted to the Maine and Maritimes atlases in 2012, whereas only 7 records were submitted in 2013.

Threats: The abundance of native elm trees, an important food source for the Question Mark, is greatly diminished from former levels in much of North America due to Dutch elm disease, likely contributing to lower populations of this butterfly in the modern era.

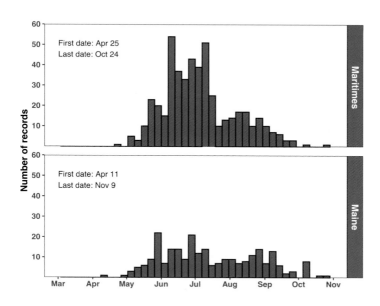

First date: Apr 25
Last date: Oct 24

Maritimes

First date: Apr 11
Last date: Nov 9

Maine

Number of records

EASTERN COMMA

Polygonia comma (T. Harris)

Subspecies: None.

Distinguishing characters:
  Size: Medium.
  Upperside: Orange with bold black spots and slight purple scaling on the margin; the innermost spot on the forewing not doubled or only faintly so; hindwing dark brown on apical two-thirds.

Female—Woodstock, NB, 12 July 2009 (Jim Edsall)

  Underside: Male heavily mottled, typically with tawny patches and marginal green scaling; female more uniformly brown; hindwing with silver comma that is clubbed or hooked at both ends.
  Additional: Wing margins lightly scalloped, hindwing with stout tail.
    Summer form described above; **overwintering form** similar, except hindwing upperside is orange with black spots.

Similar species: Other commas and the Question Mark.

Status and distribution in Maine: Resident. Common in south, uncommon in north; widespread. S4

Status and distribution in Maritimes: Resident. Uncommon in New Brunswick, rare in Nova Scotia, very rare in Prince Edward Island, where it may only be a stray. NB-S4, NS-S1?, PE-SU

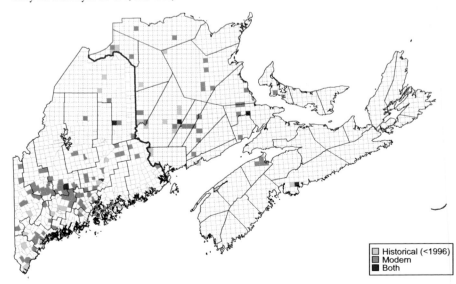

Historical (<1996)
Modern
Both

Habitat: Deciduous and mixed forests, especially in openings and along trails and forest margins.

Biology: Two generations per year. Adults of the first brood, the summer form, begin to emerge in June or July. Those of the second brood, the overwintering form, emerge in August and September, then overwinter in sheltered areas, such as tree cavities and under loose bark, and take to the wing again in early spring. Larval food plants are nettles (Urticaceae), elms (*Ulmus* spp.), and hops (*Humulus* spp.).

Adult behavior: Like other commas, this species is a strong flyer. Males are most frequently encountered because of their habit of establishing territories along woodland trails. They very rarely visit flowers and instead feed on fresh sap, rotting fruit, dung, and damp ground.

Comments: After emerging from hibernation, worn adults of the Eastern Comma can be seen basking along sunlit forested trails in early spring. The only record of this species from Prince Edward Island is a single individual collected on 25 May 2012 by Donna Martin. It is unclear if it originated from elsewhere or was from a previously undocumented resident population. It is also scarce in Nova Scotia, where several recent records from the Annapolis Valley suggest an established population exists there. Like the Question Mark and Gray Comma, the Eastern Comma has two seasonal forms, with individuals of the first brood exhibiting extensive darkening on the dorsal hindwing.

Threats: None documented in our region.

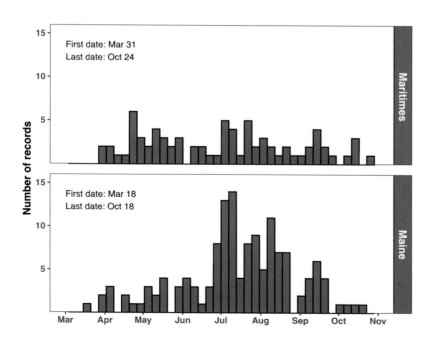

SATYR COMMA
Polygonia satyrus
(W. H. Edwards)

Subspecies: Only the sub-species *P. s. neomarsyas* dos Passos occurs in eastern North America.

Distinguishing characters:
    Size: Medium.
    Upperside: Forewing golden orange with bold black spots, the inner-most spot doubled and usually touching; margin

Kings County, PE, 6 August 2015 (Ken McKenna)

    of hindwing markedly lighter than that of forewing.
    Underside: Mottled reddish brown, with fine striations; hind wing with silver comma that is clubbed or hooked at each end.
    Additional: Wing margins scalloped, hindwing with stout tail.

Similar species: Other commas, especially the Eastern Comma.

Status and distribution in Maine: Resident. Rare; potentially widespread in the north. State Special Concern. S2S3

Status and distribution in Maritimes: Resident. Uncommon in northern and central New Brunswick and eastern Prince Edward Island; rare or absent elsewhere. NB-S4?, NS-S1?, PE-S3

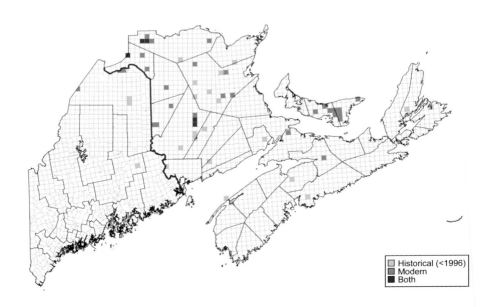

Historical (<1996)
Modern
Both

Habitats: Mixed and coniferous forests, typically near streams; most often encountered on dirt roads and along forest margins.

Biology: One generation per year. Fresh adults appear during midsummer and fly until fall. They overwinter in sheltered areas, such as tree cavities and under loose bark, and take to the wing again in early spring. Larvae feed on Stinging Nettle (*Urtica dioica*) and potentially other species in the nettle family (Urticaceae).

Adult behavior: Like other anglewings, this butterfly is a strong flyer. It is more often found feeding on carrion, sap, rotting fruit, and dung than at flowers.

Comments: The Satyr Comma is a boreal species that is much more common in western North America. The relatively large number of records from northern New Brunswick suggest that its apparent scarcity in northern Maine may be due to limited survey efforts. In Nova Scotia, this species was more common historically, and its decline there was already apparent by the 1950s (Ferguson 1954). Of the eighteen known Nova Scotia records, only four are dated after the 1940s. Adults are rarely encountered over most of our region, though larvae can sometimes be located by searching patches of the food plant (Webster and deMaynadier 2005).

Threats: None documented in our region. It is unknown why the species has declined in Nova Scotia, where continued monitoring is advisable.

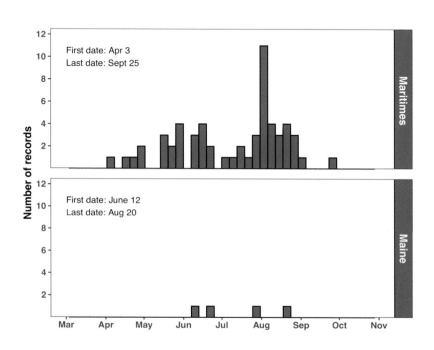

GRAY COMMA

Polygonia progne (Cramer)

Subspecies: None.

Distinguishing characters:

    Size: Medium.

    Upperside: Forewing orange with bold black spots; spots smaller than in other commas, the innermost spot not doubled; hindwing margin of overwintering form usually lighter than that of summer form.

Scoudouc, NB, 23 July 2008 (Jim Edsall)

    Underside: Gray brown with heavy striations; forewing lighter on outer half; hindwing more uniform, with a thin silver comma that tapers at the ends.

    Additional: Wing margins scalloped, hindwing with stout tail.

Similar species: Other commas, especially the Hoary Comma.

Status and distribution in Maine: Resident. Uncommon; widespread. S4

Status and distribution in Maritimes: Resident. Common in New Brunswick, uncommon in Nova Scotia and Prince Edward Island; widespread. NB-S5, NS-S4?, PE-S3

Habitat: Deciduous and mixed forests; most often encountered on dirt roads and along forest margins.

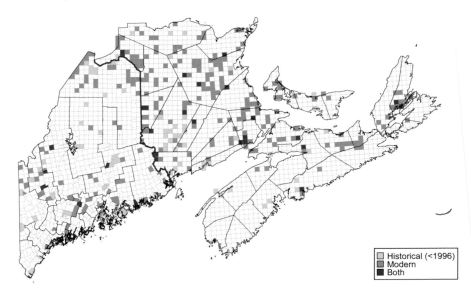

Historical (<1996)
Modern
Both

Biology: Two generations per year. Adults of the summer generation begin to emerge in June or July. Those of the overwintering generation emerge from July to September, then overwinter in sheltered areas, such as tree cavities and under loose bark, and take to the wing again in early spring. Larvae feed on currants and gooseberries (*Ribes* spp.).

Adult behavior: Like other commas, this species is a strong flyer, and it is more often found feeding on carrion, sap, rotting fruit, and dung than at flowers. Males are avid puddlers.

Comments: This species is one of the most frequently encountered *Polygonia* in our region, generally second in abundance to the Eastern Comma in Maine and the Green Comma in most of the Maritimes. Ferguson (1954) reported that the Gray Comma, while no longer the common species it had been, was still the most abundant anglewing in Nova Scotia. The situation in Nova Scotia remains similar today, with the Gray Comma being uncommon yet widespread and more abundant than resident congeners.

Threats: None documented in our region.

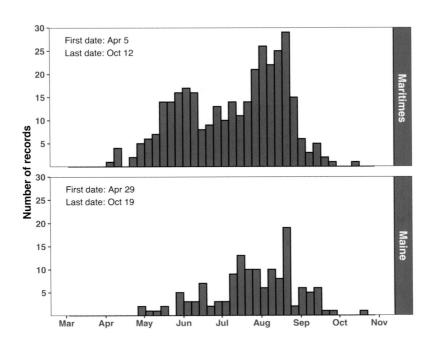

HOARY COMMA

Polygonia gracilis
(Grote & Robinson)

Allardville Parish, NB, 15 June 2018 (Jim Edsall)

Subspecies: Only the nomino-typical subspecies (*P. g. gracilis*) occurs in eastern North America.

Distinguishing characters:
> **Size:** Medium.
> **Upperside:** Orange with bold black spots; inner-most spot on forewing not doubled.
> **Underside:** Heavily striated; inner half of both wings dark brown, outer half lighter gray (hoary), giving a strong two-toned appearance; hindwing with comma relatively thick and tapered at both ends.
> **Additional:** Wing margins scalloped, hindwing with stout tail.

Similar species: Other commas, especially the Gray Comma.

Status and distribution in Maine: Resident. Rare; widespread in north. S3

Status and distribution in Maritimes: Resident. Uncommon and widespread in northern New Brunswick; known only from historical records in southern New Brunswick, northern mainland Nova Scotia, and Cape Breton. NB-S4?, NS-SH

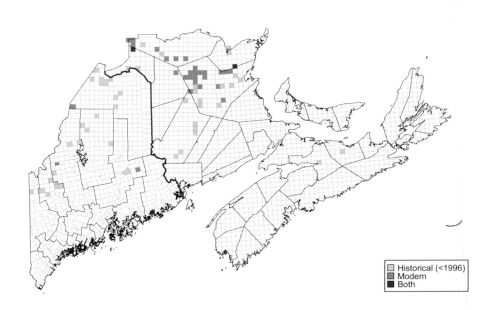

Historical (<1996)
Modern
Both

Habitat: Forested areas, often near streams; most often encountered on dirt roads and in other openings.

Biology: One generation per year. Fresh adults can be found in July and August. They overwinter in the adult stage in sheltered areas, such as tree cavities and under loose bark, and take to the wing again in early spring. The larval food plants are currants and gooseberries (*Ribes* spp.).

Adult behavior: Like other commas, this species is a strong flyer. Among our anglewings, it has the greatest propensity to visit flowers, though adults are more often found puddling along gravel roads. In our region it has been observed visiting Pearly Everlasting (*Anaphalis margaritacea*) and goldenrods (*Solidago* spp.).

Comments: It is curious that all three Nova Scotia records are from the 1920s, perhaps suggesting a temporary influx of this species into the region rather than a permanent population. Similar disjunct records exist elsewhere on this species' southern distributional boundary. For example, there are two records in central Ontario, both from 1960, which are 250 km (155 mi) south of any other Hoary Comma records in the province (Macnaughton et al. 2018).

Threats: The Hoary Comma was more widespread historically, even if partly due to population irruptions. The Acadian region is at the southern edge of this boreal species' range, and it is possible that its distribution has contracted northward as a result of climate change.

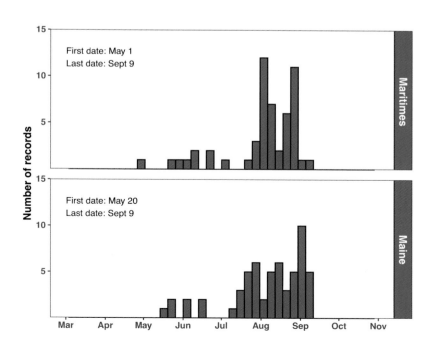

GREEN COMMA

Polygonia faunus
(W. H. Edwards)

Subspecies: Only the nomino-
typical subspecies (*P. f. faunus*)
occurs in northeastern North
America.

Distinguishing characters:

Size: Medium.

Upperside: Orange with
bold black spots; typ-
ically darker overall
than other commas; the
innermost forewing spot
sometimes doubled.

York County, NB, 27 July 2011 (John Klymko)

Underside: Mottled gray and brown; both wings with two submarginal
rows of irregular green markings (indistinct in worn individuals); hindwing with silver comma that is usually clubbed or hooked at least on one
end.

Additional: Wing margins scalloped; hindwing with stout tail.

Similar species: Other commas.

Status and distribution in Maine: Resident; common in north, uncommon in
south; widespread. S5

Status and distribution in Maritimes: Resident; common in New Brunswick,
rare in Prince Edward Island, uncommon in Nova Scotia; widespread. NB-S5,
NS-S3S4, PE-S1

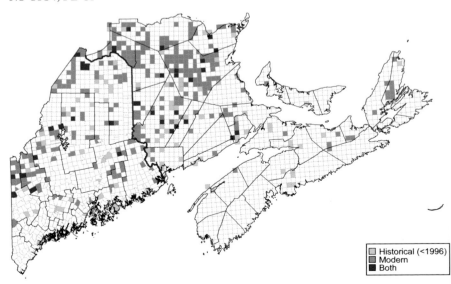

☐ Historical (<1996)
▨ Modern
■ Both

Habitats: Coniferous and mixed forests; most often encountered on dirt roads and along forest margins.

Biology: One generation per year. Fresh adults can be found in July and August. They overwinter in sheltered areas, such as in tree cavities and under loose bark, then take to the wing again in early spring. Larvae feed on a variety of woody plants, including willows (*Salix* spp.), birches (*Betula* spp.), and alders (*Alnus* spp.).

Adult behavior: The Green Comma is a strong flyer, and it is often seen sunning on dirt roads or other bare ground. Also, like other anglewings, it is more often seen feeding on carrion, sap, rotting fruit, and dung than at flowers.

Comments: A drive down a forested gravel road during September often affords a glimpse of one or more adults. When disturbed they rapidly fly off, only to land farther down the road or on a nearby tree trunk, where they often perch head downward with wings closed, resembling dead leaves. Further pursuit is often rewarded, as the Green Comma will commonly alight on clothing or skin to sip moisture (Cech and Tudor 2005; Hall et al. 2014).

Threats: None documented in our region.

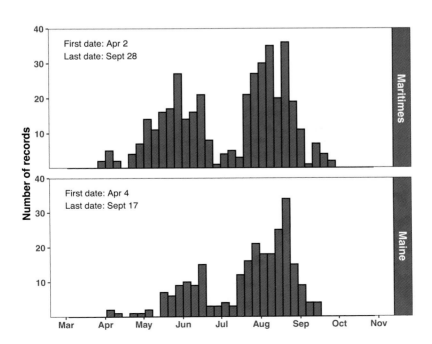

AMERICAN LADY

Vanessa virginiensis (Drury)

Subspecies: None.

Distinguishing characters:
 Size: Medium.
 Upperside: Orange with black and white markings; forewing square tipped.
 Underside: Brown with intricate cobweb-like markings; reddish-orange patch on basal half of forewing; hindwing with two bold eyespots.

Eliot, ME, 28 July 2019 (Bryan Pfeiffer)

Similar species: Painted Lady.

Status and distribution in Maine: Frequent colonist. Common most years, in some years abundant; widespread but most frequent along the coast. S5B

Status and distribution in Maritimes: Frequent colonist. Common most years, in some years abundant; widespread but most frequent along the coast. NB-S5B, NS-S5B, PE-S4B

Habitat: Nearly any open vegetated area, including old fields, gardens, and roadsides.

Biology: This migrant butterfly typically arrives by late May, and numbers peak in late summer. There are usually three generations in our region. Larvae

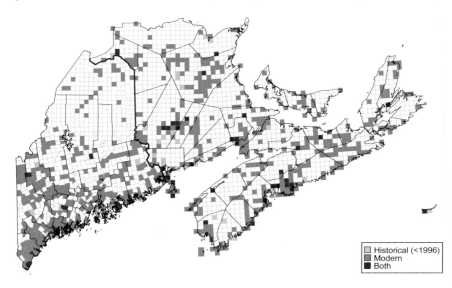

Historical (<1996)
Modern
Both

feed on species in the aster family (Asteraceae), particularly Pearly Everlasting (*Anaphalis margaritacea*), cudweeds (*Gnaphalium* spp.), and pussytoes (*Antennaria* spp.). The species does not typically overwinter anywhere in the Northeast (but see comments).

Adult behavior: Like other migrants, the American Lady is a strong flyer that can quickly cover great distances. In 2012, the American Lady and the Red Admiral invaded the Acadian region in a single day, 17 April, when strong winds were blowing from the south. The butterflies were reported from seven localities, from York County, Maine, northward to Cumberland County, Nova Scotia. Most of these records involved multiple butterflies, and at some sites a movement of northbound flyers was noted. This species uses a variety of nectar sources, including ornamental plants. During spring it frequents Common Dandelion (*Taraxacum officinale*), and later in summer it is regularly seen at goldenrods (*Solidago* spp.), asters (*Symphyotrichum* spp.), and planted coneflowers (*Echinacea* spp.).

Comments: More cold tolerant than the Painted Lady and Red Admiral, it is possible that pupae of the American Lady occasionally survive the winter in our region, especially southward. Indeed, the fresh condition of an individual found dead on 14 April 2020 in Kennebec County, Maine, may support the notion that some pupae survive mild winters in Maine. It is often claimed in the literature that the genus name *Vanessa* means "butterfly" in Greek, but this name was evidently proposed in 1713 in a poem by the Irish essayist Jonathan Swift, and the genus name was probably derived from Swift's poem (Evans 1993).

Threats: None documented in our region.

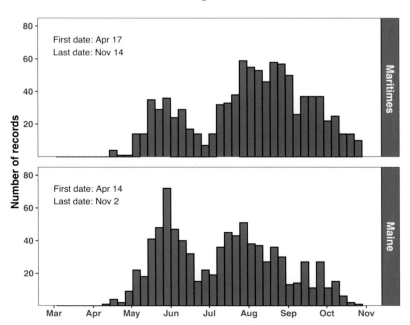

PAINTED LADY

Vanessa cardui (Linnaeus)

Subspecies: None.

Distinguishing characters:
    Size: Medium.
    Upperside: Orange
       with black and white
       markings.
    Underside: Brown with
       intricate cobweb-like
       markings; reddish-orange
       patch on basal half of
       forewing; hindwing with
       four small eyespots.

Monhegan, ME, 8 August 2019 (Bryan Pfeiffer)

Similar species: American Lady.

Status and distribution in Maine: Frequent colonist. Generally rare to uncommon but in some years abundant; widespread but most frequent along the coast. S5B

Status and distribution in Maritimes: Frequent colonist. Generally rare to uncommon but in some years abundant; widespread but most frequent along the coast. NB-S5B, NS-S4B, PE-S4B

Habitat: Nearly any open vegetated area, including old fields, gardens, and roadsides.

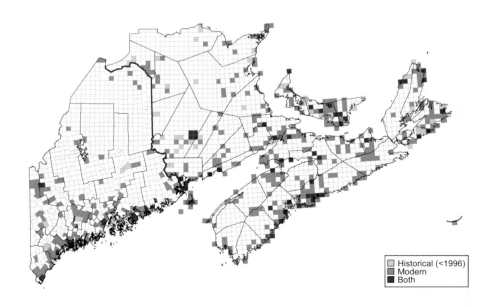

Historical (<1996)
Modern
Both

Biology: This species is a migrant in most of North America. In our region, it typically arrives by late June or early July and reaches peak abundance during late summer. It produces up to three generations per year, depending on when immigrants first arrive. Larvae are known to feed on numerous plants of several families, especially thistles (*Cirsium* spp.). There is no evidence of overwintering in our region.

Adult behavior: Like others in the genus *Vanessa*, this species is a strong flyer, and during migration they take advantage of wind currents hundreds of meters above the ground. The Painted Lady visits a variety of native and culti-vated flowers in our region, such as American-asters (*Symphyotrichum* spp.), knapweeds (*Centaurea* spp.), and thistles (*Cirsium* spp.).

Comments: This species' alternative name, the Cosmopolitan, reflects its extraordinary worldwide distribution, which includes every continent except Antarctica. The year 2012 was exceptional for all migrant butterfly species in the Northeast, especially the Painted Lady. Nearly 85% of Painted Lady records submitted to the Maine and Maritime atlases are from that year alone. During such outbreaks, this species may use a variety of alternate food plants. It can sometimes be a pest of soy crops, as in Prince Edward Island in 2017, when the species was also frequent across the region. The southward migra-tion of the Painted Lady is more fully documented in Europe than in North America, but the perennial autumn concentration of this species in coastal areas of our region, where they co-occur with other migrant butterflies such as Monarch, Mourning Cloak, Red Admiral, and American Lady (Hildreth 2011, 2013, 2016), suggests the occurrence of such a late-season movement.

Threats: None documented in our region.

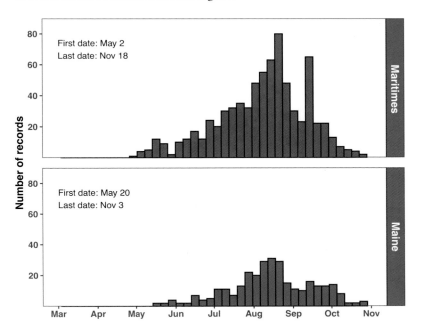

RED ADMIRAL

Vanessa atalanta (Linnaeus)

Moncton, NB, 28 July 2010 (Jim Edsall)

Subspecies: Only the subspecies *V. a. rubria* (Fruhstorfer) occurs in North America.

Distinguishing characters:
  Size: Medium.
  Upperside: Dark brown to black with bold orange bands across forewing and along hindwing border.
  Underside: Mottled brown and gray; forewing with orange-pink markings on basal half.

Similar species: None.

Status and distribution in Maine: Frequent colonist. Common most years, sometimes abundant; widespread but most frequent along the coast. S5B

Status and distribution in Maritimes: Frequent colonist. Common most years, sometimes abundant; widespread but most frequent along the coast. NB-S5B, NS-S4B, PE-S4B

Habitat: Open areas, such as old fields, gardens, roadsides, and forest edges.

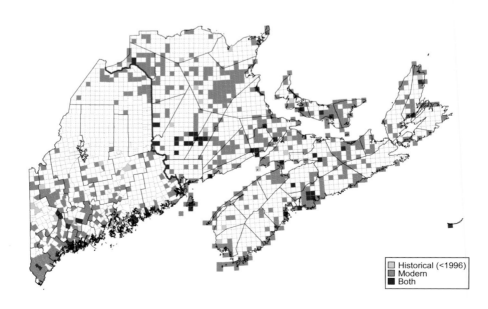

Historical (<1996)
Modern
Both

Biology: This species is a regular migrant that typically arrives during May and reaches peak abundance in late summer. There are two or three generations in our region. Larvae feed on species in the nettle family (Urticaceae), including Stinging Nettle (*Urtica dioica*) and Canada Wood-nettle (*Laportea canadensis*). There is no evidence of overwintering in our region.

Adult behavior: Like other species in the genus *Vanessa*, the Red Admiral is a strong flyer. It feeds from a variety of native and cultivated flowers in our region, including goldenrods (*Solidago* spp.), American-asters (*Symphyotrichum* spp.), knapweeds (*Centaurea* spp.), coneflowers (*Echinacea* spp.), and Orange-eye Butterfly-bush (*Buddleja davidii*).

Comments: As with other migrants, the abundance of the Red Admiral varies considerably from year to year. Data gathered in Maine and the Maritimes reflect this variability. For example, all migratory species were relatively common in 2012, and 567 Red Admiral records were submitted. In 2013, a relatively poor year for migrant species, only 25 Red Admiral records were submitted. During 2001, another big year for Red Admirals in the Northeast, Richard Hildreth (2008) counted over 600 adults in one small region of eastern coastal Maine. Notably, also in 2001, Hildreth reported that a late-summer (23 August) census of a single Stinging Nettle patch in Steuben yielded 35 Red Admiral silken tents per square yard (0.8 m²), with each tent hosting one fully grown larva, for an estimated total of 5250 mature larvae in an area of only 150 yd² (125 m²).

Threats: None documented in our region.

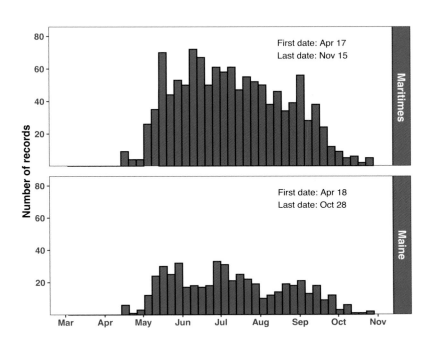

COMMON BUCKEYE

Junonia coenia Hübner

Camden, ME, 6 August 2012 (Roger Rittmaster)

Subspecies: Only the nominotypical subspecies (*J. c. coenia*) occurs in northeastern North America.

Distinguishing characters:
- **Size:** Medium.
- **Upperside:** Gray brown, with orange and white markings and bold black and blue eyespots (one on forewing, two on hindwing).
- **Underside:** Similar to upperside but lighter in color and with hindwing eyespots reduced (in late season form *rosa* underside is reddish, with dark transverse line on hindwing).

Similar species: None.

Status and distribution in Maine: Rare or frequent colonist. Uncommon most years, nearly absent in others; widespread but most records from the southern counties of York and Cumberland. S4B

Status and distribution in Maritimes: Stray. Recorded from scattered locations in New Brunswick and southern Nova Scotia; somewhat regular on Grand Manan, New Brunswick. NB-SNA, NS-SNA

Habitat: Various open spaces, including old fields, roadsides, utility corridors, and gardens; most frequent in coastal areas.

Biology: The Common Buckeye is a migrant species. In most years it is first detected in July, and its abundance increases over the summer. The number of

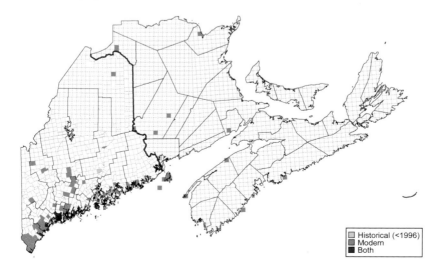

Historical (<1996)
Modern
Both

generations depends on when the first colonizers arrive; multiple overlapping generations are possible if adults arrive early in the season, and a southward migration may occur in the autumn. Larvae feed on plants of several families, including false fox-gloves (*Agalinis* spp.) and plantains (*Plantago* spp.). In Maine, it has been observed ovipositing on Slender-leaf Agalinis (*Agalinis tenuifolia*), and a larva was found on this plant. Adults overwinter in reproductive diapause but not in our region.

Adult behavior: This species is commonly seen sunning on roads or other open, disturbed areas, particularly those with low-growing vegetation interspersed with bare patches of ground. It is a strong flyer, and while adults frequently alight on the ground, they are often wary and difficult to approach. It visits a variety of flowers, including Red Clover (*Trifolium pratense*), goldenrods (*Solidago* spp.), and asters (*Symphyotrichum* spp.).

Comments: In the Acadian region, the Common Buckeye occurs much more frequently now than in the past. Scudder (1888–1889) described the species as exceedingly rare in New England, with only three known Maine records. Brower (1974) noted it had been taken at twelve localities in the state. In recent years, this butterfly has been relatively common in southern Maine (there were fifty-one records in 2011), where it occasionally establishes temporary populations (Gobeil and Gobeil 2012). In August 2020, local populations were documented in Maine as far north as Benton, Kennebec County, and Skowhegan, Somerset County; both closely associated with patches of Slender-leaf Agalinis. The oldest Maritimes record is from 1954 at Partridge Island, Cumberland County, Nova Scotia (Ferguson 1955), and it was not recorded in New Brunswick until 1984. Although it likely occurs annually on Grand Manan Island, New Brunswick, breeding has not been confirmed in the Maritimes, where few fresh adults have been found.

Threats: None documented in our region.

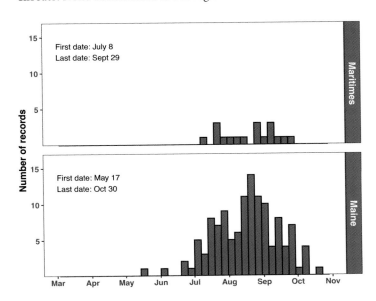

BALTIMORE CHECKERSPOT

Euphydryas phaeton (Drury)

Subspecies: We tentatively recognize only the nominotypical subspecies (*E. p. phaeton*) in our region (see comments).

Distinguishing characters:
  Size: Medium.
  Upperside: Black with bold white and orange spots.
  Underside: Similar to upperside but with more orange basally.

Dallas Plantation, ME, 6 July 2017 (Ron Butler)

Similar species: None.

Status and distribution in Maine: Resident. Uncommon; widespread but localized. S4

Status and distribution in Maritimes: Resident. Uncommon and localized in New Brunswick and Prince Edward Island, rarer in Nova Scotia; widespread. NB-S4, NS-S4?, PE-S4

Habitat: Various wet areas including marshes, riparian meadows, stream banks, and moist ditches.

Biology: One generation per year. Adults typically appear mid to late June and fly until late July. Adults are occasionally found well after the species' typical early summer flight period, but these are likely late-emerging individuals, not a partial second brood. In our region, young larvae usually feed on White

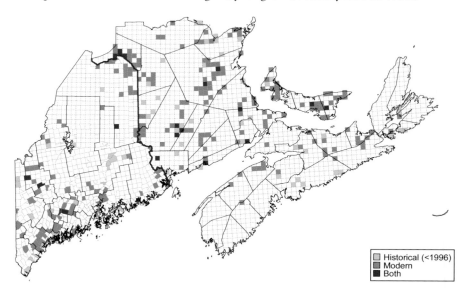

Historical (<1996)
Modern
Both

Turtlehead (*Chelone glabra*), while older larvae (the following spring) can use other plants, including plantains (*Plantago* spp.), ashes (*Fraxinus* spp.), and viburnums (*Viburnum* spp.). Larvae feed exclusively on introduced English Plantain (*Plantago lanceolata*) in some southern New England populations. Overwinters as a partially grown larva.

Adult behavior: Somewhat sluggish in flight, adults usually remain close to wet habitats that support their primary food plant, White Turtlehead. They frequent a variety of flowers, including Ox-eye Daisy (*Leucanthemum vulgare*), clovers (*Trifolium* spp.), Golden Groundsel (*Packera aurea*), Black-eyed Susan (*Rudbeckia hirta*), and Spreading Dogbane (*Apocynum androsaemifolium*). They also visit damp areas on dirt roads near wetlands.

Comments: The Baltimore Checkerspot has brightly marked larvae, pupae, and adults. This aposematic coloration serves to warn predators of the toxic iridoid glycosides that larvae sequester from their primary food plants (Bowers 1980) and to help protect slow-flying adults. Described from specimens collected at Lincoln (Penobscot County), Maine, the subspecies *E. p. borealis* F. Chermock and R. Chermock is sometimes applied to populations in Maine and the Maritimes (e.g., Pavulaan 2021), but more research is needed to validate that taxon. The sedentary nature of this species, as well as variation in local microclimate, can account for differences in adult emergence times and slight variations in wing pattern, even between nearby colonies of this highly variable species.

Threats: The filling and degradation of wet meadows and stream banks has undoubtedly led to extirpation of some Baltimore Checkerspot populations in our region. Its recent acceptance of English Plantain, a widespread and common weedy species, as a primary food source has possibly benefitted this butterfly in Massachusetts and elsewhere (Bowers and Richardson 2013; Stichter 2015).

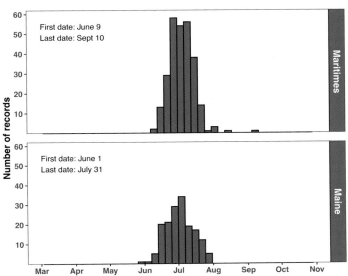

SILVERY CHECKERSPOT

Chlosyne nycteis (E. Doubleday)

Subspecies: Only the nominotypical subspecies (*C. n. nycteis*) occurs in northeastern North America.

Distinguishing characters:
   Size: Small to medium.
   Upperside: Orange with complex blackish brown markings; hindwing with series of submarginal spots, some with white centers.
   Underside: Hindwing variably tan or creamy white with brown veins on basal two-thirds, dark brown on apical third; a distinct white crescent present in the middle of a marginal brown band.

Similar species: Harris' Checkerspot (upperside) and the crescents.

Status and distribution in Maine: Resident. Rare; widespread, but few reported localities and mostly historical records in several southern counties. State Special Concern. S3

Status and distribution in Maritimes: Resident. In New Brunswick, it is widespread, rare in the south, and uncommon in the north and east. In Nova Scotia, it is known from only two historical records from a single location. Not recorded on Prince Edward Island. NB-S4, NS-SH

Habitat: A variety of open or semi-open areas, including wet meadows, moist roadsides, utility corridors, and open woodlands; almost always near streams.

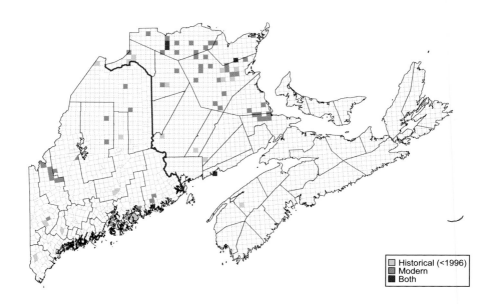

Historical (<1996)
Modern
Both

Coplin Plantation, ME, 21 June 2021 (Phillip deMaynadier)

Eustis, ME, 18 June 2021 (John V. Calhoun)

Biology: One generation per year. Adults appear early to mid-June and fly for about 1 month. Larvae feed on a variety of species in the aster family (Asteraceae). In Maine, larvae have been found feeding on Large-leaved Wood-aster (*Eurybia macrophylla*) (Calhoun 2022). Overwinters as a partially grown larva.

Adult behavior: Males patrol for females along water courses and forest edges. Adults tend to fly slowly near the ground, resembling large crescents with shallow wingbeats and brief glides. This species congregates at damp soil and dung and is an active nectar feeder. In our region, it has been documented feeding at buttercups (*Ranunculus* spp.), hawkweeds (*Hieracium* spp.), dogbanes (*Apocynum* spp.), Ox-eye Daisy (*Leucanthemum vulgare*), Bush-honeysuckle (*Diervilla lonicera*), Withe-rod (*Viburnum nudum*), and Common Yarrow (*Achillea millefolium*).

Comments: The Silvery Checkerspot is intensely local and known to establish ephemeral populations of fluctuating abundance. There are no known records from Nova Scotia after 1951, but given the numerous records from Westmorland County, New Brunswick, there is a chance that the butterfly still occurs there or could repopulate the area. A major decline in the Silvery Checkerspot has been noted in portions of its range, including New England (O'Donnell et al. 2007). A similar pattern may be occurring in Maine, where few individuals have been recorded in recent years and modern reports are lacking from the southern half of the state. It is possible that this species has always been scarce over much of our region, as Scudder (1888–1889) referred to it as a "very rare insect" in New England and Brower (1974) listed few records from Maine, all

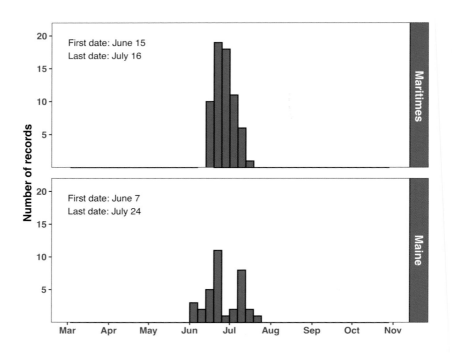

historical. In June 2021, John Calhoun found several localized colonies, some consisting of many individuals, covering an area of at least 8 km² (3 mi²) in northern Franklin County, Maine; all were closely associated with Large-leaved Wood-aster (Calhoun 2022). Additional new localities were documented in 2022 in Aroostook, Franklin, Oxford, and Penobscot Counties, suggesting that the species may be more widespread in Maine than generally believed. Long ago, the synonymous species name *oenone* Scudder was proposed for this butterfly, based partly on specimens collected by S. I. Smith at Norway, Maine.

Threats: The cause of the apparent decline of the Silvery Checkerspot in New England is unclear. Habitat loss, particularly in riparian areas, is probably a contributing factor.

HARRIS' CHECKERSPOT

Chlosyne harrisii (Scudder)

Subspecies: The nominotypical subspecies (*C. h. harrisii*) occurs in our region (see comments).

Distinguishing characters:
Size: Small to medium.
Upperside: Orange with complex blackish-brown markings; hindwing with series of submarginal spots, some with white centers.
Underside: Hindwing with bright orange and white bands divided by black lines.

Similar species: Silvery Checkerspot and the crescents (upperside).

Status and distribution in Maine: Resident. Common; widespread. S5

Status and distribution in Maritimes: Resident. Common; widespread. NB-S5, NS-S5, PE-S4

Habitat: A variety of damp, open areas, including wet meadows, marshes, and roadside ditches.

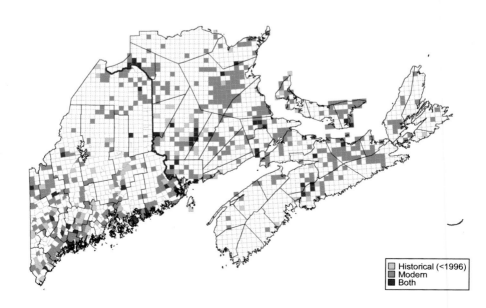

Steuben, ME, 27 June 2017 (Bryan Pfeiffer)

Hammonds Plains, NS, 20 June 2018 (Krista Melville)

Biology: One generation per year. Adults appear early to mid-June and fly for 3 to 4 weeks. Larvae feed on Tall White-aster (*Doellingeria umbellata*). Overwinters as a partially grown larva.

Adult behavior: This species is a relatively weak flyer and typically does not stray far from areas that support its food plant, though it apparently disperses to nearby areas (Williams 2002). Adults frequent damp soil and a variety of flower species, including clovers (*Trifolium* spp.), hawkweeds (*Hieracium* spp.), and Ox-eye Daisy (*Leucanthemum vulgare*) in our region.

Comments: It has been suggested that this species is a mimic of the unpalatable Baltimore Checkerspot, which often occurs in the same habitats (Bowers 1983, Williams 2002). Adults have similar ventral patterns and fly at the same time of the year with a similar sluggish manner. Even the larvae are nearly identical and construct the same type of communal webs. Some of the specimens used to describe *C. h. harrisii* came from Norway, Maine. Although the subspecies *C. h. albimontana* (Avinoff) (described from the White Mountains of New Hampshire) is usually considered to be the subspecies found in the Maritimes (e.g., Ferguson 1954; Pohl et al. 2018), Harris' Checkerspot is extremely variable, and we have found no consistent differences between

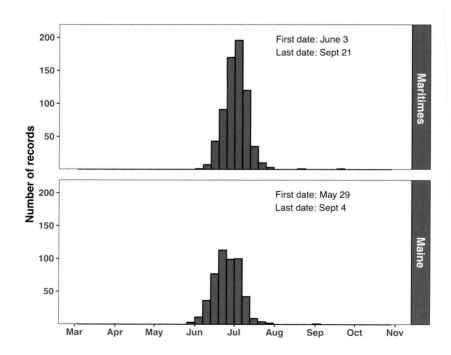

populations in the Acadian region. We therefore consider all populations in our region to represent *C. h. harrisii*.

Threats: Like the Silvery Checkerspot, the Harris' Checkerspot has declined in parts of New England. Its range seems to be contracting northward in Massachusetts, possibly because of climate change (Stichter 2015). Such a decline is not apparent in the Acadian region, but the species should be monitored here.

PEARL CRESCENT

Phyciodes tharos (Drury)

Female—Sherborn, MA, 15 May 2016 (Michael Newton)

Subspecies: Only the nomino-typical subspecies (*P. t. tharos*) occurs in eastern North America.

Distinguishing characters:

    Size: Small (slightly smaller than Northern Crescent).

    Upperside: Orange with extensive dark markings; hindwing with large orange patch divided by black veins.

    Underside: Variable; hindwing mottled orange, brown, and white with fine dark striations and marginal silvery crescent in middle of brown patch. There are two seasonal forms, with spring and autumn generations exhibiting a greater amount of white mottling on hindwing than the summer generation.

    Additional: Underside and tip of antennal club usually orange in females and black or dark brown in males (rarely orange).

Similar species: Nearly identical to the Northern Crescent.

Status and distribution in Maine: Resident. Common in the southern third of the state, mostly absent from the north. S4

Status and distribution in Maritimes: Not documented.

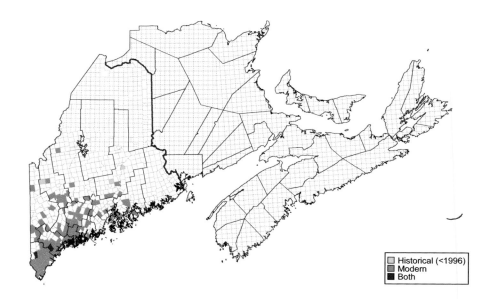

☐ Historical (<1996)
▨ Modern
■ Both

Habitat: A variety of open areas, including old fields, gardens, and roadsides; often in drier sites than the Northern Crescent.

Biology: Typically three generations per year, with adult flight from mid-May until late October. The midsummer generation, from mid-July into August, is most abundant. Larvae feed on various asters (formerly of the genus *Aster*), especially those of the genus *Symphyotrichum*. Overwinters as a partially grown larva.

Adult behavior: The Pearl Crescent has a rapid, darting, low flight. Adults frequent damp soil and a variety of flower species, including American-asters (*Symphyotrichum* spp.), clovers (*Trifolium* spp.), and goldenrods (*Solidago* spp.).

Comments: Identifying our two crescent species is challenging and is not always possible in the field or with photographs. The Pearl Crescent is far more restricted in distribution and has different peak flight periods. Crescent identification is simplified in northern Maine and the Maritimes, where only the Northern Crescent is known to occur. However, the Pearl Crescent's range may be expanding northward (Hall et al. 2014), and the species should be watched for in southwestern New Brunswick in the future. Additional cryptic species of *Phyciodes* may occur in the Acadian region.

Threats: None documented in our region.

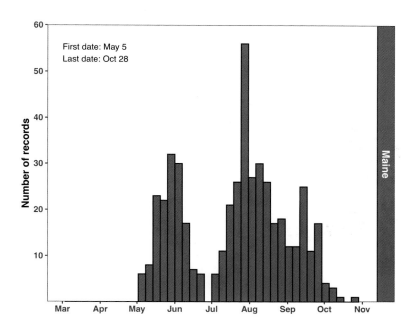

First date: May 5
Last date: Oct 28

NORTHERN CRESCENT

Phyciodes cocyta (Cramer)

Male—Steuben, ME, 30 June 2017 (Bryan Pfeiffer)

Subspecies: Only the nominotypical subspecies (*P. c. cocyta*) occurs in northeastern North America.

Distinguishing characters:
 Size: Small.
 Upperside: Orange with extensive dark markings; hindwing with large orange patch mostly undivided by black veins.
 Underside: Variable; hindwing mottled orange, brown, and white with fine dark striations and marginal silvery crescent in middle of brown patch.
 Additional: Underside and tip of antennal club usually orange in both sexes.

Similar species: Nearly identical to the Pearl Crescent.

Status and distribution in Maine: Resident. Common; widespread. S5

Status and distribution in Maritimes: Resident. Common; widespread. NB-S5, NS-S5, PE-S5

Habitat: A variety of open areas, including old fields, gardens, roadsides, and riparian meadows; often in damper sites than the Pearl Crescent.

Historical (<1996)
Modern
Both

Biology: One generation with a partial, overlapping second brood. The first generation appears in late May and peaks in early July. Adults of the second generation fly into early autumn. Larvae feed on various asters (formerly of the genus *Aster*), such as species of *Symphyotrichum*. Overwinters as a partially grown larva.

Adult behavior: This species has a rapid, low flight with periods of flapping and gliding. Adults frequent damp soil and a variety of flower species, including Ox-eye Daisy (*Leucanthemum vulgare*), hawkweeds (*Hieracium* spp.), and buttercups (*Ranunculus* spp.).

Comments: Long confused with the very similar Pearl Crescent, the Northern Crescent is abundant throughout the Acadian region and is the most frequently documented butterfly in the combined data sets of the Maine and Maritimes atlas projects. This species is frequently encountered by even the most casual butterfly enthusiast. Adults are brightly marked, avid flower visitors, and reasonably easy to approach in just about any flowery field. The designated type specimen of *Papilio cocyta* (= the name of this butterfly as originally assigned in 1777) was collected in 1983 on Cape Breton Island, Nova Scotia.

Threats: None documented in our region.

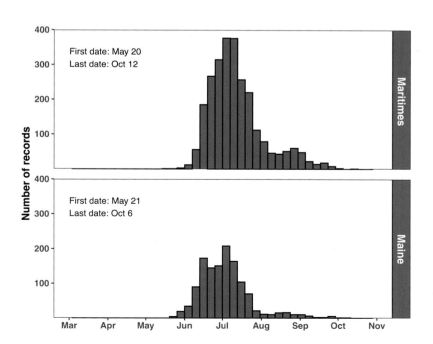

MARITIME RINGLET

Coenonympha nipisiquit
McDunnough

Subspecies: None.

Distinguishing characters:
    Size: Small to medium.
    Upperside: Orange brown; forewing lighter than hindwing (rarely seen).
    Underside: Forewing orange with weak whitish band and white scaling near apex, small eyespot present in most females and some males; hindwing brown with weak, jagged whitish band; scattered white scales present a frosted appearance.

Gloucester County, NB, 29 July 2011 (Maxim Larrivée)

Similar species: Common Ringlet is nearly identical.

Status and distribution in Maine: Not documented.

Status and distribution in Maritimes: Resident. Rare; extremely localized, found only in northeastern New Brunswick on Chaleur Bay. Nationally and provincially Endangered. NB-S1

Habitat: Salt marshes.

Biology: One generation per year. Adults are on the wing from mid-July until mid-August, with peak flight in early August. Larvae feed on Saltmeadow Cordgrass (*Spartina patens*). Overwinters as a partially grown larva.

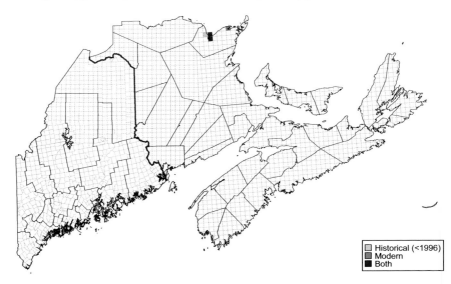

Historical (<1996)
Modern
Both

Adult behavior: A slow, weak flyer, with males patrolling low over salt marsh meadows throughout the day for females. This species nectars almost exclusively on Carolina Sea-lavender (*Limonium carolinianum*).

Comments: Described from specimens collected in New Brunswick, the Maritime Ringlet's global range is limited to salt marshes in the Chaleur Bay area in New Brunswick and Quebec, where it occupies just 455 hectares (1124 acres) of habitat (NBMRRT 2005; COSEWIC 2009). There are six known populations in New Brunswick, four of which are located at Nepisiguit Bay. The Maritime Ringlet is believed to have become genetically isolated from the *Coenonympha tullia* species complex in the past 100,000 years, and it remained physically isolated in a glacial refugium (Sei and Porter 2007). It now co-occurs with the Common Ringlet, which colonized the Northeast in the past century, but the two are mostly isolated by different flight periods and habitat preferences. The Maritime Ringlet was discovered and described as a subspecies of *Coenonympha tullia* in 1939. It was treated as a full species by Layberry et al. (1998), and this has been accepted in many (e.g., COSEWIC 2009; Pohl et al. 2018) but not all (e.g., Pelham 2022) subsequent works.

Threats: The limited distribution and fragmented nature of its habitat make this species inherently vulnerable. Sea level rise and erosion associated with increasing ice scour threaten all populations. Waterfront development and pollution from effluents are threats in the Bathurst area, where two of the largest Maritime Ringlet populations exist. Two of New Brunswick's populations were introduced during the 1990s near Bas-Caraquet and Rivière-du-Nord in salt marshes far from urban areas, in a deliberate attempt to insure the species' long-term survival (Webster 2001).

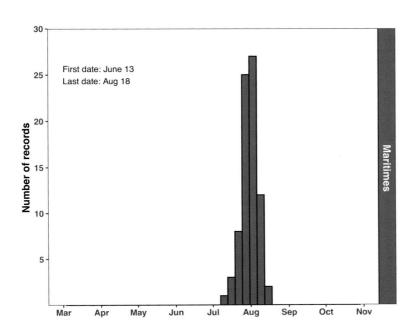

First date: June 13
Last date: Aug 18

COMMON RINGLET (INORNATE RINGLET)

Coenonympha california
Westwood

Kennebunk, ME, 31 August 2018 (Josh Lincoln)

Subspecies: Only *C. c. inornata* W. H. Edwards occurs in most of eastern North America, including the Acadian region.

Distinguishing characters:
    Size: Small to medium.
    Upperside: Ochre or orange brown; forewing usually lighter than hindwing (rarely seen).
    Underside: Forewing orange with weak pale band and extensive white scaling near apex, small eyespot present in most individuals; hindwing brown or grayish brown with weak, jagged pale band, scattered white scales present a frosted appearance.

Similar species: Maritime Ringlet is nearly identical.

Status and distribution in Maine: Resident. Common; widespread. S5

Status and distribution in Maritimes: Resident. Common; widespread. NB-S5, NS-S5, PE-S5

Habitat: Virtually any open, grassy area, including old fields, roadsides, pastures, and even yards.

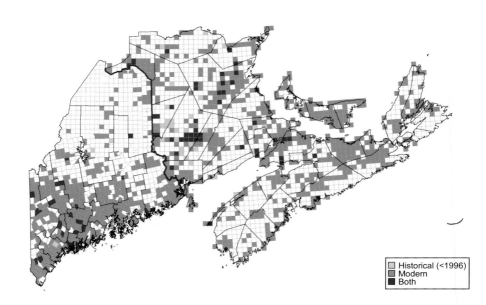

Historical (<1996)
Modern
Both

Biology: Two generations per year. Adults are on the wing from late May to mid-October, with peak flights in late June–early July and late August. Larvae feed on grasses (Poaceae), including Kentucky Bluegrass (*Poa pratensis*). Overwinters as a partially grown larva.

Adult behavior: A slow, weak flyer. Males patrol throughout the day with a low, bobbing flight over low vegetation. Although adults nectar at various flowers, including hawkweeds (*Hieracium* spp.), Ox-eye Daisy (*Leucanthemum vulgare*), clovers (*Trifolium* spp.), and goldenrods (*Solidago* spp.), they do not visit flowers as frequently as other common meadow-associated butterflies.

Comments: Today one of our most widespread and common butterflies, the Common Ringlet is a relative newcomer to the Acadian region and the eastern United States. Although a capture in 1968 was long thought to be the first from Maine (Ferris 1970), an older state specimen exists from 1948. It was first recorded in the Maritimes in 1966. The growing number of grass-lined corridors, such as railways and especially highways, likely hastened this butterfly's spread from its historical range to the north and west. This species continues to expand its range, and it is now known to occur as far south as West Virginia. It was long believed that the species *Coenonympha tullia* (Müller) was present in both North America and Eurasia and that most, if not all, New World populations belonged to that species. However, molecular and morphological studies suggested that *C. tullia* is almost exclusively Eurasian (Guppy and Shepard 2001; Kodandaramaiah and Wahlberg 2009). Recent genomics research by Zhang et al. (2020) supports this treatment; thus the name *C. california* is now applied to most North American populations.

Threats: None documented in our region.

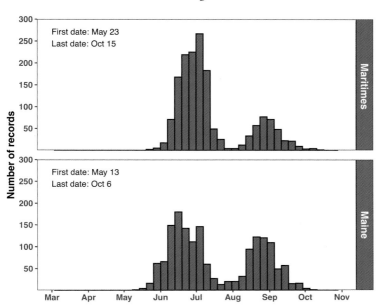

NORTHERN PEARLY-EYE

Lethe anthedon (A. Clark)

Subspecies: None
(see comments).

Distinguishing characters:
  Size: Medium.
  Upperside: Brown with
    row of black spots on
    both wings; forewing
    with three large spots
    and sometimes a fourth
    reduced spot.
  Underside: Brown with
    violet sheen when fresh,
  with dark striations and bold eyespots ringed with orange yellow; the
  second eyespot of the forewing smaller than the others.
  Additional: Hindwing margin scalloped.

Big Twenty Township, ME, 17 July 2019 (Bryan Pfeiffer)

Similar species: Eyed Brown and Appalachian Brown.

Status and distribution in Maine: Resident. Common; widespread. S5

Status and distribution in Maritimes: Resident. Common in New Brunswick, uncommon in Nova Scotia and Prince Edward Island; widespread. NB-S5, NS-S4, PE-S4S5

Habitat: Damp deciduous and mixed forests, often adjacent to streams and wetlands.

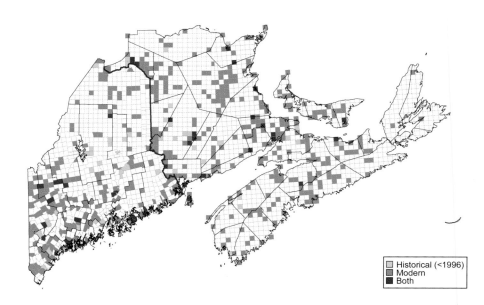

☐ Historical (<1996)
▨ Modern
■ Both

Biology: One generation per year. Adults are on the wing from late June into September, with peak flight in July. Larvae mostly feed on a variety of woodland grasses (Poaceae), but sedges (Cyperaceae) are also reported in the Northeast (Hoag 2014). Overwinters as a partially grown larva.

Adult behavior: The Northern Pearly-Eye has a rapid, erratic flight. Unlike most butterflies in our region, they are active in dimly lit forest, and individuals fly well into twilight. Adults are typically seen perched on large branches and tree trunks, often upside down. This species rarely nectars at flowers, instead preferring to feed on sap, rotting fruit, dung, and damp earth.

Comments: A forest inhabitant that is not often seen outside that environment, the Northern Pearly-Eye was once thought to be rare throughout its range. Two subspecies, *L. a. anthedon* and *L. a. borealis* (A. Clark), are recognized by some (e.g., Grkovich and Pavulaan 2003) and ignored by others (e.g., Layberry et al. 1998). Both putative subspecies occur in the Acadian region, but because perceived differences are very subtle and inconsistent, we do not apply any subspecies names to our populations.

Threats: None documented in our region.

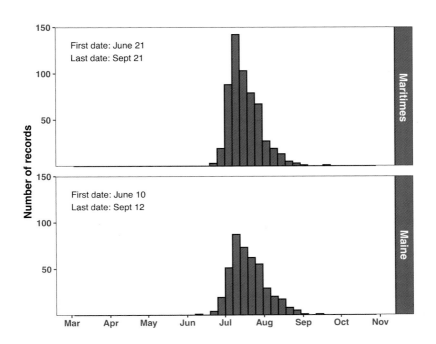

First date: June 21
Last date: Sept 21

First date: June 10
Last date: Sept 12

EYED BROWN

Lethe eurydice (Linnaeus)

Subspecies: Only the nomino-typical subspecies, *L. e. eurydice*, occurs in eastern North America.

Distinguishing characters:
    Size: Medium.
    Upperside: Pale to medium brown with dark spots near edges of wings; forewing spots smaller toward apex; posterior-most hindwing spots usually with small white pupils.

Ovipositing female—Camden, ME, 12 July 2019 (Roger Rittmaster)

    Underside: Pale to medium brown with row of bold eyespots ringed with yellow; forewing eyespots roughly the same size; outermost dark line on hindwing very jagged, especially toward the inner margin.
    Additional: Hindwing margin rounded and not scalloped.

Similar species: Northern Pearly-Eye and Appalachian Brown (nearly identical).

Status and distribution in Maine: Resident. Uncommon to common; widespread but localized. S4S5

Status and distribution in Maritimes: Resident. Uncommon to common; widespread but localized. NB-S5, NS-S5, PE-S5

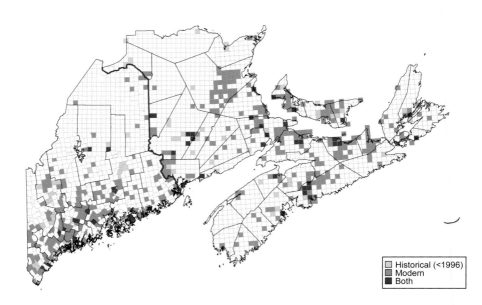

Historical (<1996)
Modern
Both

Habitat: Open, sedge-dominated wetlands, including salt marshes and occasionally roadside ditches.

Biology: One generation per year. Adults appear in late June and fly into early August, with peak activity in early July. Larvae feed on wetland sedges (*Carex* spp.). Overwinters as a partially grown larva.

Adult behavior: The Eyed Brown has a weak, bobbing flight that becomes quite erratic in alarmed individuals. They often perch on sedges or other wetland plants. Males patrol throughout the day for females just above or through wetland vegetation. Adults occasionally nectar at flowers, such as Common Milkweed (*Asclepias syriaca*). They also feed at rotting fruit, dung, and damp earth.

Comments: This species is generally restricted to open wetland habitats, where it can be quite common. Habitat alone helps distinguish the Eyed Brown from the very similar Appalachian Brown, which is typically found in the shade of wooded swamps. However, these species sometimes stray into the other's habitat, especially along edges where marshes transition into swamps. Under these circumstances, both species can sometimes be found together. Both species are variable, and some individuals appear to have intermediate markings, making identification very difficult.

Threats: None documented in our region.

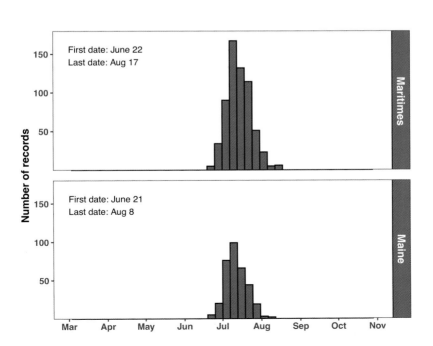

APPALACHIAN BROWN

Lethe appalachia R. Chermock

Berwick, ME, 27 July 2018 (Bryan Pfeiffer)

Subspecies: Only the nomino-typical subspecies (*L. a. appalachia*) occurs in the Acadian region.

Distinguishing characters:
   Size: Medium.
   Upperside: Grayish brown with dark spots near edges of wings, fresh specimens sometimes with purplish cast; forewing spots often very reduced; posterior-most hindwing spots with small white pupils.
   Underside: Grayish brown with bold eyespots ringed with yellow; forewing eyespots often vary in size; outermost dark line on hindwing typically less jagged than in Eyed Brown (but can be very similar).
   Additional: Hindwing margin rounded and not scalloped.

Similar species: Northern Pearly-Eye and Eyed Brown (nearly identical).

Status and distribution in Maine: Resident. Rare to uncommon; all records from the southwest. S3

Status and distribution in Maritimes: Not documented.

Habitat: Hardwood-dominated swamps with abundant sedges, often near marshes or streams.

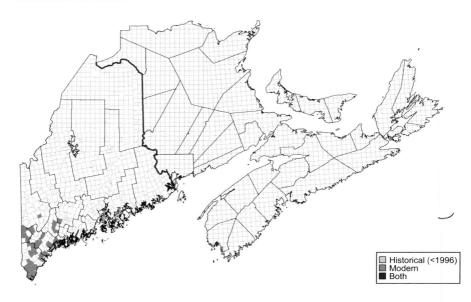

Historical (<1996)
Modern
Both

Biology: One generation per year. Adults fly from early July to late August, with a peak of activity in mid-July. Larvae feed on wetland sedges (*Carex* spp.). Overwinters as a partially grown larva.

Adult behavior: This species has an erratic, bobbing flight, but they perch often and are well camouflaged in the dappled light of the forest understory. They usually occur in close association with sedge-filled clearings in the forest but wander more widely from their breeding habitat than the Eyed Brown, often into adjacent marshes and upland forest. This butterfly rarely nectars at flowers, instead preferring to feed on rotting fruit, sap, and damp earth.

Comments: The Appalachian Brown was considered to be a subspecies of the Eyed Brown until 1970 (Cardé et al. 1970). These two species are extremely similar in appearance, but their habitat preferences differ (see Eyed Brown account). Unlike its sibling species, the Appalachian Brown rarely occurs in abundance. The paucity of historical records from our region probably reflects a lack of swamp exploration for butterflies, though a recent range expansion may be involved. A pre-1904 specimen labeled "Bangor Me," from the collection of the Pennsylvania lepidopterist Henry Engel (1873–1943), is questionable. Further surveys in central and coastal Maine are needed to better document the current northeastern range of this species. Many swamps with abundant growths of sedge do not seem to support this butterfly, suggesting that its habitat preferences are very specific.

Threats: Forested swamps are vulnerable to loss and degradation by development, pollution, and changes to wetland hydrology, especially in southern Maine.

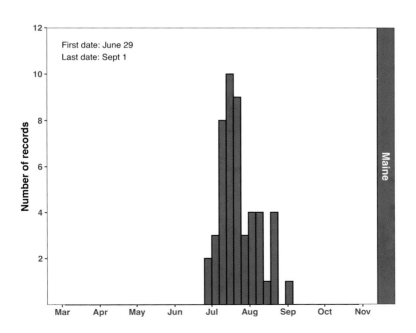

LITTLE WOOD-SATYR
Megisto cymela (Cramer)

Gloucester County, NB, 4 July 2018 (Krista Melville)

Subspecies: Only the nominotypical subspecies (*M. c. cymela*) occurs in north-eastern North America.

Distinguishing characters:
    Size: Medium.
    Upperside: Grayish brown; wings with yellow-ringed eyespots (two on forewing, one to three on hindwing).
    Underside: Tawny with two vertical, dark brown lines; forewing and hindwing each with two widely separated bold, yellow-ringed eyespots; hindwing with additional reduced eyespots and patches of silver scaling.

Similar species: Northern Pearly-Eye, Eyed Brown, Appalachian Brown, and Common Wood-Nymph.

Status and distribution in Maine: Resident. Common; widespread. S4S5

Status and distribution in Maritimes: Resident. Uncommon to common and widespread in New Brunswick and Prince Edward Island; uncommon in Nova Scotia, most records from the west. NB-S5, NS-S4, PE-S4

Habitat: Mixed and deciduous forest, generally in woodland openings or along their edges.

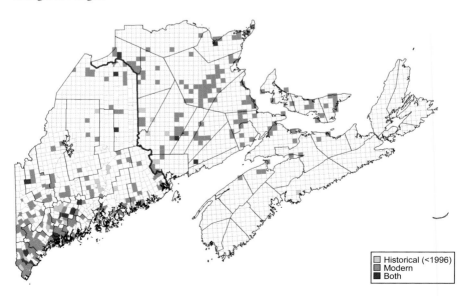

Historical (<1996)
Modern
Both

Biology: One generation per year. Adults are on the wing from late May until early August (rarely late August), with a peak flight in mid-June. Larvae feed on various grasses (Poaceae), including Kentucky Blue Grass (*Poa pratensis*), Orchard Grass (*Dactylus glomerata*), and presumably some woodland species. Overwinters as a partially grown larva.

Adult behavior: This species has a relatively slow, erratic, bouncing flight. Males patrol throughout the day for females, and both sexes often perch near the ground on sunlit vegetation. Adults visit several flowers in our region, including Spreading Dogbane (*Apocynum androsaemifolium*), Common Strawberry (*Fragaria americana*), and Common Blackberry (*Rubus allegheniensis*). Like other satyrs, it is more attracted to sap, rotting fruit, and dung.

Comments: The Little Wood-Satyr is a recent colonist in much of our region. Although Brower (1974) described it as "a species of the southwest quarter of Maine," modern records now extend statewide. In Nova Scotia, Ferguson (1954) described it as "apparently very local"; the only known specimens at that time were from Dartmouth in 1946. By the mid-1990s it was reported in New Brunswick from scattered locations in the southwest and at Bathurst (Thomas 1996). Maritime Butterfly Atlas records suggest it is still absent from most of Nova Scotia. The cause of the species' range expansion is unclear. A warming climate is a possibility, but this species has long been known from northerly latitudes in central Canada. It is interesting to note that the wings of many specimens from Maine (~3%) display damage from bird attacks, which is consistent with the hypothesis that marginal eye spots help increase survivorship (Kodandaramaiah 2011).

Threats: None documented in our region.

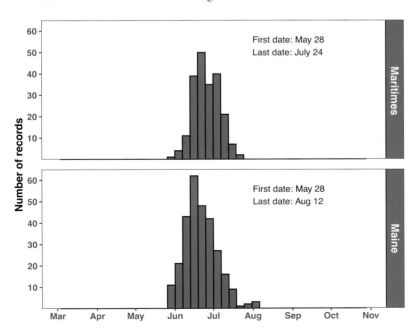

COMMON WOOD-NYMPH
Cercyonis pegala (Fabricius)

Subspecies: Two subspecies occur in our region: *C. p. alope* (Fabricius) in southern Maine and *C. p. nephele* (W. Kirby) in northern Maine and northern New Brunswick; intermediates elsewhere.

Distinguishing characters:
    Size: Medium.
    Upperside: Brown with two eyespots on the forewing and one or more small eyespots on the hindwing; forewing eyespots sometimes surrounded by creamy yellow or orange yellow (subspecies *alope* and similar phenotypes).
    Underside: Light to dark brown with fine striations; forewing with two large eyespots; hindwing with irregular series of several small eyespots; forewing eyespots sometimes surrounded by creamy yellow or orange yellow (subspecies *alope* and similar phenotypes).
    Additional: Hindwing margins scalloped.

Similar species: Jutta Arctic.

Status and distribution in Maine: Resident. Common; widespread. S5

Status and distribution in Maritimes: Resident. Common; widespread. NB-S5, NS-S5, PE-S5

Habitat: Open grassy areas, such as old fields, pastures, streamsides, and salt marshes. Sometimes found in the dappled shade of forest margins.

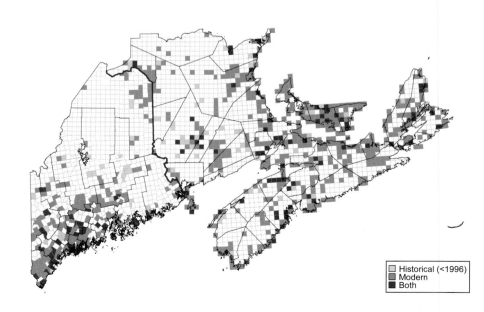

Historical (<1996)
Modern
Both

C. p. nephele—Grand Manan, NB, 9 August 2012 (Kent McFarland)

C. p. alope—Falmouth, ME, 6 August 2018 (Josh Lincoln)

Biology: One prolonged generation per year, flying from late June until late September, with peak flight in late July and early August. Larvae feed on many grass species (Poaceae). Overwinters as a newly hatched larva.

Adult behavior: This butterfly has a slow, erratic, bouncing flight, but they can be quite evasive when alarmed. Males patrol tirelessly throughout the day. Females are less active and often perch within areas of dense shrubbery. Some adults (particularly of the subspecies *alope*) roost in trees for the evening. Common Wood-Nymphs visit flowers far more often than other satyrs, particularly knapweeds (*Centaurea* spp.), goldenrods (*Solidago* spp.), thistles (*Cirsium* spp.), meadowsweets (*Spirea* spp.), and dogbanes (*Apocynum* spp.). It also frequents rotting fruit and damp soil.

Comments: Like some other summer-flying meadow species (e.g., greater fritillaries), male Common Wood-Nymphs emerge about 1 week before females, which sometimes fly very late into the season. Populations that consistently produce forewing patches ("true" *alope*) are more southerly and

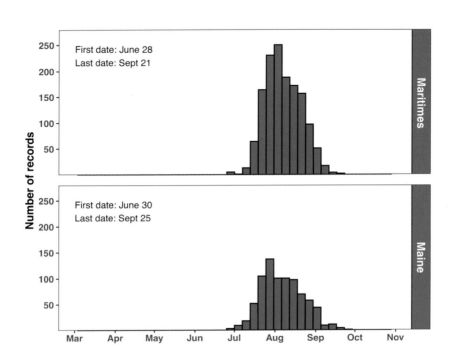

coastal in distribution. Elsewhere, most adults have reduced patches (intermediates) or no patches (*nephele*-type). Patched adults become progressively less prevalent northward, with just a few adults (mostly females) expressing this character before it disappears entirely in the northern reaches of our region ("true" *nephele*) (Calhoun 2016).

Threats: None documented in our region.

JUTTA ARCTIC

Oeneis jutta (Hübner)

North Piscataquis, ME, 15 June 2018 (Bryan Pfeiffer)

Subspecies: We tentatively consider our populations to represent *O. j. ascerta* Masters & Sorensen pending further research.

Distinguishing characters:
  Size: Medium.
  Upperside: Rarely seen; brown with series of black submarginal spots with dull orange halos (larger in female).

Underside: Forewing yellowish brown, with one to three eyespots (larger in female); hindwing mottled blackish brown with scattered gray scaling (sometimes an indistinct, dark median band) and often a small, dark-ringed submarginal eyespot; frequently also a series of pale submarginal spots.

Similar species: Katahdin Arctic and Common Wood-Nymph.

Status and distribution in Maine: Resident. Uncommon and localized; widespread but absent in the southwest. S4

Status and distribution in Maritimes: Resident. Common, localized, and widespread in New Brunswick; known from one site in Prince Edward Island; uncommon to rare in Nova Scotia, not recorded in the southwest. NB-S4, NS-S3S4, PE-S1

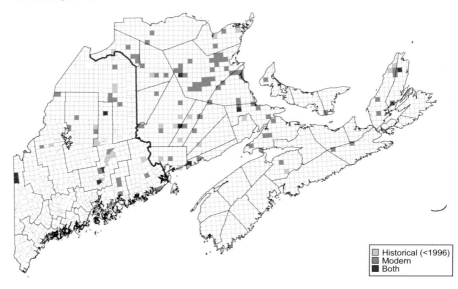

Historical (<1996)
Modern
Both

Habitat: Sphagnum bogs, mostly in the partly forested margins among Black Spruce (*Picea mariana*) and American Larch (*Larix laricina*).

Biology: One generation, but 2 years are probably required to complete development. Adults are on the wing from mid-May until early July, with a peak flight in mid-June. Larvae feed on sedges, particularly cottongrasses (*Eriophorum* spp.). Overwinters as a larva, probably over two seasons.

Adult behavior: When disturbed, adults have a very rapid, erratic flight, but they will typically alight on a nearby tree trunk, branch, or bog heath, where they are well camouflaged. Males sometimes perch on the tufts of cottongrasses. In our region, they have been observed nectaring at Labrador Tea (*Ledum groenlandicum*).

Comments: Once named the "Nova Scotian Arctic," this butterfly was for many years known in the United States only from Orono Bog, Penobscot County, Maine. As in Ontario (Hall et al. 2014), the Jutta Arctic probably produces two brood cohorts in our region: one in even numbered years and one in odd numbered years. A 2-year life cycle is an adaptation that allows some species to survive in northern climates in which a single growing season is too short for larval development. Some now believe that the Jutta Arctic is restricted to Eurasia, and the North American look-alike is a different species (Kondla and Schmidt 2010). If so, then the next available name is *Oeneis balder* (Guérin-Méneville). The similar name *balderi* (Geyer) has sometimes been used to identify the subspecies in our region, but it was proposed 5 years after the name *balder*.

Threats: The Jutta Arctic is most common in northern New Brunswick and northern Maine, an area under intensive commercial forest management. Wet forested bog margins are generally not harvested, but hydrology and habitat conditions can be affected if peatland buffers are degraded.

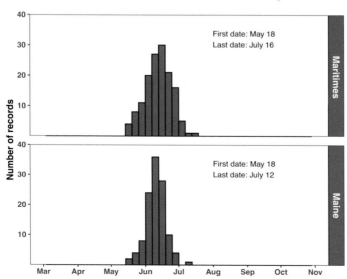

KATAHDIN ARCTIC

Oeneis polixenes (Fabricius)

Subspecies: Only *O. p. katahdin* (Newcomb) occurs in our region.

Distinguishing characters:
  Size: Medium.
  Upperside: Yellowish to orange brown with usually one (male) or multiple (female) small black eyespots on forewing; wings appear delicate and somewhat translucent.

Mount Katahdin Township, ME, 1 July 2016 (Matthew Orsie)

  Underside: Hindwing mottled dark brown and light gray; dark, scalloped central band usually flanked by narrower pale bands, especially outwardly; sometimes a series of pale submarginal spots; often a small eyespot near forewing tip.

Similar species: Jutta Arctic.

Status and distribution in Maine: Resident. Rare; restricted to Mount Katahdin. State Endangered. S1

Status and distribution in Maritimes: Not documented.

Habitat: Alpine tundra; tableland above the timberline of Mount Katahdin.

Biology: Poorly understood. One generation, but 2 years may be required to complete development. Adults fly from late June to late July, with a peak

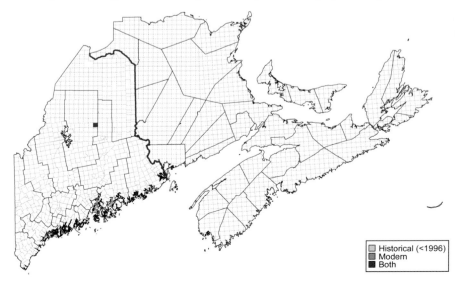

Historical (<1996)
Modern
Both

generally during early July. The food plant is unknown, but other subspecies feed on sedges (*Carex* spp.) and grasses (e.g., *Festuca* spp.). Overwinters as a larva, possibly over two winters.

Adult behavior: This elusive arctic flies low and rapidly over alpine tussocks, sometimes letting the wind carry them for considerable distances. They often rest on lichen-covered rocks or mosses, where they are well camouflaged by their mottled undersides. Adults of other subspecies are known to visit flowers occasionally, but such observations are lacking in Maine.

Comments: By far the most range-restricted butterfly in our fauna, the Katahdin Arctic is an extremely disjunct subspecies confined to the wind-swept tablelands of Mount Katahdin (1605 m; 5265 ft) within Baxter State Park (BSP). Park naturalists have conducted visual encounter surveys since the 1980s and report the species as common in some years, with single day counts of more than forty individuals in 2003 and 2005 (Jean Hoekwater, BSP naturalist, unpub. data). During a rugged expedition to the Mount Katahdin tablelands in 1901, Harry H. Newcomb reported collecting sixty-three adults over three days (Newcomb 1901). He used those specimens to describe the species *Chionobas katahdin*, which is now recognized as a subspecies of *O. polixenes*. The larval food plant and life history details of this unique subspecies requires further study.

Threats: While its mountain habitat is protected, the Katahdin Arctic remains potentially vulnerable to off-trail trampling of its food plants and illegal collection (McCollough et al. 2003). Additionally, alpine plant communities are vulnerable to climate change, particularly from woody plant succession, as documented on the summit of Bigelow Mountain in western Maine (Capers and Stone 2011). The Katahdin Arctic is therefore threatened with extinction if the treeless "sky island" it occupies is degraded in size or composition by a warming climate.

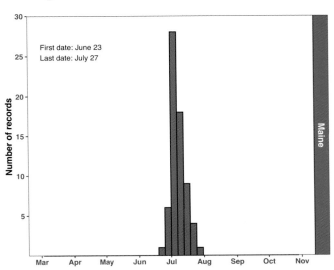

First date: June 23
Last date: July 27

BUTTERFLIES OF POSSIBLE OCCURRENCE

Despite considerable research in our region over the past century, and the discovery of several new state and provincial butterfly species records during the recent atlas era (Table 2, Methods and Biogeographical Findings), at least twenty additional species have a reasonable chance of being found in the Acadian region, either as strays or breeding residents. A recurring theme among many of these potential new species is a northward range expansion, likely in response to climate change (see climate discussions in Methods and Biogeographical Findings and Butterfly Conservation in the Acadian Region). Climate change, combined with a greater diversity of butterflies south of our region, suggests that most additions would potentially occur only in southern or coastal Maine. Notable exceptions include the Pelidne Sulphur and Freija Fritillary, species characteristic of the tundra and boreal life zones that are more likely to be found in remote northern portions of the Acadian region.

Hoary Edge—Thorybes lyciades (Geyer). This large skipper, similar in size and appearance to the Silver-spotted Skipper (*Epargyreus clarus*), was formerly included in the genus *Achalarus*. It reaches its northern range limit in New England, where it has been recorded in southern New Hampshire and Massachusetts (Opler 1995; Stichter 2015). The Hoary

Worcester, MA, 19 June 2021 (Michael Newton)

Edge is widespread but uncommon in Massachusetts and thought to be declining. The larvae feed on various legumes in and around upland forests. This species is single brooded, with adults flying from early June to early August. This skipper could occur in extreme southern Maine as a stray or temporary colonist.

Male—Northhampton, MA, 16 September 2016 (Michael Newton)

Common Checkered Skipper— *Burnsius c. communis* (Grote). This southern species regularly immigrates northward, establishing temporary populations as far north as Connecticut (O'Donnell et al. 2007). There are several late-season records (mid-August to early October) from Massachusetts and Vermont within the last two decades (McFarland and Zahendra 2010; iNaturalist 2020; Lotts and Naberhaus 2020). The Common Checkered Skipper is multiple brooded, and the larvae feed on mallows (*Malva* spp.), several of which occur in our region. This species may sporadically reach Maine and has the potential to produce at least one brood if adults arrive early enough in the season. This species was previously included in the genus *Pyrgus*.

Male—Coral Gables, FL, 12 February 2017 (Bryan Pfeiffer)

Horace's Duskywing—*Erynnis horatius* (Scudder & Burgess). A very common resident of the South, this species has been recorded as far north as southern New Hampshire (Opler 1995) and southern Vermont (McFarland and Zahendra 2010). It is known to breed in Massachusetts, where it is widespread and possibly increasing in abundance (Stichter 2015). The Horace's Duskywing has two generations in Massachusetts, with adults of the first brood flying from late April into May and the second brood flying in July and August. Because larvae feed on various oaks (*Quercus* spp.), this skipper is typically found in and around forests that support an abundance of these trees. The Horace's Duskywing is very similar to the single-brooded Juvenal's Duskywing (*Erynnis juvenalis*), but they can be confused only during spring. Any large duskywing encountered in the summer in Maine would most likely be Horace's (but see *Erynnis funeralis*, below).

Funereal Duskywing—Erynnis funeralis (Scudder & Burgess). The range of this species was originally thought to be limited to the southwestern United States and south through Mexico and Central America to Argentina. A legume feeder as a larva, the Funereal Duskywing is now known to be a regular, seasonal immigrant that establishes temporary populations in much of eastern North America (Carpenter 2015).

Male—Tuscon, AZ, 20 December 2014 (Bryan Pfeiffer)

In the last decade, it has been recorded as far northeast as New York and Rhode Island (Calhoun 2019). Strays could reach southern Maine, turning up in virtually any open, flowery habitat, most likely from late August to early October.

Columbine Duskywing—Erynnis lucilius (Scudder & Burgess). This small duskywing inhabits deciduous forests, limestone outcrops, and rocky hilltops that support an abundance of its food plant, Red Columbine (*Aquilegia canadensis*). The Columbine Duskywing has declined throughout its range and is now absent from many areas where it once occurred. It is at

Female—Lincoln County, ON, 6 June 2019 (Michael King)

least double brooded, with adults flying from May to early September. There are modern records from Vermont (McFarland and Zahendra 2010), and in New Hampshire it has been found very near the border of southern Maine (Webster and deMaynadier 2005). It is also known from Quebec, west of Maine (Handfield 2011). Habitats in Maine where Red Columbine grows should be surveyed for the presence of this skipper, especially rocky hilltops in northern York and southern Oxford Counties. Red Columbine also occurred historically in New Brunswick, where it is now considered extirpated. Sites should be checked multiple times during different times of the season, as adults may occur in low numbers.

Male—South Dartmouth, MA, 6 October 2017 (Michael Newton)

Sachem—*Atalopedes huron* (W. H. Edwards). Widespread and common in the south, this species regularly migrates northward. It erupts explosively, resulting in temporary populations in the east as far north as Massachusetts (Stichter 2015). Since 1995, reports of this skipper have dramatically increased in Massachusetts, where it produces one or more broods, depending how early adults arrive. A grass feeder as a larva, the Sachem can be found in virtually any open, flowery habitat. It probably already occurs in Maine during some years and will likely become more regular as a result of climate change, similar to what appears to be happening in the Pacific Northwest (Crozier 2002). Although it has been recorded in Massachusetts as early as June and July, it is most common later in the season. Searching clover fields in southern Maine from August to early October may reveal the presence of this highly opportunistic skipper, which may be a different species than the western *campestris*.

Zabulon Skipper—*Lon zabulon* (Boisduval & Le Conte). This widespread species is known to occur in Massachusetts, where it was originally listed as a rare vagrant following its discovery in 1988.

Milder winters have contributed to a dramatic increase of this skipper in Massachusetts, where it has become an uncommon to common resident, as indicated by Stichter (2015) and the many recent reports posted on various record-sharing platforms (e.g., iNaturalist). It was also recently recorded in southern New Hampshire. Unlike the similar Hobomok Skipper (*Lon hobomok*), which is single

Male—South Dartmouth, MA, 28 May 2017 (Michael Newton)

brooded, the Zabulon Skipper produces two generations in the north. In Massachusetts, adults have been recorded from May to mid-June and from late July to early October. Feeding on grasses as a larva, this species prefers moist woodlands and nearby fields, but it also frequents parks and gardens. Historical records from Maine are erroneous (see Butterflies Dubiously or Erroneously Reported), but it could spread into the southern part of the state or may already occur there. The Zabulon and Hobomok skippers were previously included in the genus *Poanes*.

Sleepy Orange—*Abaeis nicippe* (Cramer). A southern legume feeder in the larval stage, this species strays northward and has been recorded in Massachusetts on several occasions (Opler 1995). The Sleepy Orange could stray into our region, especially during August and September, but it would likely not establish colonies, nor would it survive the winter.

Manhatten, KS, 11 August 2013 (Bryan Pfeiffer)

Female—Chisasibi, QC, 2 August 2015
(Maxim Larrivée)

Pelidne Sulphur*—Colias p. pelidne*
Boisduval & Le Conte. This is an Arctic
species, which has been recorded as far
south as southern Newfoundland, just
across the Gulf of St. Lawrence from
Cape Breton Island, Nova Scotia (Morris 1980). Inhabiting bogs and nearby
areas, the larvae feed on blueberries
(*Vaccinium* spp.). It is single brooded,
with adults in Newfoundland on the
wing from early July to early August.
The Pelidne Sulphur is very similar to
the Pink-edged Sulphur (*Colias interior*), and some individuals share
morphological characteristics. Hammond and McCorkle (2017) went so
far as to consider these species as conspecific, but this treatment is not
widely accepted.

Polk County, FL, 1 June 2019 (Edward Perry IV)

Southern Dogface*—Zerene c. cesonia* (Stoll). Another southern legume
feeder in the larval stage, this species
irregularly wanders northward, especially during late summer and autumn.
There are historical records from
southern New Hampshire (Opler 1995),
suggesting that this butterfly could enter our region as a nonbreeding stray.

Bay Minette, AL, 1 April 2019 (Bryan Pfeiffer)

Red-banded Hairstreak*—Calycopis
cecrops* (Fabricius). Expanding northward, this southern species was first
documented in Massachusetts in 2011,
where it has since been recorded at a
number of scattered localities (Stichter
2015). The Red-banded Hairstreak is
found in semi-open habitats, such as
brushy fields and forest edges. Often
associated with sumacs (*Rhus* spp.),
larvae probably feed mostly on detritus

on the forest floor. Adults have been recorded in Massachusetts in August and September, possibly only as immigrants, though they may occasionally reproduce there. This butterfly may stray into extreme southern Maine late in the season.

'Northern' Oak Hairstreak— *Satyrium favonius ontario* (W. H. Edwards). This drab subspecies of the southeastern Oak Hairstreak (*S. favonius*) has been recorded as far north as Massachusetts, where it is rare to uncommon (Stichter 2015). The 'Northern' Oak Hairstreak primarily inhabits open oak forests and pine-oak barrens, where larvae feed on oaks (*Quercus* spp.), particularly those

Worcester, MA, 24 June 2018 (Bruce DeGraaf)

of the white oak group (subgenus *Leucobalanus*) (Gagliardi and Wagner 2016). The flight season of this single-brooded species is relatively early for a hairstreak. In Massachusetts, most records are from early June to early July, though it has rarely been recorded as late as early August. Adults probably spend the majority of their time in the forest canopy feeding on non-nectar resources, but they often descend to visit flowers (Gagliardi and Wagner 2016). This species may be resident in coastal southern Maine, where pockets of suitable habitat occur (Gagliardi et al. 2017) (see also Butterflies Dubiously or Erroneously Reported).

Hickory Hairstreak— *Satyrium caryaevorus* (McDunnough). In the Northeast, this poorly understood species ranges into Vermont and Massachusetts, where it is uncommon to rare (McFarland and Zahendra 2010; Stichter 2015). Occurring in rich, deciduous oak–hickory forests, the larvae feed mostly on hickories, such as Shagbark Hickory

West Newbury, MA, 7 July 2014 (Erik Nielsen)

(*Carya ovata*) and especially Bitternut Hickory (*C. cordiformis*). The single-brooded adults fly from June to mid-August in Massachusetts. Very similar to the resident Banded Hairstreak (*S. calanus*), the Hickory Hairstreak may occur in extreme southern Maine.

Bucks County, PA 14 April 2004 (David M. Wright)

Spring Azure—*Celastrina ladon* (Cramer). Although Wright et al. (2019) included York County, Maine, within the distribution of this species, such butterflies are now believed to represent variants of the Northern Azure (*Celastrina l. lucia*) (see Species Accounts for the Summer Azure and Butterflies Dubiously or Erroneously Reported). The Spring Azure has been recorded in Massachusetts and New Hampshire, but it is unclear if "typical" *C. ladon* could occur in Maine as many New England populations share characteristics with the Northern Azure. The primary food plant of the Spring Azure is Flowering Dogwood (*Cornus florida*), yet this does not appear to be fed upon by populations in southern New England (Schweitzer 2008). A single-brooded, woodland butterfly that flies early in spring, the New England "Spring Azure" is poorly understood and needs further study. It may be composed of an amalgamation of genetically introgressed populations where the ranges of the Northern and Spring azures intersect.

Cherry Gall Azure—*Celastrina serotina* Pavulaan & D. Wright. This species was formerly believed to occur in Maine and the Maritimes (Pavulaan and Wright 2005; Webster and deMaynadier 2005), but Schmidt and Layberry (2016) considered such butterflies in eastern Canada to represent the Northern Azure (*Celastrina l. lucia*) or the Summer Azure (*Celastrina neglecta*). Single brooded, adults of the Cherry Gall Azure fly in May and June. Larvae are primarily associated with eriophyid mite-formed galls on the leaves of Black Cherry (*Prunus serotina*). Confusing the issue, larvae of the Northern Azure have been photographed feeding on cherry galls in Nova Scotia. As we learn more about this group of butterflies, it is possible that the Cherry Gall Azure will be confirmed in our region.

Monroe County, PA, 25 May 2008 (David M. Wright)

European Common Blue—

Polyommatus i. icarus (Rottemburg). This Eurasian species was first detected in North America near Montreal, Quebec, in 2005 (Hall 2007). It is now common in the Montreal area, where it has become established around Quebec City, less than 97 km (60 mi) from the Maine border. It is continuing to expand its range and was first recorded in the United States in September 2020 in northern Vermont (DiCesare 2020; Barrington and Pfeiffer 2021), and it was recorded in New York in September 2021 (Pfeiffer 2021). The European Common Blue inhabits open areas, such as old fields and roadsides. Its larvae feed on weedy legumes, including clovers (*Trifolium* spp.) and especially Garden Bird's-foot-trefoil (*Lotus corniculatus*), an introduced Old World species that now occurs across much of North America. Multiple brooded, adults fly from late May to late October. This butterfly is well suited to further colonization, and road corridors will likely enhance its dispersal. Disturbed areas adjacent to roadways in northwestern Maine and northwestern New Brunswick should be searched for this species, particularly where Garden Bird's-foot-trefoil grows in abundance.

Clinton County, NY, 4 September 2021 (Josh Lincoln)

Big Pine Key, FL, 14 February 2017 (Bryan Pfeiffer)

Gulf Fritillary—Dione incarnata nigrior (Michener). This southern species is known to stray as far north as Massachusetts, where it was recorded in October 1997 (McFarland and Zahendra 2010). Research by Halsch et al. (2020) infers that climate change may affect the distribution of this butterfly's food plants, passion flowers (*Passiflora* spp.). As the ranges of these plants expand northward, the chances increase that strays of the Gulf Fritillary could eventually reach Maine, especially late in the season. This species was previously included in the genus *Agraulis*. Zhang et al. (2020) suggested that the species name for North American populations should be *D. incarnata*, but more research is needed.

Timiskaming District, ON, 8 June 2019 (Isabel Apkarian)

Freija Fritillary—Boloria f. freija (Thunberg). This circumpolar species has been recorded in Quebec, within 81 km (50 mi) of the northwestern border of Maine and within about 89 km (55 mi) of New Brunswick (Layberry et al. 1998; Handfield 2011). It occurs in a variety of habitats, including open, rather dry willow bogs and nearby forest clearings that support an abundance of its likely food plants, blueberries (*Vaccinium* spp.) and Red Bearberry (*Arctostaphylos uva-ursi*). Adults have also been observed ovipositing on wild cranberries (*Vaccinium* spp.) in Michigan (Nielsen 1999). Single brooded, adults fly mostly from mid-May to early June. Suitable habitats in northwestern Maine and northern New Brunswick should be searched for the presence of this butterfly, especially in view of the relatively recent discovery of the Frigga Fritillary (*B. frigga*) in the region (deMaynadier and Webster 2009).

Red-disked Alpine—Erebia discoidalis (W. Kirby). Widely distributed across much of Canada and the extreme northern United States, this species has been recorded in Quebec, about 161 km (100 mi) west of Maine and New Brunswick (Handfield 2011). Localized and rare at the eastern edge of its range, the Red-disked Alpine inhabits open peat bogs, swampy clearings, and grassy, wet meadows. It is single brooded, with adults flying in adjacent areas of Quebec from mid-May to late June. There is a remote possibility that isolated populations of this butterfly occur within northern Maine or northwestern New Brunswick.

Clearwater County, AB, 22 May 2017 (Norbert Kondla)

Chryxus Arctic—Oeneis chryxus calais (Scudder). A wide-ranging northern and montane species, the Chryxus Arctic has been recorded in Quebec, within 81 km (50 mi) of northern Maine and within about 89 km (55 mi) of New Brunswick (Layberry et al. 1998; Handfield 2011). This butterfly inhabits sandy and rocky areas, including woodlands and bogs. Single brooded, it has a 2-year life cycle, with adults in southern Quebec flying from mid-May to mid-June, most often during odd-numbered years. Food plants in the east are poorly known, but

Chippewa County, MI, 22 May 2016 (John Christensen)

in Ontario this species has been observed ovipositing on grasses, and larvae were reared on Poverty Grass (*Danthonia spicata*) (Scott 1986). Although Handfield (2011) tentatively treated populations in Quebec as *O. c. calais*, recent DNA barcode analyses could not clearly distinguish this subspecies from *O. c. strigulosa* McDunnough (Warren et al. 2016). Populations of the Chryxus Arctic may occur in remote areas of northern Maine and New Brunswick.

BUTTERFLIES DUBIOUSLY OR ERRONEOUSLY REPORTED

The following twenty-three butterfly taxa have been credited to Maine and/or the Maritime Provinces in previous publications, but the records are questionable or invalid. In some cases, applicable names have changed based on more recent taxonomic revisions.

Persius Duskywing—*Erynnis p. persius* (Scudder). This species was reported from Maine by Scudder (1868), who credited Sidney I. Smith with finding it in the vicinity of Norway, Oxford County. Fernald (1884a, 1884b) also implied that Smith found it in western Maine. In addition, Scudder (1888–1889, 1899) claimed that he personally recorded the Persius Duskywing at Moosehead Lake (probably Piscataquis County), Maine. No specimens of this species from Maine have been found, and published reports appear to be based on confusion with the Dreamy Duskywing (*Erynnis icelus*) (Calhoun 2017b). The nearest valid records of the Persius Duskywing are from southern New Hampshire.

Southern Broken-Dash—*Polites otho* (J. E. Smith). Records of this species from Maine (e.g., Fernald 1884a, 1884b) apply to the Northern Broken-Dash (*Polites egeremet*). Although *P. egeremet* was originally described as a species, it was subsequently regarded as a variety or subspecies of *P. otho*. Burns (1985) confirmed that they are separate species, with the Northern Broken-Dash occurring farther north, as its common name implies. Until recently, these species were included in the genus *Wallengrenia*.

Common Branded Skipper—*Hesperia colorado manitoba* (Scudder). Records in our region of butterflies named *manitoba* (e.g., McIntosh 1899b; dos Passos and Grey 1934a) apply to the Laurentian Skipper (*Hesperia colorado laurentina*), which was originally described as a variety of a species then known as *Pamphila manitoba*. Some later authors (e.g., MacNeill 1964; Kondla and Schmidt 2010; Handfield 2011) continued to treat *manitoba* as a species, with *laurentina* as a subspecies (i.e., *Hesperia m. laurentina*). Today, *manitoba* is usually regarded as a western North American subspecies of *H. colorado* (formerly *comma*). More research is required to fully understand the relationships within this complex group.

Zabulon Skipper—*Lon zabulon* (Boisduval & Le Conte). Reports of this species in the Acadian region (e.g., Fernald 1884a, 1884b; Bethune 1894; McIntosh 1899a) apply to the Hobomok Skipper (*Lon hobomok*), which was often regarded in older literature as a form or variety of the Zabulon Skipper. Until recently, these species were included in the genus *Poanes* (see Butterflies of Possible Occurrence).

Little Yellow—*Pyrisitia lisa euterpe* (Ménétriés). The Little Yellow (*Pyrisitia l. lisa*) was frequently identified in older literature (e.g., dos Passos and Grey 1934a) using the scientific name *euterpe*, which now defines the West Indian subspecies, *P. lisa euterpe*. This subspecies does not occur in Maine or the Maritimes.

Large Orange Sulphur—*Phoebis agarithe* (Boisduval). Several authors (dos Passos and Grey 1934a; Farquhar 1934; Brower 1974) reported that Roland Thaxter had collected a single female of this species on 3 September 1930 at Kittery Point, York County, Maine. This specimen was recently found at Boston University, where it was received many years ago from the now-defunct Boston Society of Natural History. Although an old handwritten label identifies the specimen as "*Catopsila agarithe*," it is actually a worn, pale female Orange-barred Sulphur (*Phoebis philea*), thus confirming the presence of this species in Maine as a rare stray (see Species Accounts).

Large White—*Pieris brassicae* (Linnaeus). Three individuals of this Old World species were reported by Ray and Kit Stanford on 25 July 1995 at

Lubec and Quoddy Head State Park, Washington County, Maine (Mello 1999). Opler and Warren (2006) claimed that these were sight records, explaining why these specimens were not found in Stanford's collection at Colorado State University. These sightings undoubtedly apply to the Cabbage White (*Pieris rapae*), which Stanford also reported from the same date and localities in Maine (Mello 1999). The Large White also was reported in error from Rhode Island in 2016 (Scoones 2016). This species was photographed on Staten Island, New York, in 2000 (Zirlin 2002), but it is not known to be established in North America.

Mustard White—Pieris napi (Linnaeus). Although the similar *Pieris oleracea* was originally described as a species, subsequent authors often defined it as a variety or subspecies of *P. napi*. Most researchers now regard *P. napi* as Palearctic and *P. oleracea* as a discrete North American species.

Frosted Elfin—Callophrys i. irus (Godart). This species was first reported from Maine by Scudder (1872), who later mentioned that a specimen was supposedly collected by Sidney I. Smith at Norway, Maine (Scudder 1888–1889). This specimen, deposited at the Peabody Museum of Natural History (Yale University), represents the Hoary Elfin (*Callophrys polios*) (Calhoun 2017b). Fernald (1884a, 1884b) listed "*Thecla irus*; variety *arsace*," but provided no additional information other than a description of the adult butterfly. Brower (1974) listed a male specimen of the Frosted Elfin, which was reportedly collected by Smith at Norway, Maine, on 31 May. This specimen, also at the Peabody Museum of Natural History, was actually collected by Smith in Connecticut (Calhoun 2017b). A single female Frosted Elfin, seemingly labeled "ME," was recently found at Boston University. Identified as "*Thecla Henrici*" and dated 21 June, its origin in Maine is extremely dubious (Calhoun 2017b). In addition, Perrin and Russell (1912) listed "*Incisalia irus*" and "*I. irus* var. *arsace*" from Nova Scotia, though they admitted that some of the records were applicable to the Hoary Elfin. The Frosted Elfin is not known to occur in Maine or the Maritimes; all previous records evidently apply to the Henry's Elfin (*Callophrys henrici*) and the Hoary Elfin. The nearest valid records of the Frosted Elfin are from southern New Hampshire. It was known in Canada only from Norfolk County in southern Ontario, where it was last seen in 1988 (COSEWIC 2000).

'*Northern*' *Oak Hairstreak—Satyrium favonius ontario* (W. H. Edwards). This butterfly was reported from Maine by dos Passos and Grey (1934a), who credited Charles A. Frost with finding "*Strymon ontario*" during July at Monmouth, Kennebec County. However, dos Passos and Grey (1934b) subsequently rejected the record, noting that it was based on a misidentified Acadian Hairstreak (*Satyrium acadica*). This probably corresponds to Frost's record of the Acadian Hairstreak at Monmouth on 17 July 1904, as reported by Farquhar (1934). No specimens of the 'Northern' Oak Hairstreak from Maine were found at the Museum of Comparative Zoology (Harvard University), where most of Frost's butterflies are deposited. This species has been recorded in northeastern Massachusetts (Stichter 2015) and may eventually be found in Maine (see Butterflies of Possible Occurrence).

Spring Azure—Celastrina ladon (Cramer). Previous authors (e.g., Opler 1995; Layberry et al. 1998) included Maine and the Maritimes within the range of this species, but this was based on taxonomic confusion with the Northern Azure and the Summer Azure. More recently, Wright et al. (2019) mapped *C. ladon* in York County, Maine, but such butterflies are now believed to represent variants of the Northern Azure with character introgression from *C. ladon*. (See entry for Spring Azure in Butterflies of Possible Occurrence.)

Vancouver Island Blue—Icaricia saepiolus insulanus (Blackmore). This subspecies was listed by dos Passos and Grey (1934a) in error for *Icaricia saepiolus amica* (W. H. Edwards) (Greenish Blue), which is widely distributed in North America. The name *insulanus* is currently restricted to endemic (possibly extinct) populations of *I. saepiolus* on Vancouver Island, British Columbia.

Arctic Blue—Agriades glandon labrador Schmidt, J. Scott & Kondla. Brower (1974) listed a record of "*Plebejus aquilo*" from Maine. Following Pelham (2022), this name is tentatively treated as synonymous in North America with *A. glandon*. Brower's record is based on a letter from the Pennsylvania lepidopterist Frank H. Chermock, who claimed that he and his wife collected two specimens on 1 July 1941 atop Mount Katahdin, Piscataquis County, Maine. Now deposited at the McGuire Center for Lepidoptera and Biodiversity (Florida Museum of Natural History,

Gainesville) (Calhoun 2017a), these specimens were supposedly collected within the alpine tableland that occurs above tree line on Mount Katahdin. This is the same area of the mountain where the endemic Katahdin Arctic (*Oeneis polyxenes katahdin*) occurs. For many years after its discovery in 1901, the Katahdin Arctic was the principal quarry of most lepidopterists who scaled Mount Katahdin. As a result, the tableland of the mountain has been thoroughly explored for decades by many experienced lepidopterists, yet no Arctic Blues were ever recorded. Since 1941, Mount Katahdin has been surveyed for butterflies without any additional records of this species. The Arctic Blue is not currently known to occur any closer to Mount Katahdin than central and eastern Quebec. Chermock's specimens are likely mislabeled (Calhoun 2017a).

Cranberry Blue—Agriades optilete (Knoch). Webster and deMaynadier (2005) mistakenly used the name *Agriades optilete* in reference to the dubious Maine specimens of *A. glandon*. Within North America, *A. optilete* is thought to occur only in the extreme Northwest.

Northern Blue—Plebejus argyrognomon (Bergsträsser). For many years, the name of this European species was applied to North American populations of *Plebejus idas*. Higgins (1985) emphasized that the name *argyrognomon* should be restricted to the Old World.

Karner Blue—Plebejus samuelis (Nabokov). This species was first reported from Maine by Dirig (1994), who mapped a record in Oxford County based on his discovery of a female labeled "Maine" and dated "80" (1880) at the Natural History Museum, London. Despite a lack of evidence, it was subsequently suggested that the specimen was collected by Sidney I. Smith in the vicinity of Norway, Maine, and publications thereafter included Maine within this species' historical range. The specimen deposited in London, and another labeled "Maine" in the same collection, were actually sent to a German entomologist in 1880 by the Maine naturalist Charles Fish, who probably obtained them from another correspondent (Calhoun 2017a). They were most likely collected in the vicinity of Albany, New York, which was a well-known locality for the species at that time (and the origin of specimens that were later used for the description of *samuelis*). The nearest valid records of *P. samuelis* are from southern New Hampshire, where the last native population was considered extirpated in 2000 (a reintroduction program was initiated in 2001).

Titania Fritillary—*Boloria titania* (Esper). Reports of this species apply to the Arctic Fritillary (*Boloria chariclea grandis*). The Titania Fritillary is now generally regarded as an Old World species (Shepard 1998).

Red-spotted Purple—*Limenitis arthemis astyanax* (Fabricius). This subspecies was reported from Maine by several authors (e.g., dos Passos and Grey 1934a, 1934b; Brower and Payne 1956), but such butterflies are best considered to represent "intermediate" phenotypes (form "proserpina") between the Red-spotted Purple and the more northern White Admiral (*L. a arthemis*). The Red-spotted Purple ranges from Massachusetts southward.

Hackberry Emperor—*Asterocampa c. celtis* (Boisduval & Le Conte). This species was recorded near Truro, Nova Scotia, based on a larva that was reportedly found on Yellow Birch (*Betula alleghaniensis*) (McGugan 1958). This species is known to feed only on hackberries (*Celtis* spp.; Ulmaceae), which are not native to Nova Scotia. This is undoubtedly a misidentification.

Small Tortoiseshell—*Aglais urticae* (Linnaeus). There are three specimens of this Eurasian butterfly from the Acadian region: one male (missing head and abdomen) from Saint John, New Brunswick, dated 15 August 1907 and collected by George Morrisey (New Brunswick Museum); one male that flew from a crate of books received in Halifax, Nova Scotia, from England on 7 November 1970 (Scott and Wright 1972); and one female in the Nova Scotia Museum that was collected 21 August 2019 in an urban garden by Derek Bridgehouse in Dartmouth, Nova Scotia. There is no doubt that the Halifax specimen was inadvertently imported. Given the infrequency of records, it reasonable to assume that the other two individuals also were imported and not representative of local breeding populations. This species has been documented on a number of occasions in eastern North America, where it may establish localized, temporary populations as the result of accidental introductions or releases by breeders.

Gorgone Checkerspot—Chlosyne gorgone (Hübner). A single male specimen of this species, deposited at the National Museum of Natural History (Smithsonian Institution), is labeled from Barnesville, Maine, and dated June 1902. Its printed label, however, is of modern origin, and there is no town in Maine by that name. This record probably applies to Barnesville, Clay County, Minnesota, where *C. gorgone* is known to occur. In addition, this species was listed from Nova Scotia (as *Phyciodes carlota*) by Bethune (1894), who cited Jones (1872a). This record can be traced to Belt (1864), who reported finding "*Nelitaea* [sic] *ismeria*" in July 1862 or 1863 near Lake Loon (now Loon Lake) and Lake Thomas, Halifax. Belt's specimens undoubtedly were Harris' Checkerspots (*Chlosyne harrisii*), as evidenced by the use of the name *Melitaea ismeria* for that species by Harris (1862), who Jones (1872a) cited in reference to Belt's record. Bethune (1894) simply misinterpreted Jones' (1872a) use of the name *ismeria* to mean *Phyciodes carlota* (= *C. gorgone*).

Tawny Crescent—Phyciodes b. batesii (Reakirt). This species was reported from Maine by several authors, beginning with Parlin (1922), who hesitantly listed it from Machias, Washington County. Brower (1974) mentioned a specimen that was reportedly collected by Sidney I. Smith at Norway, Maine. Schweitzer et al. (2011) referred to this specimen as the only credible record of this species in New England. This specimen, deposited at the Peabody Museum of Natural History (Yale University), is a Northern Crescent (*Phyciodes cocyta*) (Calhoun 2017a). No records of the Tawny Crescent are known from Maine or the Maritimes. The nearest valid records are from southern Quebec.

Southern Pearly-Eye—Lethe portlandia (Fabricius). This name was previously applied to butterflies now known as the Northern Pearly-Eye (*Lethe anthedon*), which was described in 1936 as a subspecies of the Southern Pearly-Eye. Heitzman and dos Passos (1974) established that the Northern Pearly-Eye is a discrete species, which ranges farther north than its southern congener.

APPENDIX A: CHECKLIST OF THE BUTTERFLIES OF MAINE AND THE MARITIME PROVINCES

The following is a list of butterflies that are reliably documented in Maine and the Maritime Provinces. The taxonomic arrangement and scientific nomenclature generally follow that of *A Catalogue of the Butterflies of the United States and Canada* (Pelham 2022), which reflects many of the changes suggested by the recent genomics research of Cong et al. (2019, 2021), Grishin (2019), Li et al. (2019), and Zhang et al. (2019a, 2019b, 2020). Continuing studies will likely reveal more about species relationships, resulting in additional changes to scientific names. With few exceptions, common names are in accordance with the North American Butterfly Association's *Checklist of North American Butterflies Occurring North of Mexico* (NABA 2018). Some common names apply to particular subspecies rather than the species as a whole. Popular alternative common names are given in parentheses.

The **residency status** of each species in Maine (**ME**), New Brunswick (**NB**), Nova Scotia (**NS**), and Prince Edward Island (**PE**) is as follows:

R = **Resident**   Known to reproduce regularly within the region.

FC = **Frequent colonist**   Not a permanent resident but frequently occurs within the region and establishes temporary breeding populations.

RC = **Rare colonist**   Occurs rarely or infrequently within the region but is capable of establishing temporary breeding populations.

ST = **Stray**   Occurs within the region rarely or with some frequency but is not known to reproduce.

EX = Extirpated   Once a resident, there are no known modern records within the region, and it is believed to be locally extinct.

SU = Status undetermined   Information is needed to determine its status.

A residency status abbreviation marked with an asterisk (*) indicates that the species is known only from historical records (pre-1996).

The **conservation status** is given for species that are listed under the Maine Endangered Species Act, New Brunswick's Endangered Species Act, Nova Scotia's Species at Risk Act, and/or Canada's Species at Risk Act. Each is prefixed by the state or province abbreviation (**CAN** for Canada), followed by its status (**SC**—Special Concern; **T**—Threatened; **E**—Endangered; **SX**—Extirpated). No species in Maine are currently listed under the U.S. Endangered Species Act. Two species are listed under Canada's Species at Risk Act, the Maritime Ringlet (Endangered) and the Monarch (Special Concern).

Scientific name	Common name	Residency status				Cons. status
		ME	NB	NS	PE	
Family Hesperiidae	**Skippers**					
Subfamily Eudaminae	**Dicot Skippers**					
Thorybes bathyllus (J. E. Smith)	Southern Cloudywing	R/RC				
Thorybes pylades pylades (Scudder)	Northern Cloudywing	R	R	R		
Urbanus proteus proteus (Linnaeus)	Long-tailed Skipper	RC/ST				
Epargyreus clarus clarus (Cramer)	Silver-spotted Skipper	R	R			
Subfamily Pyriginae	**Spread-winged Skippers**					
Pholisora catullus (Fabricius)	Common Sootywing	RC/ST*				
Erynnis icelus (Scudder & Burgess)	Dreamy Duskywing	R	R	R	R	
Erynnis brizo brizo (Boisduval & Le Conte)	Sleepy Duskywing	R				ME-T
Erynnis juvenalis juvenalis (Fabricius)	Juvenal's Duskywing	R	SU	R		
Erynnis baptisiae (W. Forbes)	Wild Indigo Duskywing	R				
Subfamily Heteropterinae	**Skipperlings**					
Carterocephalus mandan (W. H. Edwards)	Arctic Skipper	R	R	R	R	
Subfamily Hesperiinae	**Grass Skippers**					
Euphyes conspicua orono (Scudder)	Black Dash	R				
Euphyes bimacula bimacula (Grote & Robinson)	Two-spotted Skipper	R	R	R		
Euphyes vestris metacomet (T. Harris)	Dun Skipper	R	R	R	R	
Anatrytone logan logan (W. H. Edwards)	Delaware Skipper	R				
Hylephila phyleus phyleus (Drury)	Fiery Skipper	ST	ST		ST*	
Polites origenes origenes (Fabricius)	Crossline Skipper	R	R			
Polites mystic mystic (W. H. Edwards)	Long Dash	R	R	R	R	
Polites themistocles themistocles (Latreille)	Tawny-edged Skipper	R	R	R	R	

Scientific name	Common name	Residency status				Cons. status
		ME	NB	NS	PE	
Polites peckius peckius (W. Kirby)	Peck's Skipper	R	R	R	R	
Polites egeremet (Scudder)	Northern Broken-Dash	R				
Vernia verna (W. H. Edwards)	Little Glassywing	R				
Hesperia colorado laurentina (Lyman)	Common Branded Skipper (Laurentian Skipper)	R	R	R	R	
Hesperia leonardus leonardus T. Harris	Leonard's Skipper	R				ME-SC
Hesperia metea metea Scudder	Cobweb Skipper	R				ME-SC
Hesperia sassacus sassacus T. Harris	Sassacus Skipper	R	R		R	
Poanes massasoit massasoit (Scudder)	Mulberry Wing	R				
Poanes viator zizaniae Shapiro	Broad-winged Skipper	R				
Lon hobomok hobomok (T. Harris)	Hobomok Skipper	R	R	R	R	
Atrytonopsis hianna hianna (Scudder)	Dusted Skipper	R				ME-SC
Amblyscirtes hegon (Scudder)	Pepper and Salt Skipper	R	R	R		
Amblyscirtes vialis (W. H. Edwards)	Common Roadside-Skipper	R	R	R		
Thymelicus lineola lineola (Ochsenheimer)	European Skipper	R	R	R	R	
Panoquina ocola ocola (W. H. Edwards)	Ocola Skipper	ST	ST			
Ancyloxypha numitor (Fabricius)	Least Skipper	R	R	R	R	
Family Papilionidae	**Swallowtails**					
Subfamily Papilioninae	**True Swallowtails**					
Battus philenor philenor (Linnaeus)	Pipevine Swallowtail	RC				
Papilio brevicauda gaspeensis McDunnough	Short-tailed Swallowtail	R/RC	R			ME-SC
Papilio brevicauda bretonensis McDunnough	Short-tailed Swallowtail		R	R		
Papilio polyxenes asterius (Stoll)	Black Swallowtail	R	R	R	R	
Heraclides cresphontes (Cramer)	Eastern Giant Swallowtail	FC/RC	RC	ST*		
Pterourus troilus troilus (Linnaeus)	Spicebush Swallowtail	R				ME-SC

Scientific name	Common name	Residency status				Cons. status
		ME	NB	NS	PE	
Pterourus glaucus glaucus (Linnaeus)	Eastern Tiger Swallowtail	RC/ ST				
Pterourus canadensis (Roth-schild & Jordan)	Canadian Tiger Swallowtail	R	R	R	R	
Family Pieridae	**Whites & Sulphurs**					
Subfamily Coliadinae	**Sulphurs**					
Pyrisitia lisa lisa (Boisduval & Le Conte)	Little Yellow	ST	ST*	ST*		
Colias philodice Godart	Clouded Sulphur	R	R	R	R	
Colias eurytheme Boisduval	Orange Sulphur	FC	FC	FC	FC	
Colias interior interior Scudder	Pink-edged Sulphur	R	R	R	R	
Phoebis sennae eubule (Linnaeus)	Cloudless Sulphur	ST				
Phoebis philea philea (Linnaeus)	Orange-barred Sulphur	ST*	ST*			
Subfamily Pierinae	**Whites**					
Pontia protodice (Boisduval & Le Conte)	Checkered White	ST*				
Pieris rapae rapae (Linnaeus)	Cabbage White	R	R	R	R	
Pieris oleracea oleracea (T. Harris)	Mustard White	R	R	R	R	
Family Lycaenidae	**Gossamer-wings**					
Subfamily Miletinae	**Harvesters**					
Feniseca tarquinius (Fabricius)	Harvester	R	R	R	R	
Subfamily Lycaeninae	**Coppers**					
Lycaena phlaeas hypophlaeas (Boisduval)	American Copper	R	R	R	R	
Tharsalea hyllus (Cramer)	Bronze Copper	R	R	R	R	
Tharsalea dorcas claytoni (A. E. Brower)	Clayton's Copper	R	R	R		ME-T
Tharsalea dospassosi (McDunnough)	Salt Marsh Copper		R	R	R	
Tharsalea epixanthe epixanthe (Boisduval & Le Conte)	Bog Copper	R				
Tharsalea epixanthe phaedrus (G. Hall)	Bog Copper	R	R	R	R	

Scientific name	Common name	ME	NB	NS	PE	Cons. status
Subfamily Theclinae	**Hairstreaks**					
Parrhasius m-album (Boisduval & Le Conte)	White M Hairstreak	RC				
Strymon melinus humuli (T. Harris)	Gray Hairstreak	R	R	R		
Callophrys gryneus gryneus (Hübner)	Juniper Hairstreak (Olive Hairstreak)	R				ME-E
Callophrys hesseli hesseli (Rawson & Ziegler)	Hessel's Hairstreak	R				ME-E
Callophrys augustinus helenae (dos Passos)	Brown Elfin			R		
Callophrys augustinus augustinus (Westwood)	Brown Elfin	R	R	R	R	
Callophrys polios polios (Cook & F. Watson)	Hoary Elfin	R	R	R	R	ME-SC
Callophrys henrici henrici (Grote & Robinson)	Henry's Elfin	R	R	R	R	
Callophrys lanoraieensis (Sheppard)	Bog Elfin	R	R	R	R	
Callophrys niphon clarki (T. Freeman)	Eastern Pine Elfin	R	R	R	R	
Callophrys eryphon eryphon (Boisduval)	Western Pine Elfin	R	R			
Erora laeta (W. H. Edwards)	Early Hairstreak	R*	R	R*	R*	ME-SC
Satyrium titus winteri (Gatrelle)	Coral Hairstreak	R				ME-SC
Satyrium liparops strigosa (T. Harris)	Striped Hairstreak	R	R	R	R	
Satyrium calanus falacer (Godart)	Banded Hairstreak	R	R	R	R	
Satyrium edwardsii edwardsii (Grote & Robinson)	Edwards' Hairstreak	R				ME-E
Satyrium acadica acadica (W. H. Edwards)	Acadian Hairstreak	R	R	R	R*	
Subfamily Polyommatinae	**Blues**					
Glaucopsyche lygdamus mildredae F. Chermock	Silvery Blue			EX?		
Glaucopsyche lygdamus couperi Grote	Silvery Blue	R	R	R	R	
Celastrina lucia lucia (W. Kirby)	Northern Azure	R	R	R	R	
Celastrina neglecta (W. H. Edwards)	Summer Azure	R				

| Scientific name | Common name | Residency status | | | | Cons. status |
		ME	NB	NS	PE	
Cupido comyntas comyntas (Godart)	Eastern Tailed-Blue	R	R	R	SU	
Cupido amyntula maritima (Leblanc)	Western Tailed-Blue	R	R			
Icaricia saepiolus amica (W. H. Edwards)	Greenish Blue	R	R	R*		
Plebejus idas empetri T. Freeman	Crowberry Blue	R	R	R	R	ME-SC
Plebejus idas scudderii (W. H. Edwards)	Northern Blue	R				ME-SC
Family Nymphalidae	**Brushfoots**					
Subfamily Libytheinae	**Snouts**					
Libytheana carinenta bachmanii (Kirtland)	American Snout	ST	ST	ST		
Subfamily Danainae	**Milkweed Butterflies**					
Danaus plexippus plexippus (Linnaeus)	Monarch	FC	FC	FC	FC	ME-SC CAN-SC NB-SC NS-E
Subfamily Heliconiinae	**Longwings & Fritillaries**					
Euptoieta claudia (Cramer)	Variegated Fritillary	RC	RC	ST		
Boloria eunomia dawsoni (W. Barnes & McDunnough)	Bog Fritillary	R	R			
Boloria chariclea grandis (W. Barnes & McDunnough)	Arctic Fritillary (Purple Lesser Fritillary)	R	R	R		ME-T
Boloria bellona bellona (Fabricius)	Meadow Fritillary	R	R			
Boloria frigga saga (Staudinger)	Frigga Fritillary	R				ME-E
Boloria selene myrina (Cramer)	Silver-bordered Fritillary	R				
Boloria selene atrocostalis (Huard)	Silver-bordered Fritillary	R	R	R	R	
Argynnis idalia idalia (Drury)	Regal Fritillary	EX*	ST*			
Argynnis cybele cybele (Fabricius)	Great Spangled Fritillary	R				
Argynnis cybele novascotiae McDunnough	Great Spangled Fritillary	R	R	R	R	
Argynnis aphrodite aphrodite (Fabricius)	Aphrodite Fritillary	R				

| Scientific name | Common name | Residency status | | | | Cons. status |
		ME	NB	NS	PE	
Argynnis aphrodite winni Gunder	Aphrodite Fritillary	R	R	R		
Argynnis atlantis atlantis W. H. Edwards	Atlantis Fritillary	R	R	R	R	
Subfamily Limenitidinae	**Admirals & Relatives**					
Limenitis archippus archippus (Cramer)	Viceroy	R	R	R	R	
Limenitis arthemis arthemis (Drury)	White Admiral	R	R	R	R	
Subfamily Nymphalinae	**True Brushfoots**					
Aglais milberti milberti (Godart)	Milbert's Tortoiseshell	R	R	R	R	
Aglais io io (Linnaeus)	European Peacock			SU		
Nymphalis l-album j-album (Boisduval & Le Conte)	Compton Tortoiseshell	R	R	R	R	
Nymphalis antiopa antiopa (Linnaeus)	Mourning Cloak	R	R	R	R	
Polygonia interrogationis (Fabricius)	Question Mark	FC	FC	FC	FC	
Polygonia comma (T. Harris)	Eastern Comma	R	R	R	SU	
Polygonia satyrus neomarsyas dos Passos	Satyr Comma	R	R	R	R	ME-SC
Polygonia progne (Cramer)	Gray Comma	R	R	R	R	
Polygonia gracilis gracilis (Grote & Robinson)	Hoary Comma	R	R	R*		
Polygonia faunus faunus (W. H. Edwards)	Green Comma	R	R	R	R	
Vanessa virginiensis (Drury)	American Lady	FC	FC	FC	FC	
Vanessa cardui (Linnaeus)	Painted Lady	FC	FC	FC	FC	
Vanessa atalanta rubria (Fruhstorfer)	Red Admiral	FC	FC	FC	FC	
Junonia coenia coenia Hübner	Common Buckeye	RC/ FC	ST	ST		
Euphydryas phaeton phaeton (Drury)	Baltimore Checkerspot	R	R	R	R	
Chlosyne nycteis nycteis (E. Doubleday)	Silvery Checkerspot	R	R	R*		ME-SC
Chlosyne harrisii harrisii (Scudder)	Harris' Checkerspot	R	R	R	R	

Scientific name	Common name	Residency status				Cons. status
		ME	NB	NS	PE	
Phyciodes tharos tharos (Drury)	Pearl Crescent	R				
Phyciodes cocyta cocyta (Cramer)	Northern Crescent	R	R	R	R	
Subfamily Satyrinae	**Satyrs**					
Coenonympha nipisiquit McDunnough	Maritime Ringlet		R			CAN-E NB-E
Coenonympha california inornata W. H. Edwards	Common Ringlet (Inornate Ringlet)	R	R	R	R	
Lethe anthedon (A. Clark)	Northern Pearly-Eye	R	R	R	R	
Lethe eurydice eurydice (Linnaeus)	Eyed Brown	R	R	R	R	
Lethe appalachia appalachia R. Chermock	Appalachian Brown	R				
Megisto cymela cymela (Cramer)	Little Wood-Satyr	R	R	R	R	
Cercyonis pegala alope (Fabricius)	Common Wood-Nymph	R				
Cercyonis pegala nephele (W. Kirby) [including "intermediate" phenotypes]	Common Wood-Nymph	R	R	R	R	
Oeneis jutta ascerta Masters & Sorensen	Jutta Arctic	R	R	R	R	
Oeneis polixenes katahdin (Newcomb)	Katahdin Arctic	R				ME-E

APPENDIX B: CONSERVATION STATUS DEFINITIONS

NatureServe is a North American network of scientists and institutions whose mission is to track the status and trends of at-risk species and habitats based on standardized, science-based methodologies. Core to this practice is the assignment of status ranks used to indicate the threat of extinction or extirpation a species or subspecies faces at the global, national, and subnational (e.g., states and provinces) levels (G-ranks, N-ranks, and S-ranks, respectively). Conservation status ranks take into consideration factors of rarity, threats, and population trends (see Faber-Langendoen et al. 2021 for details). Following are rank suffixes and their definitions (modified from Master et al. 2012):

X   **Presumed extinct (G-rank) or extirpated (N-rank and S-rank)**—Not located despite intensive searches and virtually no likelihood of rediscovery.

H   **Possibly extinct**—Known from only historical occurrences but still some hope of rediscovery.

1   **Critically imperiled**—At very high risk of extinction or extirpation due to very restricted range, very few populations or occurrences, very steep declines, very severe threats, or other factors.

2   **Imperiled**—At high risk of extinction or extirpation due to restricted range, few populations or occurrences, steep declines, severe threats, or other factors.

3   **Vulnerable**—At moderate risk of extinction or extirpation due to a fairly restricted range, relatively few populations or occurrences, recent and widespread declines, threats, or other factors.

4   **Apparently secure**—At fairly low risk of extinction or extirpation due to an extensive range and/or many populations or occurrences but with possible cause for some concern as a result of local recent declines, threats, or other factors.

5   **Secure**—At very low risk of extinction or extirpation due to a very extensive range, abundant populations or occurrences; little to no concern from declines or threats.

U   **Unrankable**—Currently unrankable due to lack of information or substantially conflicting information about status or trends.

NR   **Unranked**—Rank not yet assessed.

NA   **Not applicable**—A conservation status rank is not applicable because the species is not a suitable target for conservation activities. Examples include nonnative species, rare colonists, and vagrants. This is typically applied only to N-ranks and S-ranks.

Several qualifiers can be applied to ranks. The "?" qualifier denotes an uncertain rank (e.g., S3?); the B qualifier is applied to species that breed in a region but are not year-round residents (e.g., S3B); and the T qualifier denotes a subspecies rank (e.g., *Papilio brevicauda bretonensis* is ranked G5T3, indicating that the species is globally secure, but the subspecies is globally vulnerable).

Range ranks (e.g., S2S3) indicate uncertainty about the exact status of a species. For an S2S3 species, data show it is imperiled or vulnerable, but data are insufficient to determine a precise rank.

LEGAL CONSERVATION STATUS

Some butterfly species in Maine and the Maritime Provinces are afforded legal protection under federal, provincial, and state laws, such as the Maine's Endangered Species Act,[1] the Canadian Species At Risk Act,[2] the New Brunswick Species At Risk Act,[3] and the Nova Scotia Endangered Species Act.[4] The purpose of these acts is to prevent the extirpation of species from the jurisdiction in question, and to recover Endangered and Threatened species through protection of individuals and populations of the species and their habitat. Each act has three primary statuses in common: Endangered, Threatened, and Special Concern or Vulnerable. While specific wording varies between acts, the general definitions of these terms are as follows:

Endangered—A species facing imminent extinction or extirpation from the nation/province/state.

Threatened—A species that is likely to become endangered if the factors affecting its vulnerability are not reversed.

Special Concern/Vulnerable[5]—A species that may become Endangered or Threatened because of characteristics that make it particularly sensitive to human activities or natural events.

1. The Maine Endangered Species Act, Inland Fisheries and Wildlife Laws, excerpts from Maine Revised Statute Authority—Title 12.
2. Species At Risk Act, Statutes of Canada 2002.
3. Species at Risk Act, Statutes of New Brunswick 2012, Chapter 6.
4. Endangered Species Act, Statutes of Nova Scotia 1998, Chapter 11.
5. The Nova Scotia Endangered Species Act uses the term Vulnerable, the other acts use the term Special Concern.

APPENDIX C: COMMUNITY SCIENTISTS WHO CONTRIBUTED DATA TO THE MAINE AND MARITIMES SURVEYS

MAINE

Jay Adams
Kevin Adams
Jeremy Adler
Donald Anderson
Maureen Anderson
Marianne Archard
Chelsea Ardle
Matt Arey
Joseph Asta-Ferrero
Amy Auchterlonie
Charles Avenengo
Hugo Avenengo
Emarie Ayala
Bruce Barker
Anne Barrett
John Bartlett
Arianne Barton
B. Baston
Rebecca Baston
Ahmet Baytas
Mathew Beaudette
Jennifer Beaven
Thomas Bedner
Allen Belden
Jane Berry
John Berry

Joe Berubi
Amanda Bessler
Kirk Betts
Louis Bevier
Susan Bickford
Christine Blais
Elijah Blais
Jordan Blais
Christina Blank
Fred Blonder
Marge Blonder
Ayala Blyther
Kelly Boland
Richard Boscoe
Paul Bourget
Bets Brown
Carla Brown
Ethan Brown
Lea Brown
Owen Brown
Raymond Brown
Ron Butler
Kevin Byron
John Calhoun
Donald Cameron
Stephanie Camp

Sarah Caputo
Gerald Carter
Tyler Case
Brian Cassie
Lindsey Chapin
Terrance Chick
Jon Chin
Thomas Ciarlante
J. Paul Ciarrocchi
Ann Clark
Katie Clarke
Dustin Colbry
Kristin Collins
Ross Conway
Maggie Cook
Hunter Corson
Zach Cray
Edward Crew
Trevor Criallos
Kalman Csigi XIV
Kristin Curay
Ethan Darling
Peter Darling
Danielle D'Auria
Ella D'Auria
Lisa Davis

Derek Dawson
Lisa Dellwo
Ernest Deluca
Elisabeth deMaynadier
Emmett deMaynadier
Phillip deMaynadier
Treva deMaynadier
Sally Demeter
Jody Despres
Phil Dilacc
Molly Docherty
Charlene Donahue
Serena Doose
Judith Dorsey
Courtney Dotterweich
Grainne Dougherty
Caleb Doyle
Grace Doyle
Sarah Drahovzal
Natalie Dumont
Paul Duncan
Stephen Dunham
Nathan Durant
Patricia Durkin
Anna Dyer
Bruce Dyer
Sara Eck
Pamela Edwards
Michael Enos
Susan Ernst
Chantelle Ervin
Gail Everett
Joan Farnsworth
Sierra Ferland
Tom Ferrari
June Ficker
Sarah Flanagan
Jean Forbes
R. M. Francis
Tulle Frazer
Elaine Gaudette
Kirk Gentalen

Nadav Gilad-Muth
Robert Gobeil
Rose Marie Gobeil
Hannah Goodman
Nathan Goodman
Tina Goodman
Nick Gordon
Adrianna Gorsky
Laura Gould
Fred Gralenski
Linda Gralenski
Pamela Green
Alex Grkovich
Barbara Grunden
Charlie Grunden
Rachel Grundl
Jonathan Guba
Susan Gurney
Geordi Hall
Georgia Hall
Anne Hammond
Hannah Hammond
Nathan Hammond
Steve Hammond
Robert Hanish
Nathan Hannah
Spencer Hardy
Kathleen Harkins
Adeline Harris
Karen Harter
Alicia Haskell
Shawn Haskell
Al Haury
Allan Haury
Annette Hauser
Bart Hersey
Janalyn Hersey
William Hersey
Richard Hildreth
Tarn Hildreth
Nicole Hinman
Doug Hitchcox

Chris Hoffman
Karen Holmes
Sherry Hooker
Olivia Hooper
Ellie Hopkins
Karen Hopkins
Skylar Hopkins
Dan Horne
Paula House
Ken Janes
Glenn Jenks
C. W. Johnson
Devin Johnstone
Melody Joliat
David Juers
Martin Junco
WNERR Junior
    Researchers
Lee Kantar
Susan Keefer
Scott Keniston
Susan Kistenmacher
Martha Kitchen
Peter Knipper
Teresa Knipper
Emily Knurek
Georgeann Kuhl
Wingyi Kung
Megan Laferriere
Bill Lane
Chelsea Lathrop
Mike Lawson
Melissa Lesh
John Leso
Tracey Levasseur
Amanda Lighteap
Emily Littlefield
Chris Livesay
Juanita Longwell
Chelsea Lopez
Zachary Lopez
Nick Lund

James MacDougall
Miranda MacFadsen
William Mackowski
Andrew Maguire
Marina Mann
Stacie Mann
Warren Mann
Susan Mansir
Judy Markowsky
Teresa Martel
Tim Martel
Kenneth Masloski
Fox Maxwell
Jonathan Mays
Patrick McBride
K. McCurdy
Kristin McCurdy
Jill McElderry-Maxwell
Corinne Michaud
D. Michaud
Tom Mickewich
Kathryn Miles
Dorcas Miller
M. Mills
Pat Milner
C. S. Minot
James Minott
Martha Mixon
Natalya Mohoff
Tatiana Mohoff
Pamela Moore
Ronald Moore
T. Moser
David Mozzoni
Sarah Mueller
Carol Muth
Walter Muth
Ziv Nelson
Jennifer Nomura
Stephen Nomura
Danielle Norsworthy
Haleigh O'Donnell

Susan Palumbo
Jim Paruk
Olivia Paruk
Mila Paul
Brian Pauly
Madilyn Peay
John Perry
Ruby Persons
Trevor Persons
Bryan Pfeiffer
Christopher Pillsbury
Margaret Pinsky
Debra Piot
John Piot
Sarah Pokorny
Erin Porter
David S. Potter
L.M. Potter
Daniel Poulin
Emily Poulin
James Poulin
Russell Poulin
John Pratte
Michael Pruyn
Gwenda Pryor
Loreli Pryor
David Putnam
Robert Pyle
Amanda Rand
Alysa Remsburg
Hope Richards
Scott Richardson
Kendall Ricker
Roger Rittmaster
Linda Rivard
Karen Robbins
Ellen Robertson
Milkala Robinson
Mike Rol
Catherine Rowe
Kristisu Sader
Kelly Safford

Anne Sarantos
Kobe Saunders
Hillary Scannell
Kerry Schlosser
Allan Seamans
Leyton Sewell
Roberta Sharp
Connor Shaw
William Sheehan
Allison Sheffield
Matt Simas
Megan Sine
Brett Skelly
Ricky Skiba
Shawn Skillern
Devin Sklar
Dmitriy Skoog
Ben Skvorak
Jackie Skvorak
John Skvorak
Katie Skvorak
Nathan Skvorak
Jerry Smith
Lee Snyder
Patricia Snyder
Steve Spear
Terry Sprague
Zackery St. Pierre
Cory Stearns
Carolyn Sternberge
Elias Steward
David Stickler
Beth Swartz
L.W. Swett
Betty Thompson
Donald Thompson
Bill Thorne
D. Tierney
Grace Toles
Judith Tollefson
Caroline Tolstad
Stephen Traften

Dain Trafton
Elizaeth Trafton
Frances Trafton
Mary Trafton
Melissa Trafton
Sandra Trafton
Steve Trafton
Vera Trafton
David Trask
Rosemart Trudel
Ross Turcotte
Kirstin Underwood

Kenniston Vicki
Margaret Viens
Barbara Volkie
Steve Walker
Stella Walsh
Mark Ward
Marina Warren
Reginald Webster
Andrew Weik
Jeffrey Wells
David White
Nancy White

Fred Wigham
Ernest Williams
P. Williams
Terry Williamson
Herb Wilson
Polly Wilson
Tom Winslow
Brooke Winter-Potter
Laura Wood
B. Woods
W. Yates

MARITIMES

Marie Abgrall
Darrell Abolit
Christopher I. G. Adam
Sarah Adams
Chantelle Alderson
Don Anderson
Jennifer Arsenault
Mark Arsenault
Mel Arsenault
Ronald G. Arsenault
Lana Ashby
Michelle Baker
Rick Ballard
Dan Banks
Ginny Barrett
Greg Barrett
Steve Barrett
Avery Bartels
Sean Basquill
Duncan M. Bayne
Alban Beaulieu
Richard Beazely
David Bell
Jim Bell
Alain Belliveau
Gilles Belliveau
Normand Belliveau

Marilyn Benjamin
Tyler Bernard
Juliet Bewley
Doris Bishop
Jo Bishop
Richard Blacquiere
Bennett Blades
Madeline Blades
Bob Blake
Sean Blaney
Sherman Boates
Alison Bogan
Larry Bogan
Marla Bojarski
Marlene Bolger
Jeanette Boudreau
Vivienne Bourgoin
William Bowerbank
Jessica Bradford
Debbie Brekelmans
Derek W. Bridgehouse
Meredith Brison-Brown
Anthony Brooks
Cecil Brown
Jean Brown
Linda Brown
John Brownlie

Marion Brunel
Michel Bruse
Don Bullock
Mike Burrell
Roger Burrows
Kelsey J. Busson
Hennie de Caluwe
Liz Cameron
Maureen Cameron-
   MacMillan
Greg Campbell
Margaret Campbell
K. Cann
Carl Canning
Vernon Carrier
Maggie Caskanette
Rick Cavasin
Cody Chapman
Roland Chiasson
Zac Chipper
David Christie
Ann Chudleigh
James Churchill
Chelsey Clem
Jeff C. Clements
James Clifford
Nelson Cloud

Christopher Clunas
Elwood Coakes
Bruce Coates
Murray Colbo
David Colville
Brian Comeau
Kevin J. Connor
Sandra Cooper
Louis-Emile Cormier
Merv Cormier
V. Cormier
Clive Cosham
Jessica Cosham
Laura Cosham
Barry Cottam
John Crabtree
Kevin Craig
L. Crain
Megan Crowley
Adam Cruickshank
Rosemary Curley
Dave Currie
Joan Czapalay
Vicki Daley
Even Dankowicz
Andrew Danylewich
Daphne Davey
Robert Davis
Lyas Dawe
Andrew Dean
Tracey Dean
Fiep DeBie
Arielle DeMerchant
Mark Dennis
Alix d'Entremont
Gisele d'Entremont
Ronnie d'Entremont
David D'Eon
Jerome D'Eon
Ted D'Eon
Keith Dewar
June DeWolfe

Mary Dicks
Sabine Dietz
Patricia Dix
Clayton D'Orsay
Catherine Doucet
Denis Doucet
Janet Doucet
Archie Doucette
Danielle Dowling
David Dowling
Irene Doyle
Derek Durston
Samara Eaton
Jim Edsall
Mark Elderkin
Ralph Eldridge
Claire Elliott
Jim Elliott
Christopher Engelhart
Cathy Etter
Aaron D. Fairweather
Sandra Feetham
Rick Ferguson
Nicole Ferron
Susan Feurtado
Danielle Fife
Edie Fillmore
Sophie Finlayson-
    Schueler
Jeanne Finn-Allen
Karen Flannagan
Graham Forbes
Celia Ann Forsyth
George Forsyth
Harold Forsyth
Paige Forsythe
Tammi Forsythe
Rick Fournier
Jesse Francis
Jamie Fraser
Holly Frazer
Troy Frech

Ken Freeman
Gabrielle Fry
Lillianne Fry
Steven Furino
Deana Gadd
Peter Gadd
Léon Gagnon
Gabriel Gallant
Sal Gallant
Gisele Gaudet
Maxim Giasson
Mischa Giasson
Pascal Giasson
Don Gibson
Marie-Andrée Giguère
Lisette Godin
Michel Godin
Jim Goltz
Steve Gordon
Bill Gough
Paul Gould
Derek Grant
Garry Gregory
Diane Griffin
William Grover
Dominique Gusset
Matt Hackett
Sam Hackett
Ross Hall
Caleb F. Harding
Clayton A. Harding
Jacob W. Harding
Jordan S. Harding
Robert W. Harding
Guillaume Harmange
Tara Harris
William Hartford
Sharon Hawboldt
Doug Hennigar
Henry Hensel
Catherine Higgins
Verna J. Higgins

Steven Hiltz

James R. Hirtle

Beth Hoar

Ardeth Holmes

Danielle Horne

Cory Hughes

Steve Hupman

Rachel Imhoff

Christine Ingeborg

Fenton Isenor

Judith James

Mitchell Jay

Margaret Jennings

Brian Johnstone

Devin Johnstone

Colin Jones

David Kaposi

Robert Keereweer

Correna Kelly

Grace Kelly

Paulette Kelly

Charlie Kendell

Chris Kennedy

Dan Kennedy

Roberta Kenny

Dorothy Keough

Sandra Keough

Leila Killen

Andrew King

Elizabeth Kingsland

Jeff Klein

John Klymko

Eric Knopf

Sheldon Lambert

Merville Landry

Rosita Lanteigne

Kevin Lantz

Charlotte LaPointe

Roy LaPointe

Maxim Larrivée

Diane LaRue

LeeAnn Latremouille

Ann Leadbeater

Danielle P. LeBlanc

Jane LeBlanc

Verica LeBlanc

Yolande Leblanc

Angela Léger

Jacob Legere

Brian Levenson

Tim Levy

Donna Lewis

Morgan Leyte

Holly Lightfoot

Judy Lincoln

Joshua Lindsay

Josh Logel

Henry Long

Keith Lowe

J.C. Lucier

Mary Macaulay

Elizabeth MacCormick

Christopher
   MacDonald

Gerald MacDonald

Gerry MacDonald

Mike MacDonald

Nicole MacDonald

Sam MacDonald

Andrew Macfarlane

Rebecca MacFarlane

Ken MacIntosh

Colin MacKay

Kathy MacKay

Frank MacKenzie

Angus MacLean

Gayle MacLean

Allan MacMillan

Harriet MacMillan

Janet MacMillan

Ellen MacNearney

Michael MacNeill

Sarah Makepeace

Scott Makepeace

Allison Manthorne

Hughie Manthorne

Anne Marsch

Don Marsch

Cathy Martin

Donna Martin

Evelyn Martin

Paul Martin

Sonya Martin

Richard G. Mash

Gillian Mastromatteo

Kitty Maurey

Blake Maybank

David Mazerolle

Charles McAleenan

Donald F. McAlpine

Dan McAskill

Leslie McClair

Tim McCluskey

Max McCosham

Fritz McEvoy

Kent McFarland

Mark McGarrigle

Harold McGee

Lois McGibbon

Matt McIver

Pat McKay

Karen McKendry

Ken McKenna

Rachel McKinley

Ian McLaren

Helen McLaughlin

Robin McLeod

Jeffie McNeil

Ian Middleton

Richard Migneault

Marg Millard

Mary Millard-Smith

Michael Milligan

Garrett
   Mombourquette

Jean Mombourquette

Bernice Moores
Anne Morrison
Nancy Mullin
Pat Murphy
Richard Murphy
Gary Myers
Kenneth A. Neil
Antoinette Neily
Larry Neily
Charles Neveu
Louise Nichols
Christine Noronha
Josh Noseworthy
Ruth Nymand
Dwaine Oakley
Judy O'Brien
Jeffrey Ogden
Michael Olsen
Donald Ostaff
Judy K. Panchuk
Julie Paquet
Johanne Paquette
Mike Parks
David G. Patriquin
Nathalie Paulin
Linda Payzant
Peter Payzant
Alicia Penney
Vince Perkins
Susan Petrie
Dennis Pitts
Kimberley Pitts
Laura Pitts
Erik Plante
Nelson Poirier
Riley Pollom
Mark Ponikvar
Dorothy Poole
Anita Pouliot
Terrance Power
Harold Preston
Chuck Priestley

Sheila Pugsley
Mark Pulsifer
Chantal Raillard
Martin Raillard
Coleen Ramsay
Tim Rawlings
Tristan Rawlings
Donna Redden
Ira Reinhart-Smith
Jean Renton
Diane Richard
Leonel Richard
Louise Richard
Lewnanny Richardson
Sarah Richer
Roger Rittmaster
John Robart
Lynda Robertson
Sue Robertson
Dawn Robichaud
Étienne Robichaud
Kimberly Robichaud-
    Leblanc
Evelynn Robinson
Kate Robinson
Norma Robinson
Sarah L. Robinson
Scott R. Robinson
Ellis Roddick
Kailum Rogers
Ruth Rogers
Abe Ross
Hannah Ross
Josh Ross
Sue Ross
Karen S. Roy
Dwayne L. Sabine
Mary E. J. Sabine
Vielka Salazar
Ma Sanford
Jesse Saroli
Donna Savoy

Peggy Scanlan
Phil Schappert
Chris Schmidt
Tim Schreckengost
Ted Sears
Marjorie Sharpe
Colin Silver
Julie Singleton
Alison Smith
Donald Smith
Mary Jane Smith
Matthew Smith
Nigel Smith
Valerie Smith
Judy Smits
John Sollows
Mary Sollows
Cindy Spicer
Kathleen Spicer
Gary Stairs
Grace Standen
Jennifer Stephen
Richard Stern
Becky Stewart
Ed Stewart
Pam Stewart
Judy Stockdale
Annette Stone
Brian Stone
Guy St-Pierre
Jeffrey St-Pierre
Noemie St-Pierre
Ed Sulis
Eric Sullivan
Kevin M. Sullivan
Michael J. Sullivan
Wendy Sullivan
Hilary Taylor
James Taylor
Marilyn Taylor
John Te Raa
Gisele Thibiodeau

Anthony W. Thomas
Peter Thomas
Cliff Thornley
Stuart Tingley
Brad Toms
Maureen Toner
Brian Townsend
Tina Trask
Cameron Trenholm
Judy Tufts
Martin Turgeon
Suzanne Turgeon
Doug van Hemmessen
Katelyn A. A.
    Vandenbroeck

Joshua Vandermeulen
Karen Vanderwolf
Michel Viau
Rita Viau
Kathleen Vinson
Neil Vinson
Marcy Wagner
Ron Walsh
Ann Walter-Fromson
Owen Washburn
Peggy Weatherson
Reginald P. Webster
Vincent Webster
Sybil Wentzell
Ann Wheatley

Rick Whitman
Robert Whitney
Bev Wigney
Heather Wilkes
Angela Wilson
Jim Wilson
Ron Wilson
Bill Winsor
Barry Wright
Terry Wyman
Jessy Wysmyk
Bev Yollem
Chris Zinck

REFERENCES

Aaron, E.M. 1907. [Butterfly captures.] Canadian Entomologist 39:104.

Acorn, J., and I. Sheldon. 2016. Butterflies of Ontario and eastern Canada. Edmonton, AB: Partners Publishing.

Alexander, C.P. 1951. Doctor William Proctor (1872–1951). Entomological News 62:237–241.

Amano, T., R.J. Smithers, T.H. Sparks, and W.J. Sutherland. 2010. A 250-year index of first flowering dates and its response to temperature changes. Proceedings of the Royal Society B 277:2451–2457.

Anderson, M.G., B. Vickery, M. Gorman, L. Gratton, M. Morrison, J. Maillet, A. Olivero, C. Ferree, D. Morse, G. Kehm, K. Rosalska, S. Khanna, and S. Bernstein. 2006. The Northern Appalachian / Acadian Ecoregion: Ecoregional assessment, conservation status and resource CD. The Nature Conservancy, Eastern Conservation Science, and the Nature Conservancy of Canada, Atlantic and Quebec regions. Boston: Nature Conservancy.

Andrew, M.E., M.A. Wulder, and N.C. Coops. 2011. How do butterflies define ecosystems? A comparison of ecological regionalization schemes. Biological Conservation 144:1409–1418.

Angelo, R., and D.E. Boufford. 2016. Atlas of the flora of New England. http://neatlas.org/.

Anonymous. 1750. [Nova Scotia plants, fruits and Animal.] Gentleman's Magazine (February): contents page.

Anonymous. 1837. Donations. Boston Journal of Natural History 1:515–527.

Anonymous. 1886. [Charles Fish collection.] Catalogue of the officers and students of Bowdoin College and the Medical School of Maine, for the year 1886–1887:38.

Anonymous. 1888. Thomas Belt, F.G.S. Monthly Chronicle of North Country Lore and Legend 2:262–263.

Anonymous. 1911. Charles Henry Fernald. In Anonymous, ed., Entomology and zoology at the Massachusetts Agricultural College, pp. 3–10. Amherst: Massachusetts Agricultural College.

Anonymous. 1938. Among our authors. Maine Forester Annual Edition 1938:46.

Anonymous. 1941. [Obituary of Philip Laurent.] Entomological News 53:277.

Anonymous. 1995. [Obituary of Auburn E. Brower.] *In* R. Dearborn and C. Grainger, eds., Forest & shade tree insect & disease conditions for Maine: A summary of the 1994 situation. Insect and Disease Management Division Summary Report 9:2. Augusta: Maine Department of Conservation, Maine Forest Service.

Armstrong, J.A, and C.A. Cook. 1993. Aerial spray applications on Canadian forests: 1945–1990. Information report. Ottawa: Forestry Canada.

Audubon, M.R., and E. Coues. 1897. Audubon and his journals, vol. 1. New York: Charles Scribner's Sons.

Bain, F. 1885a. Butterflies of Prince Edward Island. Canadian Science Monthly 3:72.

Bain, F. 1885b. [Notes on the natural history of Prince Edward Island.] Canadian Science Monthly 3:118–119.

Bain, F. 1890. The natural history of Prince Edward Island. Charlottetown, PE: G. Herbert Haszard.

Barnes, W., and J. McDunnough. 1917. Check list of the Lepidoptera of boreal America. Decatur, IL: Herald Press.

Barrington, D., and B. Pfeiffer. 2021. *Polyommatus icarus* (European Common Blue) expands into the United States. News of the Lepidopterists' Society 63:108–109.

Barton, A.M., A.S. White, and C.V. Cogbill. 2012. The changing nature of the Maine woods. Durham: University of New Hampshire Press.

Bean, R.C. 1948. John Crawford Parlin. Rhodora 50:130–131.

Belt, T. 1864. List of butterflies observed in the neighbourhood of Halifax, Nova Scotia. Transactions of the Nova-Scotian Institute of Natural Science 1:87–92.

Belt, T. 1874. The naturalist in Nicaragua: A narrative of a residence at the gold mines of Chontales; journeys in the savannahs and forests. With observations on animals and plants in reference to the theory of evolution of living forms. London: John Murray.

Belt, T. 1888. The naturalist in Nicaragua: A narrative of a residence at the gold mines of Chontales; journeys in the savannahs and forests. With observations on animals and plants in reference to the theory of evolution of living forms. Second edition, revised and corrected. London: Edward Bumpus.

Bethune, C.J.S. 1894. The butterflies of the eastern provinces of Canada. Annual Report of the Entomological Society of Ontario 25:29–44.

Bethune, C.J.S. 1897. The butterflies of the eastern provinces of Canada. Annual Report of the Entomological Society of Ontario 27:106–110.

Bethune, C.J.S. 1900. Henry Herbert Lymam, M.A. Annual Report of the Entomological Society of Ontario 30:123.

Bethune, C.J.S. 1914. Henry Herbert Lyman, M.A. Canadian Entomologist 46:221–225.

Bethune, C.J.S., and J.M. Jones. 1870. Nova Scotian Lepidoptera. Proceedings and Transactions of the Nova Scotian Institute of Science 2:78–87.

Black, S.H. 2018. Insects and climate change: Variable responses will lead to climate winners and losers. *In* D.A. Dellasala and M.I. Goldstein, eds., Encyclopedia of the Anthropocene pp. 95–101. New York: Elsevier.

Black, S.H., and S. Jepsen. 2007. The Xerces Society: 36 years of butterfly conservation. News of the Lepidopterists Society 49(4): 112–115.

Boggs, C.L., and K.D. Freeman. 2005. Larval food limitation in butterflies: Effects on adult resource allocation and fitness. Oecologia 144:353–361.

Bonmatin, J.-M., C. Giorio, V. Girolami, D. Goulson, D.P. Kreutzweiser, C. Krupke, M. Liess, E. Long, M. Marzaro, E.A.D. Mitchell, D.A. Noome, N. Simon-Delso, and A. Tapparo. 2014. Environmental fate and exposure: Neonicotinoids and fipronil. Environmental Science and Pollution Research 22:35–67.

Bowers, M.D. 1980. Unpalatability as a defense strategy of *Euphydryas phaeton* (Lepidoptera: Nymphalidae). Evolution 4:586–600.

Bowers, M.D. 1983. Mimicry in North American checkerspot butterflies: *Euphydryas phaeton* and *Chlosyne harrisii* (Nymphalidae). Ecological Entomology 8:1–8.

Bowers, M.D., and L.L. Richardson. 2013. Use of two oviposition plants in populations of *Euphydryas phaeton* Drury (Nymphalidae). Journal of the Lepidopterists' Society 67:299–300.

Braun, A.F. 1921. Charles Henry Fernald. Entomological News 32:129–133.

Breed, G.A., S. Stichter, and E.E. Crone. 2013. Climate-driven changes in northeastern US butterfly communities. Nature Climate Change 3:142–145.

Brittain, W.H. 1918. The insect collections of the Maritime Provinces. Canadian Entomologist 50:117–122.

Brower, A.E. 1932. *Eurymus eurytheme* in Maine. Journal of the New York Entomological Society 40:510.

Brower, A.E. 1940. Descriptions of some new macrolepidoptera from eastern America. Bulletin of the Brooklyn Entomological Society 35:138–140.

Brower, A.E. 1974. A list of the Lepidoptera of Maine. Part 1: The macrolepidoptera. Life Sciences and Agriculture Experiment Station Technical Bulletin 66. Orono: University of Maine.

Brower, A.E. 1983. A list of the Lepidoptera of Maine. Part 2: The macrolepidoptera. Section 1. Limacodidae through Cossidae. Maine Department of Conservation, Maine Forest Service, Division of Entomology, Augusta, and the Department of Entomology, Maine Agricultural Experiment Station Technical Bulletin 109. Orono: University of Maine.

Brower, A.E. 1984. A list of the Lepidoptera of Maine. Part 2. The macrolepidoptera. Section 2. Cosmopterigidae through Hepialidae. Maine Department of Conservation, Maine Forest Service, Division of Entomology, Augusta, and the Department of Entomology, Maine Agricultural Experiment Station Technical Bulletin 114. Orono: University of Maine.

Brower, A.E., and R.-M. Payne. 1956. Check list of Maine butterflies. Maine Field Naturalist 12:42–44.

Brower, L.P., O.R. Taylor, E.H. Williams, D.A. Slayback, R.R. Zubieta, and M.I. Ramirez. 2012. Decline of monarch butterflies overwintering in

Mexico: Is the migratory phenomenon at risk? Insect Conservation and Diversity 5:95–100.

Brown, J.W. 1987. The peninsular effect in Baja California: An entomological assessment. Journal of Biogeography 14:359–365.

Brown, J.W., and P.A. Opler. 1990. Patterns of butterfly species density in peninsular Florida. Journal of Biogeography 17:615–622.

Brues, C.T. 1933. Charles Willison Johnson, 1863–1932. Entomological News 44:113–116.

Buggey, S., and G. Davies. 1988. McCulloch, Thomas. Dictionary of Canadian Biography 7:529–541.

Burns, J.M. 1985. *Wallengrenia otho* and *W. egeremet* in eastern North America (Lepidoptera: Hesperiidae: Hesperiinae). Smithsonian Contributions to Zoology 423. Washington, DC: Smithsonian Institution Press.

Bushell, S., J.S. Carlson, and D.W. Davies, eds. 2020. Romantic cartographies: Mapping, literature, culture, 1789–1832. Cambridge: Cambridge University Press.

Cale, J.A., M.T. Garrison-Johnston, S.A. Teale, and J.D. Castello. 2017. Beech bark disease in North America: Over a century of research revisited. Forest Ecology and Management 394:86–103.

Calhoun, J.V. 2016. A reevaluation of *Papilio pegala* F. and *Papilio alope* F., with a lectotype designation and a review of *Cercyonis pegala* (Nymphalidae: Satyridae) in eastern North America. Journal of the Lepidopterists' Society 70:20–46.

Calhoun, J.V. 2017a. Notes on historical butterfly records from Maine. Part 1. News of the Lepidopterists' Society 59:77–83.

Calhoun, J.V. 2017b. Notes on historical butterfly records from Maine. Part 2. With the designation of a lectotype. News of the Lepidopterists' Society 59:128–133.

Calhoun, J.V. 2019. Watch for *Erynnis funeralis* (Hesperiidae) in the east. Southern Lepidopterists' News 41:285–287.

Calhoun, J.V. 2022. A local irruption of *Chlosyne nycteis* (Nymphalidae) in Maine, with an important new food plant record. News of the Lepidopterists' Society. 64:26–33.

Canadian Endangered Species Conservation Council. 2016. Wild species 2015: The general status of species in Canada. Ottawa: National General Status Working Group.

Capers, R.S., and A.D. Stone. 2011. After 33 years, trees more frequent and shrubs more abundant in northeast U.S. alpine community. Arctic, Antarctic, and Alpine Research 43:495–502.

Cardé, R.T., A.M. Shapiro, and H.K. Clench. 1970. Sibling species in the *eurydice* group of *Lethe* (Lepidoptera: Satyridae). Psyche 77:70–103.

Carpenter, R.W. 2015. The Funereal Duskywing, *Erynnis funeralis* (Hesperiidae): Seasonal range expansion into eastern North America. Journal of the Lepidopterists' Society 69:114–124.

Catling, P.M., R.A. Layberry, J.P. Crolla, and P.W. Hall. 1998. Increase in populations of Henry's Elfin, *Callophrys henrici*, (Lepidoptera: Lycaenidae) in Ottawa-Carleton, Ontario, associated with man-made habitats and glossy Buckthorn, *Rhamnus frangula*, thickets. Canadian Field-Naturalist 112: 335–337.

Cech, R., and G. Tudor. 2005. Butterflies of the east coast: An observer's guide. Princeton, NJ: Princeton University Press.

Champlain, A.B. 1945. Classified collections of insects. Proceedings of the Pennsylvania Academy of Science 19:26–30.

Clinton, G.P. 1935. Biographical memoir of Roland Thaxter 1858–1932. Biographical Memoirs of the National Academy of Sciences 17:55–68.

Cockerell, T.D.A. 1920. Biographical memoir of Alpheus Spring Packard 1839–1905. National Academy of Sciences Biographical Memoirs 9:181–236.

Coe, W.R. 1929a. Biographical memoir of Addison Emery Verrill 1839–1926. National Academy of Sciences Biographical Memoirs 14:19–66.

Coe, W.R. 1929b. Biographical memoir of Sidney Irving Smith 1843–1926. National Academy of Sciences Biographical Memoirs 14:5–16.

Cong, Q., J. Shen, A.D. Warren, D. Borek, Z. Otwinowski, and N.V. Grishin. 2016. Speciation in cloudless sulphurs gleaned from complete genomes. Genome Biology and Evolution 8:915–931.

Cong, Q., J. Shen, J. Zhang, W. Li, L.N. Kinch, J.V. Calhoun, A.D. Warren, and N.V. Grishin. 2021. Genomics reveals the origins of historical specimens. Molecular Biology and Evolution 38:2166–2176.

Cong, Q., J. Zhang, J. Shen, and N.V. Grishin. 2019. Fifty new genera of Hesperiidae (Lepidoptera). Insect Mundi 0731:1–56.

Copenheaver, C.A., A.S. White, and W.A. Patterson. 2000. Vegetation development in a southern Maine pitch pine–scrub oak barren. Journal of the Torrey Botanical Society 127(1): 19–32.

COSEWIC (Committee on the Status of Endangered Wildlife in Canada). 2000. COSEWIC assessment and status report on the Frosted Elfin *Callophrys [Incisalia] irus* in Canada. Ottawa: Committee on the Status of Endangered Wildlife in Canada.

COSEWIC (Committee on the Status of Endangered Wildlife in Canada). 2009. COSEWIC assessment and update status report on the Maritime Ringlet *Coenonympha nipisiquit* in Canada. Ottawa: Committee on the Status of Endangered Wildlife in Canada.

COSEWIC (Committee on the Status of Endangered Wildlife in Canada). 2016. COSEWIC assessment and status report on the Monarch *Danaus plexippus* in Canada. Ottawa: Committee on the Status of Endangered Wildlife in Canada.

Cox, K.W. 1993. Wetlands: A celebration of life. Final report of the Canadian Wetlands Conservation Task Force. Sustaining Wetlands Issues Paper 1993–1. Ottawa: North American Wetlands Conservation Council (Canada).

Creed, P.R. 1930. The Boston Society of Natural History 1830–1930. Boston: Boston Society of Natural History.

Creese, M.R.S., and T.M. Creese. 2010. Ladies in the laboratory III. South African, Australian, New Zealand, and Canadian women in science: Nineteenth and early twentieth centuries: a survey of their contributions. Lanham, MD: Scarecrow Press.

Crossley, M.S., T.D. Meehan, M.D. Moran, J. Glassberg, W.E. Snyder, and A.K. Davis. 2022. Opposing global change drivers counterbalance trends in breeding North American monarch butterflies. Global Change Biology 28:4726–4735.

Crozier, L. 2002. Climate change and its effects on species range boundaries: A case study of the Sachem skipper butterfly, *Atalopedes campestris*. In S.H. Schneider and T.L. Root, eds., Wildlife responses to climate change, pp. 57–91. Washington, DC: Island Press.

Dahl, T.E. 1990. Wetlands losses in the United States, 1780's to 1980's. Washington, DC: U.S. Department of the Interior, U.S. Fish and Wildlife Service.

Davis, B.N.K., K.H. Lakhani, and T.J. Yates. 1991. The hazards of insecticides to butterflies of old field margins. Agriculture Ecosystems and Environment 36:151–161.

Davis, D.R., and G.F. Hevel. 1995. Donation of the Auburn E. Brower collection to the Smithsonian Institution. Journal of the Lepidopterists' Society 49:253–254.

deMaynadier, P.G. 2009. Spicebush Swallowtail: Rare butterfly rediscovered in Maine. Inland Fisheries and Wildlife Insider (November): 1–2.

deMaynadier, P.G., and R.P. Webster. 2009. *Boloria frigga saga* (Nymphalidae): A significant new record for Maine and northeastern North America. Journal of the Lepidopterist's Society 63:177–178.

Devictor, V., C. van Swaay, T. Brereton, L. Brotons, D. Chamberlain, J. Heliölä, S. Herrando, R. Julliard, M. Kuussaari, Å. Lindström, J. Reif, D.B. Roy, O. Schweiger, J. Settele, C. Stefanescu, A. Van Strien, C. Van Turnhout, Z. Vermouzek, M.W. DeVries, I. Wynhoff, and F. Jiguet. 2012. Differences in the climatic debts of birds and butterflies at a continental scale. Nature Climate Change 2:121–124.

DiCesare, L. 2020. Un-common blue discovery in St. Albans, VT. Vermont Entomological Society News 109:4–5.

Dirig, R. 1994. Historical notes on wild lupine and the Karner Blue butterfly at the Albany Pine Bush, New York. *In* D.A. Andow, R.J. Baker, and C.P. Lane, eds., Karner Blue butterfly: A symbol of a vanishing landscape. Minnesota Agricultural Experiment Station Miscellaneous Publication 84(1994): 23–36. St. Paul: University of Minnesota.

Dirzo, R., H.S. Young, M. Galetti, G. Ceballos, N.J.B. Issac, and B. Collen. 2014. Defaunation in the Anthropocene. Science 345:401–406.

dos Passos, C.F. 1964. A synonymic list of the Nearctic Rhopalocera. Lepidopterists' Society Memoir 1:1–145.

dos Passos, C.F., and L.P. Grey. 1934a. A list of the butterflies of Maine with notes concerning some of them. Canadian Entomologist 66:188–192.

dos Passos, C.F., and L.P. Grey. 1934b. Additions and corrections to "A list of the butterflies of Maine." Canadian Entomologist 66:278.

dos Passos, C.F., and L.P. Grey. 1947. Systematic catalogue of *Speyeria* (Lepidoptera Nymphalidae) with designation of types and fixations of type localities. American Museum Novitates 1370:1–30.

Doubleday, E. 1844. List of the specimens of lepidopterous insects in the collection of the British Museum. Part I. London: Trustees of the British Museum.

Doubleday, E. 1847. List of the specimens of lepidopterous insects in the collection of the British Museum. Part II. London: Trustees of the British Museum.

Doucet, D. 2009. Census of globally rare, endemic butterflies of Nova Scotia Gulf of St Lawrence salt marshes. NS Species at Risk Conservation Fund Final Report 2008. https://novascotia.ca/natr/wildlife/conservationfund /final08/NSSARCF07_04RareButterfly.pdf.

Drummond, M.A., and T.R. Loveland. 2010. Land-use pressure and a transition to forest-cover loss in the eastern United States. Bioscience 60(4): 286–298.

Eaton, L.C. (with H. Piers). 1896. The butterflies of Truro, N. S. Proceedings and Transactions of the Nova Scotian Institute of Science 9:xvii–xxi.

eButterfly. 2020. eButterfly: An online database of butterfly distribution and abundance. http://www.e-butterfly.org (accessed 29 September 2020).

Edwards, G. 1743. A natural history of uncommon birds, and of some other rare and undescribed animals, quadrupeds, reptiles, fishes, insects, etc. exhibited in two hundred and ten copper-plates from designs copied immediately from nature, and curiously coloured after life. With a full and accurate description off each figure. Part I. London: published by the author.

Edwards, W.H. 1862. Descriptions of certain species of diurnal Lepidoptera found within the limits of the United States and British America. No. 2. Proceedings of the Academy of Natural Sciences of Philadelphia 14:54–55.

Environment Canada. 2012. Recovery strategy for the Maritime Ringlet (*Coenonympha nipisiquit*) in Canada. Species at Risk Act Recovery Strategy Series. Ottawa: Environment Canada.

Evans, C.K. 1993. How Vanessa became a butterfly: A psychologist's adventure in entomological etymology. Names: A Journal of Onomastics 41: 276–281.

Faber-Langendoen, D., J. Nichols, L. Master, K. Snow, A. Tomaino, R. Bittman, G. Hammerson, B. Heidel, L. Ramsay, A. Teucher, and B. Young. 2012. NatureServe conservation status assessments: Methodology for assigning ranks. Arlington, VA: NatureServe.

Fairweather, A.D., and D.F. McAlpine. 2011. History and status of the Natural History Society of New Brunswick entomology collection: 1897–1931. Journal of the Acadian Entomological Society 7:14–19.

Farquhar, D.W. 1934. The Lepidoptera of New England. PhD diss., Harvard University, Cambridge, MA.

Ferguson, D.C. 1954. The Lepidoptera of Nova Scotia. Part I, macrolepidoptera. Proceedings of the Nova Scotian Institute of Science 23:161–375.

Ferguson, D.C. 1955. The Lepidoptera of Nova Scotia. Part I (macrolepidoptera). Nova Scotia Museum of Science Bulletin 2:161–375.

Ferguson, D.C. 1962. James Halliday McDunnough (1877–1962): A biographical obituary and bibliography. Journal of the Lepidopterists' Society 16:209–228.

Fernald, C.H. 1884a. The butterflies of Maine. Annual reports of the trustees, president, farm superintendent and treasurer of the State College of Agriculture and the Mechanic Arts, Orono, Me, 1883. Appendix. 1–106.

Fernald, C.H. 1884b. The butterflies of Maine. Designed for the use of the students in the Maine State College, and the farmers of the state. Augusta: Sprague and Son.

Ferris, C.D. 1970. Occurrence of *Coenonympha inornata* (Satyridae) in Maine. Journal of the Lepidopterists' Society 24:202.

Field, W.D., C.F. dos Passos, and J.H. Masters. 1974. A bibliography of the catalogs, lists, faunal and other papers on the butterflies of North America north of Mexico arranged by state and province (Lepidoptera: Rhopalocera). Smithsonian Contributions to Zoology 157:1–104.

Fisher, R., A. Corbet, and C. Williams. 1943. The relation between the number of species and the number of individuals in a random sample of an animal population. Journal of Animal Ecology 12:42–58.

Fletcher, J. 1897. *Argynnis idalia* in New Brunswick. Canadian Entomologist 29:93.

Fletcher, J. 1907. Entomological record, 1906. Annual report of the Entomological Society of Ontario 37:86–106.

Forbes, D.L., G.S. Parkes, and L.A. Ketch. 2006. Sea-level rise and regional subsidence. *In* The impacts of sea level rise and climate change on the coastal zone of southeastern New Brunswick, pp. 34–94. New Brunswick: Environment Canada.

Forister, M.L., and A.M. Shapiro. 2003. Climatic trends and advancing spring flight of butterflies in southern California. Global Change Biology 9:1130–1135.

Foster, D. 1995. Land use history and forest transformations in central New England. *In* M.J. McDonnell and S.T.A. Pickett, eds., Humans as components of ecosystems, pp. 91–110. New York: Springer-Verlag.

Foster, D.R., B.M. Donahue, D.B. Kittredge, K.F. Lambert, M.L. Hunter, B.R. Hall, L.C. Irland, R.J. Lilieholm, D.A. Orwig, A.W. D'Amato, E.A. Colburn, J.R. Thompson, J.N. Levitt, A.M. Ellison, W.S. Keeton, J.D. Aber, C.V. Cogbill, C.T. Driscoll, T.J. Fahey, and C.M. Hart. 2010. Wildlands and woodlands: A vision for the New England landscape. Petersham, MA: Harvard Forest.

Freeman, T.N. 1962. James Halliday McDunnough, 1877–1962. Canadian Entomologist 94:1094–1102.

Gagliardi, B.L., and D.L. Wagner. 2016. 'Northern' oak hairstreak (*Satyrium favonius ontario*) (Lepidoptera: Lycaenidae): Status survey in Massachusetts, false rarity, and its use of non-nectar sugar resources. Annals of the Entomological Society of America 109:503–512.

Gagliardi, B.L., D.L. Wagner, and J.M. Allen. 2017. Species distribution model for the 'Northern' Oak hairstreak (*Satyrium favonius ontario*) with comments on its conservation status in the northeastern United States. Journal of Insect Conservation 21:781–790.

Gallinat, A.S., R.B. Primack, and D.L. Wagner. 2015. Autumn, the neglected season in climate change research. Trends in Ecology and Evolution 30:169–176.

Gawler, S., and A. Cutko. 2018. Natural landscapes of Maine: A guide to natural communities and ecosystems. Augusta:Maine Department of Agriculture, Conservation, and Forestry, Maine Natural Areas Program.

Gawler, S.C., J.J. Albright, P.D. Vickery, and F.C. Smith. 1996. Biological diversity in Maine: An assessment of status and trends in the terrestrial and freshwater landscape. Report prepared for the Maine Forest Biodiversity Project. Augusta: Maine Department of Conservation, Maine Natural Areas Program.

Gibson, A. 1924. The development of applied entomology in Canada. Annual Report of the Quebec Society for the Protection of Plants 16:24–56.

Glassburg, J. 1999. Butterflies through binoculars: The East. New York: Oxford University Press.

Glick, P., B.A. Stein, and N.A. Edelson, eds. 2011. Scanning the conservation horizon: A guide to climate change vulnerability assessment. Washington, DC: National Wildlife Federation.

Gobeil, R.E., and R.M. Gobeil. 2012. Status of the Common Buckeye in Maine. Newsletter of the Maine Butterfly Survey. Limenitis 5:6–8.

Gobeil, R.E., and R.M. Gobeil. 2014. The importance of power transmission line right-of-ways as habitat for butterflies in Maine. News of the Lepidopterists' Society 56:24–27.

Gobeil, R.E., and R.M. Gobeil. 2016. Notes on the status and distribution of the Wild Indigo Duskywing, *Erynnis baptisiae* (Forbes), in Maine. News of the Lepidopterists' Society 58:142–144.

Gobeil, R.E., and R.M. Gobeil. 2020. Notes on the status of the Fiery Skipper (*Hylephila phyleus*) in Maine. Maine Entomologist 24(1): 5–6.

Goltz, J.P. 2014. How the new forest strategy threatens plant species at risk in New Brunswick. New Brunswick Naturalist 41(1): 8–10.

Gochfeld, M., and J. Burger. 1997. Butterflies of New Jersey: A guide to their status, distribution, conservation and appreciation. New Brunswick, NJ: Rutgers University Press.

Grey, L.P. 1932. A good butterfly transition form (Lepid. Nymphalidae). Entomological News 43:241–242.

Grey, L.P. 1934. Lepidoptera from a Maine bog. Entomological News 45: 35–36.

Grey, L.P. 1956. Bog collecting in central Maine. Lepidopterists' News 10:119–120.

Grey, L.P. 1965. The flight period of *Boloria eunomia*. Journal of the Lepidopterists' Society 19:184–185.

Grey, L.P. 1988. Memories of Cyril F. dos Passos (1887–1986). Journal of the Lepidopterists' Society 42:164–167.

Grishin, N.V. 2019. Expanded phenotypic diagnoses for 24 recently named new taxa of Hesperiidae (Lepidoptera). Taxonomic Report 8(1): 1–15.

Grkovich, A., and H. Pavulaan. 2003. The case for taxonomic recognition of the taxon *Enodia anthedon borealis* A. H. Clark (Satyridae). Taxonomic Report of the International Lepidoptera Survey 4(5): 1–15.

Guppy, C.S., and J.H. Shepard. 2001. Butterflies of British Columbia: Including western Alberta, southern Yukon, the Alaska panhandle, Washington, northern Oregon, northern Idaho, and northern Montana. Vancouver: Royal British Columbia Museum and University of British Columbia Press.

Hagen, R.H., R.C. Lederhouse, J.L. Bossart, and J.M. Scriber. 1991. *Papilio canadensis* and *P. glaucus* (Papilionidae) are distinct species. Journal of the Lepidopterists' Society 45:245–258.

Hall, P.W. 2007. The European Common Blue *Polyommatus icarus*: New alien butterfly to Canada and North America. News of the Lepidopterists' Society 49:111.

Hall, P.W. 2009 Sentinels on the wing: The status and conservation of butterflies in Canada. Ottawa: NatureServe Canada.

Hall, P.W., C.D. Jones, A. Guidotti, and B. Hubley. 2014. Butterflies of Ontario. Toronto: Royal Ontario Museum.

Halsch, C.A., A.M. Shapiro, J.H. Thorne, D.P. Waetjen, and M.L. Forister. 2020. A winner in the Anthropocene: Changing host plant distribution explains geographical range expansion in the Gulf Fritillary butterfly. Ecological Entomology 45:652–662.

Hammond, P.C., and D.V. McCorkle. 2017. Taxonomy, ecology, and evolutionary theory of the genus *Colias* (Lepidoptera: Pieridae: Coliadinae). Corvallis: published by the authors.

Handfield, L. 2011. Le guide des papillons du Québec. Version scientifique. Vol. 1. Partie 1. Saint-Constant, QC: Broquet.

Harris, T.W. 1862. A treatise on some of the insects injurious to vegetation, 3rd ed. Boston: William White.

Haskin, J.R. 1931. Some unusual occurrences of butterflies in Connecticut (Lepidoptera: Pieridae, Nymphalidae). Entomological News 42:201–202.

Hawes, B., ed. 1835. Report from the select committee on the condition, management and affairs of the British Museum; together with the minutes of evidence, appendix and index. London: Hansard.

Healthy Forest Partnership. 2021. Available online at https://healthyforest partnership.ca/.

Heitzman, J.R., and C.F. dos Passos. 1974. *Lethe portlandia* (Fabricius) and *L. anthedon* (Clark), sibling species, with descriptions of new subspecies of the former (Lepidoptera: Satyridae). Transactions of the American Entomological Society 100:52–99.

Hessel, S.A., ed. 1951. The field season summary of North American Lepidoptera for 1951. Northeast—Delaware and Pennsylvania north to southern Quebec. Lepidopterists' News 5:108–109.

Heustis, C.E. 1879a. [*Papilio thoas* in New Brunswick]. Canadian Entomologist 11:239–240.

Heustis, C.E. 1879b. Scarcity of Papilionidae in Nova Scotia and New Brunswick. Canadian Entomologist 11:39.

Heustis, C.E. 1880. [Note on butterflies]. Canadian Entomologist 13:19–20.

Higgins, L.G. 1985. The correct name for what has been called *Lycaeides argyrognomon* in North America. Journal of the Lepidopterists' Society 39:145–146.

Hildreth, R.W. 2008. 2007—The year of the Red Admiral. The Maine Entomologist 12(2): 8–10.

Hildreth, R.W. 2011. Observations of migrant Insects in Downeast coastal Maine during the 2010 field season. The Maine Entomologist 15(4): 8–10.

Hildreth, R.W. 2013. Butterfly migration in Downeast coastal Maine (Hancock and Washington Counties) during the 2012 field season. Maine Entomologist 17(1): 3–6.

Hildreth, R.W. 2016. Sightings of migrant insects in Downeast coastal Maine during September 2015. The Maine Entomologist 20(1): 5–9.

Hoag, D.J. 2014. Larval host plants of *Enodia anthedon, Satyrodes appalachia* and *S. eurydice* in Vermont, USA. Taxonomic Report of the International Lepidoptera Survey 7(6): 1–6.

Hodges, R.W. 2003. Alexander Douglas Campbell Ferguson (1926–2002). Journal of the Lepidopterists' Society 57:299–303.

Horsfall, J.G. 1979. Roland Thaxter. Annual Review of Phytopathology 19:29–35.

Hunter, M.L., Jr. 1991. Coping with ignorance: The coarse filter strategy for maintaining biological diversity. *In* K. Kohm, ed., Balancing on the brink of extinction, pp. 266–281. Washington, DC: Island Press.

Iftner, D.C., J.A. Shuey, and J.V. Calhoun. 1992. Butterflies and skippers of Ohio. Ohio Biological Survey Bulletin (n.s.) 9.

iNaturalist. 2020. iNaturalist. https://www.inaturalist.org (accessed 30 December 2020).

IPCC (Intergovernmental Panel on Climate Change). 2019. Climate change and land: An IPCC special report on climate change, desertification, land degradation, sustainable land management, food security, and greenhouse gas fluxes in terrestrial ecosystems. https://www.ipcc.ch/srccl/ (accessed 23 December 2019).

Jacobson, G.L., I.J. Fernandez, P.A. Mayewski, and C.V. Schmitt, eds. 2009. Maine's climate future: An initial assessment. Orono: University of Maine. http://www.climatechange.umaine.edu/mainesclimatefuture/.

Jaffe, S. 2009. The caterpillars of Massachusetts 2009: A summary of my notes and findings. https://www.naba.org/chapters/nabambc/downloads/jaffe -caterpillar-notes-and-guide-2009.pdf.

Johns, R.C., J.J. Bowden, D.R. Carleton, B.J. Cooke, S. Edwards, E.J.S. Emilson, P.M.A. James, D. Kneeshaw, D.A. MacLean, V. Martel, E.R.D. Moise, G.D. Mott, C.J. Norfolk, E. Owens, D.S. Pureswaran, D.T. Quiring, J. Régnière, B. Richard, and M. Stastny. 2019. A conceptual framework for the spruce budworm early intervention strategy: Can outbreaks be stopped? Forests 10(10): 910.

Johnson, C.W. 1927. Biological survey of the Mount Desert region. Part I. The insect fauna with reference to the flora and other biological features. Philadelphia: Wistar Institute of Anatomy and Biology.

Jones, J.M. 1859. The naturalist in Bermuda; a sketch of the geology, zoology, and botany, of that remarkable group of islands; together with meteorological observations. London: Reeves and Turner.

Jones, J.M. 1870. Nova Scotian Lepidoptera. Canadian Entomologist 2:157.

Jones, J.M. 1872a. Review of Nova Scotian diurnal Lepidoptera. Proceedings and Transactions of the Nova Scotian Institute of Natural Science 3:18–27.

Jones, J.M. 1872b. Review of Nova Scotian diurnal Lepidoptera. Proceedings and Transactions of the Nova Scotian Institute of Science 3:100–103.

Keeler, M.S., and F.S. Chew. 2008. Escaping an evolutionary trap: Preference and performance of a native insect on an exotic invasive host. Oecologia 156:559–568.

Kendall, R.O., compiler. 1977. The Lepidopterists' Society commemorative volume 1945–1973. New Haven, CT: Lepidopterists' Society.

Kershaw, A.K. 1993. Early printed maps of Canada. Vol. 3. Ancaster, ON: Kershaw Publishing.

Kiel, W.J. 1976. *Callophrys eryphon* (Lycaenidae) in Maine. Journal of the Lepidopterists' Society 30:16–18.

Kiel, W.J. 2003. The butterflies of the White Mountains of New Hampshire. Concord: Audubon Society of New Hampshire.

Kirby, W. 1837. Order Lepidoptera. *In* J. Richardson, ed., Fauna boreali-americana; or the zoology of the northern parts of British America: containing descriptions of the objects of natural history collected on the late northern land expeditions, under command of Captain Sir John Franklin, R.N., pp. 286–309. Norwich, UK: Josiah Fletcher.

Klassen, G., and A. Locke. 2010. Inland waters and aquatic habitats of the Atlantic Maritime Ecozone. *In* D.F. McAlpine and I.M. Smith, eds., Assessment of species diversity in the Atlantic Maritime Ecozone, pp. 43–62. Ottawa: NRC Research Press.

Klots, A.B. 1939. *Brenthis aphirape* (Huebner) in North America, with a new record of the species from Maine (Lepidoptera, Nymphalidae). Bulletin of the Brooklyn Entomological Society 34:259–264.

Klots, A.B. 1951. A field guide to the butterflies of North America, east of the Great Plains. Boston: Houghton Mifflin.

Klymko, J. and K. Anderson. 2022. First records of the invasive beech leaf-mining weevil (*Orchestes fagi*) from New Brunswick and Prince Edward Island, Canada. Journal of the Acadian Entomological Society 18:23–25.

Klymko, J., C.S. Blaney, and D.G. Anderson. 2012. The first record of Dorcas Copper (*Lycaena dorcas*) in Nova Scotia. Journal of the Acadian Entomological Society 8:41–42.

Knurek, E.S. 2009. Taxonomic and population status of the Clayton's Copper butterfly (*Lycaena dorcas claytoni*). Master's thesis, University of Maine, Orono.

Kodandaramaiah, U. 2011. The evolutionary significance of butterfly eyespots. Behavioral Ecology 22:1264–1271.

Kodandaramaiah, U., and N. Wahlberg. 2009. Phylogeny and biogeography of *Coenonympha* butterflies (Nymphalidae: Satyrinae)—patterns of colonization in the Holarctic. Systematic Entomology 34:315–323.

Kondla, N.G., and B.C. Schmidt. 2010. Appendix. Taxonomic changes to Lepidoptera. Butterflies. *In* G.R. Pohl, G.G. Anweiler, B.C. Schmidt, and N.G. Kondla, eds., An annotated list of the Lepidoptera of Alberta, Canada. ZooKeys 38:495–497.

Krohn, W. B., R.B. Boone, and S.L. Painton. 1999. Quantitative delineation and characterization of hierarchical biophysical regions of Maine. Northeastern Naturalist 6(2): 139–164.

Latham, R.E. 2003. Shrubland longevity and rare plant species in the northeastern United States. Forest Ecology and Management 185:21–39.

Laurent, P. 1895. Notes on the insect fauna of Somerset Co., Maine. Canadian Entomologist 27:322–324.

Layberry, R.A., P.W. Hall, and J.D. Lafontaine. 1998. The Butterflies of Canada. Toronto: University of Toronto Press.

Lennox, J. 2007. An empire on paper: The founding of Halifax and conceptions of imperial space, 1744–55. Canadian Historical Review 88:373–412.

Li, W., Q. Cong, J. Shen, J. Zhang, W. Hallwachs, D.H. Janzen, and N.V. Grishin. 2019. Genomes of skipper butterflies reveal extensive convergence of wing patterns. Proceedings of the National Academy of Sciences of the USA 116:6232–6237.

Lloyd, C.G. 1917. Professor Roland Thaxter. Mycological Notes 45:622.

Loo, J., L. Cwynar, B. Freedman, and N. Ives. 2010. Changing forest landscapes in the Atlantic Maritime Ecozone. *In* D.F. McAlpine and I.M. Smith, eds., Assessment of species diversity in the Atlantic Maritime Ecozone, pp. 35–42. Ottawa: NRC Research Press.

Losey, J.E., and M. Vaughan. 2006. The economic value of ecological services provided by insects. Bioscience 56(4): 311–323.

Lotts, K., and T. Naberhaus, coordinators. 2020. Butterflies and moths of North America. https://www.butterfliesandmoths.org (accessed 30 December 2020).

Lyman, H.H. 1874. [Notes about butterflies]. Canadian Entomologist 6:38.

Lyman, H.H. 1876. Notes on the occurrence of *Argynnis idalia* Drury. Canadian Entomologist 8:208–210.

Lyman, H.H. 1880. List of diurnal Lepidoptera taken in the vicinity of Portland, Maine. Canadian Entomologist 12:7–9.

Lyman, H.H. 1910. Rare butterflies of Maine. Entomological News 21:84.

Macgregor, C.J., C.D. Thomas, D.B. Roy, M.A. Beaumont, J.R. Bell, T. Brereton, J.R. Bridle, C. Dytham, R. Fox, K. Gotthard, A.A. Hoffmann, G. Martin, I. Middlebrook, S. Nylin, P.J. Platts, R. Rasteiro, I.J. Saccheri, R. Villoutreix, C.W. Wheat, and J.K. Hill. 2019. Climate-induced phenology shifts linked to range expansions in species with multiple reproductive cycles per year. Nature Communications 10:1–10.

Macnaughton, A., R. Layberry, R. Cavasin, B. Edwards, and C. Jones. 2018. Ontario butterfly atlas. https://www.ontarioinsects.org/atlas_online.htm (accessed 30 May 2018).

MacNeill, C.D. 1964. The skippers of the genus *Hesperia* in western North America with special reference to California (Lepidoptera: Hesperiidae). Berkeley: University of California Press.

Mader, E., M. Shepherd, M. Vaughan, S.H. Black, and G. LeBuhn. 2011. Attracting native pollinators: Protecting North America's bees and butterflies. The Xerces Society guide. North Adams, MA: Storey Publishing.

Martin, K. 1979. Francis Bain, farmer naturalist. The Island Magazine 6:3–8.

Martin, K. 1983. Inventory of natural science specimens of Prince Edward Island. Charlottetown: University of Prince Edward Island Department of Extension and Prince Edward Island Department of Community and Cultural Affairs.

Martin, K. 1990. Bain, Francis. Dictionary of Canadian Biography 12:47–49.

Master, L.L., D. Faber-Langendoen, R. Bittman, G.A. Hammerson, B. Heidel, L. Ramsay, K. Snow, A. Teucher, and A. Tomaino. 2012. NatureServe conservation status assessments: Factors for evaluating species and ecosystem risk. Arlington, VA: NatureServe.

McAlpine, D.F. 2018. The gaudy sphinx: McIntosh, Leavitt and entomology in Maritime Canada, 1897–1922. *In* G. Davies, P. Larocque, and C. Verduyn, eds., The creative city of Saint John, pp. 32–135. Halifax: Formac Publishing.

McAlpine, D.F., and I.M. Smith. 2010. The Atlantic Maritime Ecozone: Old mountains tumble into the sea. *In* D.F. McAlpine and I.M. Smith, eds., Assessment of species diversity in the Atlantic Maritime Ecozone, pp. 1–12. Ottawa: NRC Research Press.

McCaskill, G.L., T. Albright, C.J. Barnett, B.J. Butler, S.J. Crocker, C.M. Kutz, W.H. McWilliams, P.D. Miles, R.S. Morin, M.D. Nelson, R.H. Widmann, and C.W. Woodall. 2016. Maine forests 2013. Resource Bulletin NRS-103. Newtown Square, PA: U.S. Department of Agriculture, Forest Service, Northern Research Station.

McCollough, M., C. Todd, B. Swartz, P. deMaynadier, and H. Givens. 2003. Maine's endangered and threatened wildlife. Augusta: Maine Department of Inland Fisheries and Wildlife.

McCollough, M.A. 1997. Conservation of invertebrates in Maine and New England: Perspectives and prognoses. Northeastern Naturalist 4(4): 261–278.

McCulloch, T. 1920. Life of Thomas McCulloch, D.D. Truro, Nova Scotia: published by the author.

McDermott, F.A. 1963. Frank Morton Jones (1869–1962). Entomological News 74:29–36.

McDunnough, J.H. 1935a. A new race of *Argynnis cybele* from Nova Scotia. Canadian Entomologist 67:18–19.

McDunnough, J.H. 1935b. Notes on two Nova Scotia butterflies. Canadian Entomologist 67:211–212.

McDunnough, J.H. 1938. Check list of the Lepidoptera of Canada and the United States of America. Part 1. Macrolepidoptera. Memoirs of the Southern California Academy of Sciences 1.

McDunnough, J.H. 1939a. Check list of the Lepidoptera of Canada and the United States of America. Part II. Microlepidoptera. Memoirs of the Southern California Academy of Sciences 2.

McDunnough, J.H. 1939b. *Papilio brevicauda* in Cape Breton Island, N.S. Canadian Entomologist 71:153–158, pl. 22.

McDunnough, J.H. 1940. A new race of *Lycaena dorcas* from northeastern New Brunswick. Canadian Entomologist 72:130–131.

McFarland, K.P., and B. Pfeiffer. 2022. Vermont butterfly survey. Vermont

Center for Ecostudies—Vermont Atlas of Life. http://val.vtecostudies.org (accessed 1 February 2022).

McFarland, K., and S. Zahendra. 2010. Vermont butterfly survey 2002–2007. Final report to the Natural Heritage Information Project of the Vermont Department of Fish and Wildlife. Norwich: Vermont Center for Ecostudies.

McGugan, B.M., ed. 1958. Forest Lepidoptera of Canada recorded by the Forest Insect Survey. Vol. I—Papilionidae to Arctiidae. Ottawa: Canada Department of Agriculture, Forest Biology Division Publication 1034.

McIntosh, W. 1899a. The butterflies of New Brunswick. Bulletin of the Natural History Society of New Brunswick 4:114–121.

McIntosh, W. 1899b. The butterflies of New Brunswick. Bulletin of the Natural History Society of New Brunswick 4:223–225.

McIntosh, W. 1904. Supplementary list of the Lepidoptera of New Brunswick. Bulletin of the Natural History Society of New Brunswick 5:355–357.

McIntosh, W. 1907. Report of zoology. Bulletin of the Natural History Society of New Brunswick 5:570–571.

McLaine, L.S. 1937. Joseph Perrin. Canadian Entomologist 69:19–20.

McMahon, J.S. 1990. The biophysical regions of Maine: Patterns in the landscape and vegetation. Master's thesis, University of Maine, Orono.

McTavish, L. 2008. Strategic donations: Women and museums in New Brunswick, 1862–1930. Journal of Canadian Studies 42:93–116.

McTavish, L., and J. Dickison. 2007. William MacIntosh, natural history and the professionalization of the New Brunswick Museum, 1898–1940. Acadiensis 36:72–90.

MDEP (Maine Department of Environmental Protection). 2003. Wetlands protection: A federal, state and local partnership. Augusta: Maine Department of Environmental Protection. https://www.maine.gov/dep/land/nrpa/ip-wet-protectionl.html.

MDIFW (Maine Department of Inland Fisheries and Wildlife). 2015. Maine's wildlife USFWS. Augusta: Maine Department of Inland Fisheries and Wildlife.

Meier, A.J., S.P. Bratton, and D.C. Duffy. 1995. Possible ecological mechanisms for loss of vernal-herb diversity in logged eastern deciduous forests. Ecological Applications 5(4): 935–946.

Melander, A.L. 1932. The entomological publications of C. W. Johnson. Psyche 39:87–99.

Melero, Y., C. Stefanescu, and J. Pino. 2016. General declines in Mediterranean butterflies over the last two decades are modulated by species traits. Biological Conservation 201:336–342.

Mello, M.J., coordinator. 1999.1998 season summary. Zone 10 northeast. Maine. News of the Lepidopterists' Society 41 (supplement S1): 68–69.

Miller, S.J., and D.B. Lyons. 1979. An entomological survey of Kouchibouguac National Park. Ottawa: Agriculture Canada, Biosystematics Research Institute.

Miller-Rushing, A.J., and R.B. Primack. 2008. Global warming and flowering times in Thoreau's Concord: A community perspective. Ecology 89:332–341.

Mittelhauser, G., M. Barr, and A. Swann. 2014. Catalog of the butterflies and moths (Lepidoptera) of Mount Desert Island, Maine from early- to mid-1900s: Based on the catalog of William Procter's insect collection. Gouldsboro: Maine Natural History Observatory.

Monarch Watch. 2022. Total area occupied by monarch colonies at overwintering sites in Mexico. https://monarchwatch.org/blog/uploads/2021/02/monarch-population-figure-monarchwatch-2021.png.

Monroe, J.L., and D.M. Wright. 2017. Butterflies of Pennsylvania: A field guide. Pittsburgh, PA: University of Pittsburgh Press.

Moola, F.M., and L. Vasseur. 2008. The maintenance of understory residual flora with even-aged management: A review of temperate forests in northeastern North America. Environmental Reviews: 16:141–155.

Morris, R.F. 1980. Butterflies and moths of Newfoundland and Labrador: The macrolepoptera. Ottawa: Agriculture Canada.

Mulé, R., G. Sabella, L. Robba, and B. Manachini. 2017. Systematic review of the effects of chemical insecticides on four common butterfly families. Frontiers in Environmental Science. 5:1–5.

NABA (North American Butterfly Association). 2018. Checklist of North American butterflies occurring North of Mexico—edition 2.4. https://www.naba.org/pubs/enames2_4.html (accessed 2 July 2020).

Nakazawa, T., and H. Doi. 2012. A perspective on match/mismatch of phenology in community contexts. Oikos 121:489–495.

NatureServe. 2002. Element occurrence data standard. https://www.natureserve.org/conservation-tools/element-occurrence-data-standard (accessed 15 February 2018).

NatureServe Explorer. 2021. Online encyclopedia of life. http://explorer.natureserve.org/index.htm (accessed 4 March 2021).

Nazari, V., L. Handfield, and D. Handfield. 2018. The European Peacock butterfly, *Aglais io* (Linnaeus, 1758) in North America (Lepidoptera: Nymphalidae). News of the Lepidopterists' Society 60:128–129.

NBMRRT (New Brunswick Maritime Ringlet Recovery Team). 2005. Recovery strategy and action plan for the Maritime Ringlet (*Coenonympha nipisiquit*) in New Brunswick. Fredericton: New Brunswick Department of Natural Resources.

Neily, P., S. Basquill, E. Quigley, B. Stewart, and K. Keys. 2010. Forest ecosystem classification for Nova Scotia. Part I: Vegetation types. Nova Scotia Department of Natural Resources, Renewable Resources Branch, Report FOR 2011-1.

New Brunswick Department of Energy and Mines. 2014. Peat mining policy. https://www2.gnb.ca/content/dam/gnb/Departments/en/pdf/Minerals-Minerales/Peat_Mining_Policy-e.pdf.

Nielsen M.C. 1999. Michigan butterflies and skippers. East Lansing: Michigan State University Extension.

O'Donnell, J.E., L.F. Gall, and D. L. Wagner, eds. 2007. The Connecticut butterfly atlas. State Geological and Natural History Survey Bulletin 118. Hartford, CT: Department of Environmental Protection.

Olmstead, N., and M. Yurlina. 2019. Maine invasive plants field guide. Augusta: Maine Department of Agriculture, Conservation, and Forestry, Maine Natural Areas Program.

Olofsson, P., C.E. Holden, E.L. Bullock, and C.E. Woodcock. 2016. Time series analysis of satellite data reveals continuous deforestation of New England since the 1980s. Environmental Research Letters 11. https://doi.org/10.1088/1748-9326/11/6/064002.

Omernik, J.M., and G.E. Griffith. 2014. Ecoregions of the conterminous United States: Evolution of a hierarchical spatial framework. Environmental Management 54:1249–1266.

Opler, P.A. 1995. Lepidoptera of North America. 2. Distribution of the butterflies (Papilionoidea and Hesperioidea) of the eastern United States. Contributions of the C. P. Gillette Museum of Arthropod Diversity. Fort Collins: Colorado State University.

Opler, P.A., and G.O. Krizek. 1984. Butterflies east of the Great Plains. Baltimore: The Johns Hopkins University Press.

Opler, P.A., and A.D. Warren. 2006. Butterflies of North America. 4. Scientific names list for butterfly species of North America, north of Mexico. Contributions of the C. P. Gillette Museum of Arthropod Diversity. Fort Collins: Colorado State University.

Packard, A.S. 1862. Entomological report. Annual report of the secretary of the Maine Board of Agriculture 7:143–219.

Packard, A.S. 1869. Guide to the study of insects, and treatise on those injurious and beneficial to crops: For the use of colleges, farm-schools, and agriculturists. Salem, MA: Naturalist's Book Agency.

Parlin, J.C. 1922. Butterflies collected by J. C. Parlin. Maine Naturalist 2: 73–74.

Parlin, J.C. 1923. Additional Lepidoptera collected in 1922. Maine Naturalist 3:51.

Parmesan, C., N. Ryrholm, C. Stefanescu, J.K. Hill, C.D. Thomas, H. Descimon, B. Huntley, L. Kaila, J. Kullberg, T. Tammaru, J. Tennent, J.A. Thomas, and M. Warren. 1999. Poleward shifts in geographical ranges of butterfly species associated with regional warming. Nature 399:579–583.

Pavulaan, H. 2021. Reevaluation of the described subspecies of *Euphydryas phaeton* (Drury, 1773) with a replacement name for *Melitaea phaeton schausi* (Clark, 1927). Taxonomic Report 9(10): 1–20.

Pavulaan, H., and D.M. Wright. 2005. *Celastrina serotina* (Lycaenidae: Polyommatinae): A new butterfly species from the northeastern United States and eastern Canada. Taxonomic Report 6(6): 1–18.

Peatland Ecology Research Group. 2021. Peat industry. https://www.gret-perg.ulaval.ca/industrial-research-partnership/peat-industry/ (accessed 12 March 2021).

Pelham, J.P. 2022. A catalogue of the butterflies of the United States and Canada. https://www.butterfliesofamerica.com/US-Can-Cat.htm (revised 2 February 2022; accessed 17 March 2022).

Perrin, J. 1919. Additions to the catalogue of butterflies and moths collected in the neighbourhood of Halifax, etc., Nova Scotia. Proceedings and Transactions of the Nova Scotian Institute of Science 14:49–56.

Perrin, J., and J. Russell. 1912. Catalogue of butterflies and moths, mostly collected in the neighbourhood of Halifax and Digby, Nova Scotia. Proceedings and Transactions of the Nova Scotian Institute of Science 12:258–290.

Pfeiffer, B. 2021. Follow up sightings of *Polyomatus icarus* in the United States. News of the Lepidopterists' Society 63:155

Piers, H. 1903. Sketch of the life of J. M. Jones. Proceedings and Transactions of the Nova Scotian Institute of Science 10:lxxx–lxxxii.

Piers, H. 1906. Provincial Museum and science library. *In* E. Gilpin, ed., Report of the Department of Mines, Nova Scotia, for the year ended 30th September, 1905, pp. 132–142. Halifax: Commissioner of Public Works and Mines.

Piers, H. 1915. A brief historical account of the Nova Scotian Institute of Science, and the events leading up to its formation; with biographical sketches of its deceased presidents and other prominent members. Proceedings and Transactions of the Nova Scotian Institute of Science 13:liii–cix.

Piers, H. 1918. The Orthoptera (cockroaches, locusts, grasshoppers and crickets) of Nova Scotia; with descriptions of the species and notes on their occurrence and habits. Proceedings and Transactions of the Nova Scotian Institute of Science 14:201–354.

Platt, A.P., G.W. Rawson, and G. Balogh. 1978. Inter-specific hybridization involving *Limentitis archippus* and its congeneric species (Nymphalidae). Journal of the Lepidopterists' Society 32:289–303.

Pohl, G.R., G. Anweiler, C. Schmidt, and N. Kondla. 2010. An annotated list of the Lepidoptera of Alberta, Canada. ZooKeys 38.

Pohl, G.R., J.-F. Landry, B.C. Schmidt, J.D. Lafontaine, A.D. Macaulay, E.J. van Nieukerken, J.R. deWaard, J.J. Dombroskie, J. Klymko, V. Nazari, and K. Stead. 2018. Annotated checklist of the moths and butterflies (Lepidoptera) of Canada and Alaska. Series Faunistica 118. Sofia, Bulgaria: Pensoft.

Polgar, C.A., R.B. Primack, E.H. Williams, S. Stichter, and C. Hitchcock. 2013. Climate effects on the flight period of lycaenid butterflies in Massachusetts. Biological Conservation 160:25–31.

Porter, C.J.M., S.P. Basquill, and J.T. Lundholm. 2020. Barrens ecosystems in Nova Scotia: Classification of heathlands and related plant communities. Joint publication of Nova Scotia government and Saint Mary's University. Nova Scotia Department of Lands and Forestry, Biodiversity Reference Guide 2020–001. https://novascotia.ca/natr/wildlife/pdf/Barrens -Classification.pdf.

Pöyry, J., M. Luoto, R. Heikkinen, R.M. Kuussaari, and K. Saarinen. 2009. Species traits explain recent range shifts of Finnish butterflies. Global Change Biology 15:732–743.

Pratt, G.F., G.R. Ballmer, and D.M. Wright. 2011. Allozyme-based phylogeny of North American *Callophrys* (s. l.) (Lycaenidae). Journal of the Lepidopterists' Society 65:205–222.

Pratt, G.F., and D.M. Wright. 2002. Allozyme phylogeny of North American coppers (Lycaeninae: Lycaenidae). Pan-Pacific Entomologist 78:219–229.

Procter, W. 1938. Biological survey of the Mount Desert region. Part VI. The insect fauna with references to methods of capture, food plants, the flora and other biological features. Philadelphia: Wistar Institute of Anatomy and Biology.

Procter, W. 1946. Biological survey of the Mount Desert region. Part VII. The insect fauna with references to methods of capture, food plants, the flora and other biological features. Philadelphia: Wistar Institute of Anatomy and Biology.

Raine, J. 1857. A memoir of the Rev. John Hodgson. Vol. 1. London: Longman, Brown, Green, Longmans, and Roberts.

Raymond, C., and J.S. Mankin. 2019. Assessing present and future coastal moderation of extreme heat in the eastern United States. Environmental Research Letters 14. https://doi.org/10.1088/1748-9326/ab495d.

Remington, C.L. 1954. Frank Morton Jones collection presented to Yale. Lepidopterists' News 8:47.

Renaud, A. 2010. Into the mist: The story of the Empress of Ireland. Toronto: Dundurn Press.

Renwick, J.A.A. 2002. The chemical world of crucivores: Lures, treats and traps. Entomologia Experimentalis et Applicata 104:35–42.

Ritland, D.B. 1995. Comparative unpalatability of mimetic viceroy butterflies (*Limenitis archippus*) from four southeastern United States populations. Oecologia 103:327–336.

Ritland, D.B., and L.P. Brower. 1991. The viceroy butterfly is not a Batesian mimic. Nature 350: 497–498.

Roy, D.B., and T.H. Sparks. 2000. Phenology of British butterflies and climate change. Global Change Biology 6:407–416.

Russell, C., and C.B. Schultz. 2010. Investigating the use of herbicides to control invasive grasses: Effects on at-risk butterflies. Journal of Insect Conservation 14:53–63.

Russell, K.N., H. Ikerd, and S. Droege. 2005. The potential conservation value of unmowed powerline strips for native bees. Biological Conservation 124:133–148.

Sánchez-Bayo, F. and K.A.G. Wyckhuys. 2019. Worldwide decline of the entomofauna: A review of its drivers. Biological Conservation 232:8–27.

Sánchez-Bayo, F., and K.A.G. Wyckhuys. 2021. Further evidence for a global decline of the entomofauna. Australian Entomology 60:9–26.

Sardinas, H., J. Hopwood, J.K. Cruz, J. Eckberg, K. Gill, R. Powers, S.F. Jordan, M. Vaughan, N.L. Adamson, and E. Lee-Mader. 2018. Maintaining diverse stands of wildflowers planted for pollinators: Long-term management of planted pollinator habitat for pollinator conservation. Portland, OR: The Xerces Society for Invertebrate Conservation.

Schlawin, J., K. Puryear, D. Circo, A. Cutko, S. Demers, and M. Docherty. 2021. An assessment of accomplishments and gaps in Maine land conservation. A technical report submitted to the Maine Department of Agriculture, Conservation, and Forestry, Augusta, ME.

Schmidt, B.C. 2020. More on Ontario Tiger Swallowtails. *In* L. Hockley and

A. Macnaughton, eds., Ontario Lepidoptera 2019. Occasional Publication 51(2020): 3–11. Toronto: Toronto Entomologists' Association.

Schmidt, B.C., and R.A. Layberry. 2016. What azure blues occur in Canada? A re-assessment of *Celastrina* Tutt species (Lepidoptera, Lycaenidae). ZooKeys 584:135–164.

Schuerman, T.P., and K. Puryear. 2001. Waterboro Barrens 2001 Lepidoptera survey final report. Technical report for The Nature Conservancy, Brunswick, ME.

Schweitzer, D.F. 2008. *Celastrina ladon*: Spring Azure. NatureServe Explorer. https://explorer.natureserve.org (accessed 19 May 2020).

Schweitzer, D.F., M.C. Minno, and D.L. Wagner. 2011. Rare, declining, and poorly known butterflies and moths (Lepidoptera) of forests and woodlands in the eastern United States. Morgantown, WV: U.S. Forest Service, Forest Health Technology Enterprise Team, FHTET-2011-01.

Scoones, J. 2016. 2016 RI butterfly count report. Smithfield: Audubon Society of Rhode Island.

Scott, F.W., and B. Wright. 1972. Accidental occurrence of *Aglais urticae* (Nymphalidae) in Nova Scotia. Journal of the Lepidopterists' Society 26:116.

Scott, J.A. 1986. The butterflies of North America: A natural history and field guide. Stanford, CA: Stanford University Press.

Scriber, J.M, M. Deering, and A. Stump. 2008. Hybrid zone ecology and Tiger Swallowtail trait clines in North America. *In* C.R. Boggs, W. Watt, and P.R. Ehrlich, eds., Butterflies: Ecology and evolution taking flight, pp. 367–391. Chicago: University of Chicago Press.

Scudder, S.H. 1863. A list of the butterflies of New England. Proceedings of the Essex Institute. 3:161–179.

Scudder, S.H. 1868. Supplement to a list of the butterflies of New England. Proceedings of the Boston Society of Natural History 11:375–384.

Scudder, S.H. 1872. A systematic revision of some of the American butterflies; with brief notes on those known to occur in Essex County, Mass. Annual Report of the Trustees of the Peabody Academy of Sciences 4:24–83.

Scudder, S.H. 1887a. The introduction and spread of *Pieris rapae* in North America, 1860–1885. Memoirs of the Boston Society of Natural History 4:53–69.

Scudder, S.H. 1877b. Notice of a small collection of butterflies made by Mr. Roland Thaxter, on Cape Breton Island. Proceedings of the Boston Society of Natural History 18:188–190.

Scudder, S.H. 1888–1889. The butterflies of the eastern United States and Canada with special reference to New England. 3 vols. Cambridge, MA: published by the author.

Scudder, S.H. 1899. Every-day butterflies: A group of biographies. Boston: Houghton Mifflin.

Sei, M., and D.H. Porter. 2007. Delimiting species boundaries and the conservation genetics of the endangered Maritime Ringlet butterfly (*Coenonympha nipisiquit* McDunnough). Molecular Ecology 16:3313–3325.

Semmens, B.X., D.J. Semmens, W.E. Thogmartin, R. Wiederholt, L.

López-Hoffman, J.E. Diffendorfer, J.M. Pleasants, K.S. Oberhauser, and O.R. Taylor. 2016. Quasi-extinction risk and population targets for the eastern, migratory population of Monarch butterflies (*Danaus plexippus*). Science Reports 6:23265.

Sferra, N., and T. Patton. 1996. Preliminary survey of invertebrates for Waterboro Barrens Preserve. Technical report for The Nature Conservancy, Brunswick, ME.

Shapiro, A.M. 1966. Butterflies of the Delaware Valley. Philadelphia: American Entomological Society.

Shepard, J.H. 1998. The correct name for the *Boloria chariclea/titania* complex in North America (Lepidoptera: Nymphalidae). *In* T. C. Emmel, ed., Systematics of western North American butterflies, pp. 727–730. Gainesville, FL: Mariposa Press.

Slayback, D.A., L.P. Brower, M.I. Ramirez, and L.S. Fink. 2007. Establishing the presence and absence of overwintering colonies of the monarch butterfly in Mexico by the use of small aircraft. American Entomologist 53:28–39.

Smith, M.P., and L. Hayden, eds. 2009. Conserving freshwater and coastal resources in a changing climate. A report prepared for The Nature Conservancy by E. Grubin, A. Hardy, R. Lyons, A. Schmale, and T. Sugii, Tufts University, Medford, MA.

Sourakov, A. 2013. Two heads are better than one: False head allows *Calycopis cecrops* (Lycaenidae) to escape predation by a jumping spider, *Phidippus pulcherrimus* (Salticidae). Journal of Natural History 47:1047–1054.

Stark, J.D., X.D. Chena, and C.S. Johnson. 2012. Effects of herbicides on Behr's Metalmark butterfly, a surrogate species for the endangered butterfly, Lange's Metalmark. Environmental Pollution 164:24–27.

Stein B.A., L.S. Kutner, and J.S. Adams, eds. 2000. Precious heritage: The status of biodiversity in the United States. New York: Oxford University Press.

Stichter, S. 2015. The butterflies of Massachusetts: Their history and future. Middletown, DE: published by the author.

Summerville, K.S. 2011. Managing the forest for more than the trees: Effects of experimental timber harvest on forest Lepidoptera. Ecological Applications 21(3): 806–816.

Summerville, K.S. 2013. Forest lepidopteran communities are more resilient to shelterwood harvests compared to more intensive logging regimes. Ecological Applications 23(5): 1101–1112.

Summerville, K.S., and T.O. Crist. 2002. Effects of timber harvest on forest Lepidoptera: Community, guild, and species responses. Ecological Applications 12(3): 820–835.

Sweeney, J., R.S. Anderson, R.P. Webster, and R. Neville. 2012. First records of *Orchestes fagi* (L.) Coleoptera: Curculionidae: Curculioninae) in North America, with a checklist of the North American Rhamphini. Coleopterist's Bulletin 66: 297–304.

Sweeney, J.D., C. Hughes, H. Zhang, N.K. Hillier, A. Morrison, and R. Johns. 2020. Impact of the invasive beech leaf-mining weevil, *Orchestes fagi*, on

American Beech in Nova Scotia, Canada. Frontiers in Forest and Global Change 3:1–11.

Tallamy, D.W. 2019. Nature's best hope: A new approach to conservation that starts in your yard. Portland, OR: Timber Press.

Tang, G., and B. Beckage. 2010. Projecting the distribution of forests in New England in response to climate change. Diversity and Distributions 16:144–158.

Theobald, D.M. 2010. Estimating natural landscape changes from 1992 to 2030 in the conterminous US. Landscape Ecology 25:999–1011.

Thomas, A.W. 1991. Life across the border. Journal of the Lepidopterists' Society 45:180.

Thomas, A.W. 1996. A preliminary atlas of the butterflies of New Brunswick. New Brunswick Museum Publications in Natural Science 11.

Thompson, J.R., D.N. Carpenter, C.V. Cogbill, and D.R. Foster. 2013. Four centuries of change in northeastern United States forests. PLoS ONE 8(9): e72540.

Toftegaard, T., D. Posledovich, J.A. Navarro-Cano, C. Wiklund, K. Gotthard, and J. Ehrlén. 2019. Butterfly–host plant synchrony determines patterns of host use across years and regions. Oikos 128:493–502.

Uhler, P.R. 1865. Report on the collection of insects. In W. Gray, ed., Annual report of the trustees of the Museum of Comparative Zoology, at Harvard College, in Cambridge, together with the report of the director, 1864, pp. 29–37. Boston: Wright and Potter.

USFWS (U.S. Fish and Wildlife Service). 2020. Endangered and threatened wildlife and plants; 12-month finding for the monarch butterfly. Federal Register 85(243): 81813–81822.

van Klink, R., D.E. Bowler, K.B. Gongalsky, A.B. Swengel, A. Gentile, and J.M. Chase. 2020. Meta-analysis reveals declines in terrestrial but increases in freshwater insect abundances. Science 368:417–420.

Vanderweide, H. 1977. Maine's foremost naturalist. Kennebec Journal, 28 January 1977, 6.

Venter, O., N.N. Brodeur, L. Nemiroff, B. Belland, I.J. Dolinsek, and J.A. Grant. 2006. Threats to endangered species in Canada. Bioscience 56(11): 903–910.

Verrill, A.E. 1914. Monograph of the shallow-water starfishes of the North Pacific Coast from the Arctic Ocean to California. Part 2. Plates. Harriman Alaska Series, vol. 14. Washington, D.C.: Smithsonian Institution.

Verrill, A.E. 1926. Sidney Irving Smith. Science (n.s.) 66:57–58.

Vickery, V., and G. Moore. 1964. The Lyman Entomological Museum, 1914–1964. Canadian Entomologist 96:1489–1494.

Volenec, Z.M., and A.P. Dobson. 2019. Conservation value of small reserves. Conservation Biology 34(1): 66–79.

Wagner, D.L. 2000. The biodiversity of moths. In S.A. Levin, editor in chief, The encyclopedia of biodiversity, pp. 249–270. San Diego, CA: Academic Press.

Wagner, D.L. 2005. Caterpillars of eastern North America. Princeton, NJ: Princeton University Press.

Wagner, D.L. 2007. Emerald Ash Borer threatens ash-feeding Lepidoptera. News of the Lepidopterists' Society 49(1): 10–11.

Wagner, D.L. 2020. Insect declines in the Anthropocene. Annual Review of Entomology 65:457–480.

Wagner, D.L., and B.L. Gagliardi. 2015. Hairstreaks (and other insects) feeding at galls, honeydew, extrafloral nectaries, sugar bait, cars, and other routine substrates. American Entomologist 61:160–167.

Wagner, D.L., M.W. Nelson, and D.F. Schweitzer. 2003. Shrubland Lepidoptera of southern New England and southeastern New York: Ecology, conservation, and management. Forest Ecology and Management 185:95–112.

Wagner, D.L., M.S. Wallace, G.H. Boettner, and J.S. Elkinton. 1995. Status update and life history studies on the Regal Fritillary (Lepidoptera: Nymphalidae). American Butterflies 3:261–274.

Ward, M., and P. deMaynadier. 2012. Distribution, abundance, and status of the Spicebush Swallowtail (*Papilio troilus*) in Maine, 2010–2012. Report to Maine Department of Inland Fisheries and Wildlife and U.S. Fish and Wildlife Service, Augusta, ME.

Warren, A.D., S. Nakahaara, V.A. Lukhtanov, K.M. Daly, C.D. Ferris, N.V. Grishin, M. Cesanek, and J.P. Pelham. 2016. A new species of *Oeneis* from Alaska, United States, with notes on the *Oeneis chryxus* complex (Lepidoptera: Nymphalidae: Satyrinae). Journal of Research on the Lepidoptera 49:1–20.

Warren, M.S., D. Maes, C.A.M. van Swaay, P. Goffart, H. Van Dyck, N.A.D. Bourn, I. Wynhoff, D. Hoare, and S. Ellis. 2021. The decline of butterflies in Europe: Problems, significance, and possible solutions. Proceedings of the National Academy of Sciences of the USA 118. https://doi.org/10.1073/pnas.2002551117.

Webster, R.P. 2001. The establishment of new populations of the Maritime Ringlet, *Coenonympha nipisiquit*, in New Brunswick. 2001 interim report to the New Brunswick Wildlife Council Trust Fund Committee. Frederickton, NB.

Webster, R.P., and P.G. deMaynadier. 2005. A baseline atlas and conservation assessment of the butterflies of Maine. Technical report submitted to the Maine Department of Inland Fisheries and Wildlife, Augusta, ME. https://digitalmaine.com/cgi/viewcontent.cgi?article=1082&context=ifw_docs.

Webster, R.P., and M.C. Nielsen. 1984. Myrmecophily in the Edward's Hairstreak butterfly *Satyrium edwardsii* (Lycaenidae). Journal of the Lepidopterists' Society 38:124–133.

Webster, R.P., and B. Swartz. 2006. Population studies of the Clayton's Copper at Dwinal Pond Wildlife Management Area. Technical report submitted to the Maine Department of Inland Fisheries and Wildlife, Augusta, ME.

Wepprich T., J.R. Adrion, L. Ries, J. Wiedmann, and N.M. Haddad. 2019. Butterfly abundance declines over 20 years of systematic monitoring in Ohio, USA. PLoS ONE 14. https://doi.org/10.1371/journal.pone.0216270.

White, H.B., Jr., and J.V. Calhoun. 2009. Miss Mattie Wadsworth (1862–1943): Early woman author in *Entomological News*. Transactions of the American Entomological Society 135:413–429.

Whitman, A., A. Cutko, P. deMaynadier, S. Walker, B. Vickery, S. Stockwell, and R. Houston. 2013. Climate change and biodiversity in Maine: Vulnerability of habitats and priority species. Brunswick, ME: Manomet Center for Conservation Sciences (in collaboration with Maine Beginning with Habitat Climate Change Working Group) Report SEI-2013–03.

Widoff, L. 1987. Pitch Pine–Scrub Oak barrens in Maine. Augusta: Maine State Planning Office, Critical Areas Program Planning Report 86.

Wilkinson, R.S. 1988. Cyril Franklin dos Passos (1887–1986). Journal of the Lepidopterists' Society 42:155–163.

Williams, E.H. 2002. Harris' Checkerspot: A very particular butterfly. American Butterflies 10:18–25.

Williams, E.H. 2019. The butterflies (Lepidoptera) of an isolated island: Monhegan, Maine. Northeastern Naturalist. 26(3): 537–544.

Williamson, W.D. 1832. The history of the State of Maine; from its first discovery, A.D. 1602, to the separation, A.D. 1820, inclusive. Vol. 1. Hallowell, ME: Glazier, Masters.

Wilson, E.O. 1987. The little things that run the world: The importance and conservation of invertebrates. Conservation Biology 1:344–346.

Wilson, J.K., N. Casajus, R.A. Hutchinson, K.P. McFarland, J.T. Kerr, D. Berteaux, M. Larrivée, and K.L. Prudic. 2021. Climate change and local host availability drive the northern range boundary in the rapid expansion of a specialist insect herbivore, *Papilio cresphontes*. Frontiers in Ecology and Evolution 9. htpps://doi.org/ 10.3389/fevo.2021.579230.

Winter, D., coordinator. 1986. 1985 season summary. Zone 10 northeast. Massachusetts. News of the Lepidopterists' Society 2:35.

Wintle, B.A., H. Kujala, A. Whitehead, A. Cameron, S. Veloz, A. Kukkala, A. Moilanen, A. Gordon, P.E. Lentini, N.C.R. Cadenhead, and S. A. Bekessy. 2019. Global synthesis of conservation studies reveals the importance of small habitat patches for biodiversity. Proceedings of the National Academy of Sciences of the USA 116(3): 909–914.

Wojcik, V.A., and S. Buchmann. 2012. Pollinator conservation and management on electrical transmission and roadside rights-of-way: A review. Journal of Pollination Ecology 7(3): 16–26.

Wright, D.M. 1983. Life history and morphology of the immature stages of the bog copper butterfly *Lycaena epixanthe* (Bsd. & Le C.) (Lepidoptera: Lycaenidae). Journal of Research on the Lepidoptera 22:47–100.

Wright, D.M., M.A. Friedman, M.C. Minno, J.V. Calhoun, H. Pavulaan, and J.L. Monroe. 2019. New thoughts on *Celastrina* in Florida. News of the Lepidopterists' Society 61:26–29.

Wright, J. 1878. Short memoir of the life of Thomas Belt, F.G.S. Natural History Transactions of Northumberland, Durham, and Newcastle-on-Tyne 7:235–240.

Young, J. 1824. Extract from "minutes of an agricultural tour." New England Farmer 2:308.

Young, P. 1952. Eight military miniatures. Journal of the Society for Army Historical Research 30:172–173.

Youngsteadt, E., and P.J. Devries. 2005. The effects of ants on the entomophagous butterfly caterpillar *Feniseca tarquinius*, and the putative role of chemical camouflage in the *Feniseca*–ant interaction. Journal of Chemical Ecology 31:2091–2109.

Zhang, J., Q. Cong, J. Shen, P.A. Opler, and N.V. Grishin. 2019a. Changes to North American butterfly names. Taxonomic Report of the International Lepidoptera Survey 8(2): 1–11.

Zhang, J., Q. Cong, J. Shen, P.A. Opler, and N.V. Grishin. 2019b. Genomics of a complete butterfly continent. bioRxiv 829887.

Zhang, J., Q. Cong, J. Shen, P.A. Opler, and N.V. Grishin. 2020. Genomic evidence suggests further changes of butterfly names. Taxonomic Report 8(7): 1–40.

Zirlin, H. 2002. Strangers in a strange land. American Butterflies 10:4–11.

INDEX

Page numbers in *italics* refer to photographs; those followed by *t* and *fig* refer to tables and figures, respectively.

ABOUT THE AUTHORS

Phillip G. deMaynadier is a wildlife biologist at the Maine Department of Inland Fisheries and Wildlife, where his focus is nongame and endangered species biology and management. He received his doctorate in wildlife ecology from the University of Maine, where he now serves on the graduate faculty. Phillip has authored over fifty scientific publications and is committed to helping narrow the gap between conservation science and policy.

John Klymko is the zoologist at the Atlantic Canada Conservation Data Centre and was the director of the Maritimes Butterfly Atlas. He researches and surveys for rare and imperiled species throughout the Maritime Provinces to better understand their conservation statuses and co-chairs the terrestrial arthropod subcommittee of COSEWIC, the body that identifies species at risk in Canada.

Ronald G. Butler retired in 2021 as emeritus professor at the University of Maine at Farmington after 40 years of teaching. He has published papers on the ecology and behavior of mice, beavers, gulls, petrels, guillemots, skuas, penguins, damselflies, dragonflies, and bumble bees. For over 20 years, Ron has helped plan and coordinate statewide community-science projects focused on Maine insects, and he remains active in research and conservation initiatives.

W. Herbert Wilson Jr. is an emeritus professor of biology at Colby College, Waterville, Maine. He retired in 2019 after 30 years of service. He received his master of science from the University of North Carolina and his doctorate from the Johns Hopkins University. A population and community ecologist, Herb has published over fifty papers on marine intertidal invertebrates, odonates, and birds. Many of these papers have resulted from community-science projects.

John V. Calhoun is a research associate of the McGuire Center for Lepidoptera and Biodiversity (Florida Museum of Natural History). He has authored and coauthored over 100 publications on butterflies and moths, including the books *Butterflies and Skippers of Ohio* and *Butterflies through Binoculars: A Field, Finding, and Gardening Guide to Butterflies in Florida*. John has studied the butterflies of Maine for over 20 years.